CAMBRIDGE STUDIES IN ECOLOGY

Multivariate analysis in community ecology

Multivariate analysis in community ecology

HUGH G. GAUCH, Jr.
Cornell University

CAMBRIDGE UNIVERSITY PRESS
Cambridge
London New York New Rochelle
Melbourne Sydney

Published by the Press Syndicate of the University of Cambridge
The Pitt Building, Trumpington Street, Cambridge CB2 1RP
32 East 57th Street, New York, NY 10022, USA
10 Stamford Road, Oakleigh, Melbourne 3166, Australia

First published 1982
Reprinted 1983, 1984, 1985, 1986

Printed in the United States of America

Library of Congress Cataloging in Publication Data

Gauch, Hugh G., Jr., 1942–

Multivariate analysis in community ecology.

(Cambridge studies in ecology; 1)

Bibliography: p.

Includes index.
1. Biotic communities–Mathematics.
2. Ecology–Mathematics. 3. Multivariate
analysis. I. Title. II. Series.
QH541.15.M34G38 574.5′247′0151935 81-9974
ISBN 0 521 23820 X hard covers AACR2
ISBN 0 521 28240 3 paperback

To my sisters, Susan and Penny, and my brothers, James, Jonathan, and Robert

Contents

Preface

A review of multivariate analysis in community ecology is timely for several reasons. This field has been extensively developed recently, with many preferred methods only a few years old. Also, multivariate analysis is receiving wider usage in applied ecology and allied fields. Finally, this topic is timely because of the urgency of making wise choices concerning ecological problems. In tackling these problems, multivariate analysis is a major research approach because many ecological problems involve numerous variables and numerous individuals or samples (and hence involve multivariate data), and many problems cannot be investigated experimentally because of practical restraints. Multivariate analysis of community data cannot replace experimental manipulation, for example, but neither can experimentation replace multivariate analysis. Each research methodology has unique advantages, and the strongest research strategy employs all methodologies.

Two decisions were fundamental to the planning of this book. First, this book is relatively brief and emphasizes the preferred techniques of multivariate analysis. Such a strategy is possible because comparative tests have demonstrated the superiority of relatively few of the many proposed techniques. Furthermore, the primary concerns of most readers are with ecological or related subject matter, not with methods. This book is complementary to longer and more technical treatments by Orlóci (1978*a*) and Whittaker (1978*a,c*). Direct gradient analysis, ordination, and classification are presented as a methodological triad, and an integrated approach is described for solving research problems. Second, this book stresses both principles and applications. The foundations are ecological and biological principles concerning the structure of plant and animal communities, and mathematical principles concerning the intent and function of multivariate analyses. A preliminary grasp of principles is the basis for a subsequent grasp of practical research recommendations.

The intended audience includes advanced undergraduate students, graduate students, and professional researchers, primarily in ecology but also in coordinate fields (such as taxonomy, paleoecology, agronomy, forestry, and environmental management). Attention is given to plant, animal, and microbial communities; to present and fossil communities; and to terrestrial, aquatic, and marine communities. Present terrestrial plant communities receive special attention, reflecting partly the author's background and partly the primary development of the field in this context. Because multivariate analysis is ubiquitous, brief mention is also given to distant fields including business, social sciences, and medicine; specialists in these fields may also find topics of interest here because several multivariate techniques recently developed in ecology appear to be exceptionally robust and exceptionally sparing of computer resources. This book should also serve as a guide to the literature. An appendix identifies sources of relevant computer programs.

Our decade of research in the multivariate analysis of community data at Cornell University owes much to the creativity, energy, and perceptiveness of the late Robert H. Whittaker. I gratefully acknowledge his leadership and encouragment. The Section of Ecology and Systematics at Cornell University has been very helpful, as have the computer and library resources of the university. A sabbatical visit by Mark O. Hill contributed greatly to the ideas presented here. John M. Bernard, H. John B. Birks, David J. Hicks, Mark O. Hill, John T. Kunz, Peter L. Marks, and Robert K. Peet gave valuable comments on the manuscript; Robert H. Whittaker commented on the first three chapters. Steven B. Singer and John T. Kunz provided capable technical assistance, and Beth H. Marks and June Stein facilitated preparation of the manuscript. Authors too numerous to mention have contributed to the body of knowledge reviewed here; the citations to their work give some indication of the credit due them. I appreciate the invitation from H. John B. Birks and the Cambridge University Press to write this text.

HUGH G. GAUCH, JR.

Ithaca, New York
August 1981

1

Introduction

Community ecology concerns assemblages of plants and animals living together and the environmental and historical factors with which they interact. The domain of community ecology within biology as a whole may be characterized better by the spatial and temporal scales of principal interest rather than by other criteria such as conceptual viewpoint or methodology (Osmond, Björkman, & Anderson 1980). Community ecologists study living things principally on a spatial scale of meters to kilometers and a time scale of weeks to centuries. Phenomena beyond these spatial and temporal scales although relevant to community ecology and its interdisciplinary character, are studied mainly by biologists who would not be classified as community ecologists.

The community-level viewpoint is a major sector of ecology, receiving attention for many theoretical and applied purposes. Even when an ecologist's principal interest is more specific, however, such as the distribution of a particular species or the impact of a particular pollutant, much insight may result from preliminary or complementary community studies (Poore 1956, 1962; Foin & Jain 1977).

Multivariate analysis is the branch of mathematics that deals with the examination of numerous variables simultaneously: "The need for multivariate analysis arises whenever more than one characteristic is measured on a number of individuals, and relationships among the characteristics make it necessary for them to be studied simultaneously" (Krzanowski 1972). Community data are multivariate because each sample site is described by the abundances of a number of species, because numerous environmental factors affect communities, and so on. The purpose of multivariate analysis is to treat multivariate data as a whole, summarizing the data and revealing their structure. By contrast, statistical methods treating only one to several variables at a time are generally tedious, impractical, and ineffective for analysis of such data (Everitt 1978:5; Williams 1976:130–6). The application of multivariate analysis to community ecology is natural, routine, and fruitful.

Historical perspective

Plant and animal communities have been analyzed by ecologists worldwide for several decades. These studies are motivated by the importance and appeal of communities in their natural state and by the need to make intelligent management decisions. Even when interests are distinctively environmental rather than biological, the most important environmental impact is often that on plant and animal communities, so that environmental factors are best scaled and integrated through analysis of the communities. For example, the visual and economic impacts of serpentine soils are, not directly from their high magnesium and low calcium levels, but from the consequent impoverished and fragile vegetation.

Worldwide ecological interest in communities led to a diversity of schools, each with distinctive methodologies and emphases (Whittaker 1962, 1978*a,c;* Shimwell 1971; Maarel 1979*a;* Westhoff 1979). One consistent trend in community ecology, however, has been the progression, as in all sciences, toward increasingly quantitative methods (Cain & Castro 1959:1, 105; Greig-Smith 1964:210–19; Daubenmire 1968:39, 268; Orlóci 1978*a*:2). This trend has been accelerated over the past two decades by the ecologists' increasing access to computers, coupled with developments in multivariate methods. Doubtless this growth in quantitative methods has awkward and inappropriate elements, and many methods do not stand the test of time or do not prove realistic because of complexities and limitations inherent in the study of communities. Although original research cannot be expected to progress without trial and error, it is imperative to compare and evaluate methods. Community ecology is a rather young science, with almost no work prior to 1900 (except for occasional, informal descriptions of various communities) and little theoretical or quantitative development until around 1950 (as reviewed by Greig-Smith 1980). Present achievements are remarkable, considering that most of the work spans only the past three decades (Goodall 1962; Egerton 1976; McIntosh 1976, 1980).

A quantitative approach to community ecology has several advantages. (1) Quantitative methods "require the user to be precise and explicit in the formulation of the problem, foreseeing the implications for the type of data to be collected, sampling design used, and method of data analysis and statistical inference chosen" (Orlóci 1978*a*:2; also see Greig-Smith 1964:212). (2) Quantitative methods "require from the user uniformity and consistency in the implementation of the operational rules and decisions throughout the entire process of data collection, analysis, and inference" (Orlóci 1978*a*:2). This consistency encourages cooperation and exchange among ecologists, as well as with scientists in other disciplines (Orlóci 1978*a*:3). (3) "The quantitative approach allows the detection and appreciation of smaller differences" (Greig-Smith 1964:212). (4) Because of their uniformity and consistency, formal methods "naturally lend themselves to computer processing," which offers certain logistic advantages (Orlóci 1978*a*:2).

For these reasons, quantitative methods can be used to find answers to questions quite unanswerable by informal methods (Greig-Smith 1964:212; Williams 1976). Despite these advantages, quantitative methods do not necessarily require greater effort than informal methods. Indeed, "the habit of doing a thing a certain way can be efficient" (Cain & Castro 1959:105) and computer processing can save much time otherwise spent in tedious and error-prone tasks. The advantages are increasingly felt as a study grows in complexity or amount of data.

Only a few years ago, the multivariate methods available to ecologists were markedly less powerful than at present. There were three problems. (1) Proper comparison of the increasingly numerous multivariate methods required the application of a number of methods to a number of the same data sets, but this was lacking. Conflicting claims of the merits of various methods were confusing and frequently resulted in rather unfortunate choices, despite the existence of better methods. (2) Mathematical properties of communities needed to be modeled explicitly and accurately and the resulting community models compared with the underlying models of multivariate methods so that failures could be understood and more appropriate methods could be developed (Noy-Meir & Austin 1970; Swan 1970; Austin & Noy-Meir 1971; Gauch & Whittaker 1972*a,b,* 1976; Gauch, Whittaker, & Wentworth 1977; Austin 1980*b;* Hill & Gauch 1980). The goal of data analysis itself required clarification in order to foster progress toward it (Dale 1975). (3) Most multivariate methods had computer requirements rising with the square or cube (or more) of the amount of data (at least as implemented in available computer programs). Consequently, many methods, even with computers, were limited to only hundreds or even tens of samples.

Now, effective, efficient, and appropriate multivariate methods are available for ordination, hierarchical classification, and nonhierarchical classification of community data. The newer methods are also more reliable, and, consequently, recommendations for their use are more straightforward than were required a few years ago. Theoretical evaluation of the inherent limitations of community data and of the limits in computing efficiency indicates that some of these methods may be near theoretical limits, so that neither performance nor speed should improve markedly. Consequently, a review of multivariate methods in community ecology is now timely.

The two-way data matrix

Community studies involve observation of the abundances of a number of species in a number of samples. The resulting data may be tabulated in a matrix termed a community table, a primary data matrix, or simply a *data matrix* (such as in Table 1.1 in the section on multivariate methods). The data matrix has a two-way structure, with species on one side and samples on the other; most commonly, species are rows and samples are columns in the two-way samples-by-species data matrix.

The samples-by-species matrix is one specific kind of matrix from a large class of individuals-by-attributes matrices. Indeed, scientists and technicians are interested most frequently in more than one attribute, so most data are multivariate. Furthermore, as more than one individual is most frequently of interest, more scientific data are in the two-way individuals-by-attributes matrix format than in any other. The explosive, cosmopolitan development of multivariate methods is a response to the commonness of such data. In Chapter 6 the wider context of multivariate analysis will be reviewed, but here the following representative list should indicate something of the commonness of two-way data in ecology and other sciences and fields:

- Ecology
    - Samples by species
    - Sites by environmental parameters
    - Species by niches
    - Species by behavioral patterns
- Taxonomy
    - Specimens by characteristics
- Genetics
    - Varieties by genetic traits
- Agronomy
    - Soils by characteristics
- Paleontology and geology

- Samples by species
- Samples by geochemical parameters

Archeology

- Sites by artifacts

Linguistics

- Persons by vocabulary
- Languages by sounds

Psychology

- Persons by characteristics
- Nationalities by traits

Business

- Persons by consumer preferences
- Companies by products

Because applications of multivariate methods are diverse, the literature is large and widespread. Many methods commonly used in ecology were originated or developed by workers in taxonomy, psychology, archeology, and business. Because data from these various fields have much in common, the exchange of methods has proved to be profitable (Simon 1962; Crovello 1970; Blashfield & Aldenderfer 1978). Each field, however, also has its special data properties and analysis requirements, so that each field needs to test the appropriateness of different methods for its purposes. The multivariate methods used most commonly in each field are greatly influenced by the early introduction of particular methods. For example, principal components analysis has been popular in ecology and taxonomy, and nonmetric multidimensional scaling in psychology and business. Unfortunately many preferences reflect custom rather than performance, but it takes time for a given field to assess the usefulness of the many methods available.

Sometimes the most natural way to view a data set is as a three-way (or many-way) matrix, instead of the usual two-way matrix. For example, in a marine benthos study including seasonality, Williams & Stephenson (1973) have a samples-by-species-by-times data set. Similarly, from permanent plots in a montane grassland, Swaine & Greig-Smith (1980) analyze a sites-by-species-by-times-by-treatments-by-replicates data set. Ordination and classification algorithms, however, must receive a two-way data matrix as input (with the exception of some three-way nonmetric multidimensional scaling models, which are likely to be ineffective in most ecological applications (Young & Lewyckyj 1979:11, 13). Many-way data must be reorganized, summarized, or separated into one or more two-way matrices if ordination and classification are to be used (Crovello 1970; Williams & Stephenson 1973; Gillard 1976; Williams & Edye 1976; Jöreskog, Klovan, & Reyment 1976:117–18; Austin 1977; Green 1979:69; Maarel 1980*b;* Nishisato 1980:149–63; Swaine & Greig-Smith 1980). It is difficult to say whether important methodological advances could be made to handle many-way data better or whether the intrinsic complexities of such data, together with the human need to be presented with no more than two or three dimensions of variation at once, pose unavoidable problems that are tackled about as well as possible by methods used in the references just listed. In any case, as the complexity of a data set increases, the amount of data required to define a stable solution rises rapidly, so large data bases are required even if analytical techniques are effective.

The threshold at which one considers the number of individuals or attributes sufficient to justify a multivariate approach is arbitrary and varies somewhat with the properties of data sets; it may, however, be taken to be about 10 to 15. With fewer variables, especially five or fewer, multivariate approaches are not appropriate, and statistical analysis of variance and scatter plots are likely to be more fruitful. An exception, however, concerns the usefulness of multivariate analyses for some matrices that, although short in one dimension, are very long in the other dimension, such as 1000×5.

The multivariate analyses presented in this book are applicable to a data set that (1) is organized into a two-way matrix and (2) has a size of at least about 10×10 or 15×15. These two requirements must be met. The primary case treated in this book is the samples-by-species community data matrix, but the structure and analysis of numerous kinds of two-way data have many common features, giving most multivariate methods broad relevance to a variety of problems. Of course, useful results are expected only when data set properties and the assumptions of the multivariate technique match, at least in part. If the datum values are essentially random numbers or if their structure is quite different from the technique's model, multivariate analysis will be unfruitful. Reasonable match between data properties and analysis model may be considered a third requirement for results to be useful. The first two requirements are trivial to assess; the third may be trivial in some cases, but in other cases the simplest approach may be to try a multivariate analysis and see whether the results are useful.

Aspects of community data

Once the data matrix is established, analysis of a samples-by-species data matrix involves four aspects of the data: noise, redundancy, relationships, and outliers.

Noise

Replicate community samples (different samples from the same small homogeneous community) are rarely identical. Rather, they have average percentage similarities of 50 to 90%, depending on the sample size, the precision of measurements or estimates, the number of species, and so on (Dahl 1960; Maarel, Janssen, & Louppen 1978; Gauch 1981). The variation in species composition of the samples of a data set is due, in part, to interesting variation in environmental factors and, additionally, to uninteresting noise. The data are "noisy" in that samples from identical environmental conditions are not identical in species composition. Consequently, the data reflect partly interesting structure and partly noise.

The biological causes of noise are complex and include chance distribution and establishment of individuals, animal activity, local disturbances, and environmental heterogeneity at scales below that of the sample area. Noise also results from statistical limitations of finite samples and from limitations in measuring or estimating species abundances.

An operational definition of *noise* in the context of multivariate analysis of communities is variation in a species's abundances not coordinated with variation in other species' abundances (Poore 1956; Gauch 1981; also see Green 1977). "Coordinated" is used instead of "correlated" to mean related in a systematic manner detected by a multivariate analysis, not necessarily in a simple linear fashion, which would give rise to a large correlation coefficient. By contrast, data structure involves coordinated changes in a number of species simultaneously (Pignatti 1980). Coordination implies that a number of species are responding to an environmental (or other) factor in a significant and potentially interpretable way; noise implies uncoordinated and apparently chance differences in species' abundances. Coordination comes in a continuous spectrum of degrees, so the preceding definition may be refined as follows: noise is variation in a species's abundances coordinated markedly less with variation in other species' abundances than the larger coordinations observed. The boundary discriminating larger coordinations involves a subjective decision, which is sharpened by field experience.

This definition of noise is imprecise, but noise, by its very nature, tends to defy precise treatment. For routine purposes, this definition is adequate, however, being commensurate with the low level of interest attached to noise. Note that the concept of noise arises from the interaction between data and analysis. Apart from some particular analysis and research purpose, there is no definition of interest and hence no distinction between structure and noise (Simberloff 1980; Wimsatt 1980; also see Webb, Laseski, & Bernabo 1978).

The total information of a data set may be considered to be $X\%$ noise and $(100 - X)\%$ structure, where X is typically 10 to 50% (with the amount of information expressed by bits, variance, or whatever). The goal of multivariate analysis is to express data structure as faithfully as possible, with minimal expression of noise. Hence the ideal is to summarize $(100 - X)\%$ of the original information, but information reflecting data structure selectively, eliminating noise (Greig-Smith 1971; Prentice 1980*a*). If 100% of the original information were recovered, analysis would have failed completely regarding noise reduction.

Frequently, for practical reasons, ordination results are restricted to the range from one to three dimensions and classification results to 10 to 50 clusters, so it may not be possible to recover all of the interesting data structure. In this case, the goal is to express as much of the significant coordinated data structure as possible within the chosen format for results, while disregarding noise as much as possible.

Redundancy

A samples-by-species data matrix usually has a fair to large degree of redundancy in that many samples are much like other samples in their species composition and many species resemble other species in their occurrences in samples (Moore, Fitzsimons, Lambe, & White 1970; Orlóci 1974*a,* 1975, 1978*a*:17–19; Gauch 1980; also see Simon 1962). *Redundancy* involves coordinated species' responses and similar samples; it is the opposite of noise.

Figure 1.1 shows the percentage covers of *Agrostis* and *Festuca* in 128 chalk grassland samples (Kershaw 1973). There is sample redundancy in that the original 128 samples represent no more than several significantly different communities (because smaller differences are

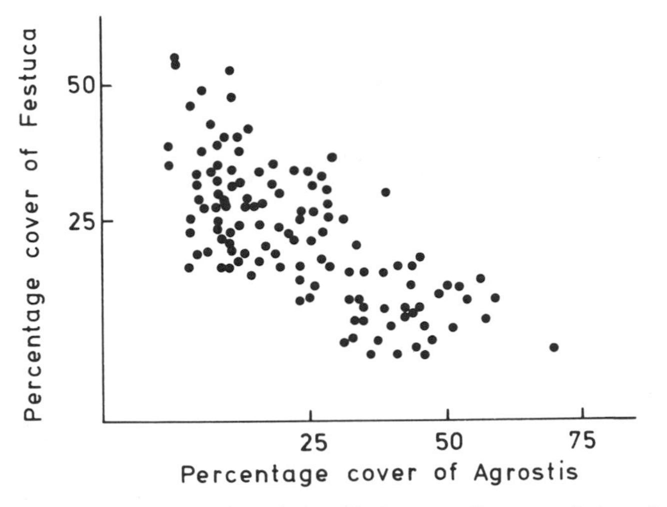

Figure 1.1. The relationship between *Festuca* and *Agrostis* covers in a chalk grassland. (After Kershaw 1973:29)

merely noise). There is also species redundancy in that coverages of *Agrostis* and *Festuca* are negatively correlated, and, consequently, knowledge of one coverage permits estimation of the other. Figure 1.2 likewise shows redundancy, in this case involving two environmental parameters, pH and base saturation, in waterlogged habitats in England (Gorham 1953; Kershaw 1973).

The detection of redundancy requires, quite obviously, a number of the same (or nearly the same) instances. Ecologists collect field data sets with several times as many samples as there are significantly different community types in order that recurrent species combinations can be detected. Noise can be distinguished from recurrent patterns only by having redundant data.

The necessity for collecting redundant field data is evident, but it must be emphasized that the desired final results from a community study are not to be redundant. Final results must be brief and economical in order to facilitate thinking, remembering, and communicating. The raw data are far too bulky and unorganized to serve these purposes.

Noise cannot be summarized because it involves uncoordinated species differences, but coordinated species responses can be summarized. A major role of multivariate analyses is to receive bulky raw data and, by summarization of redundant structure, to produce economical, brief results.

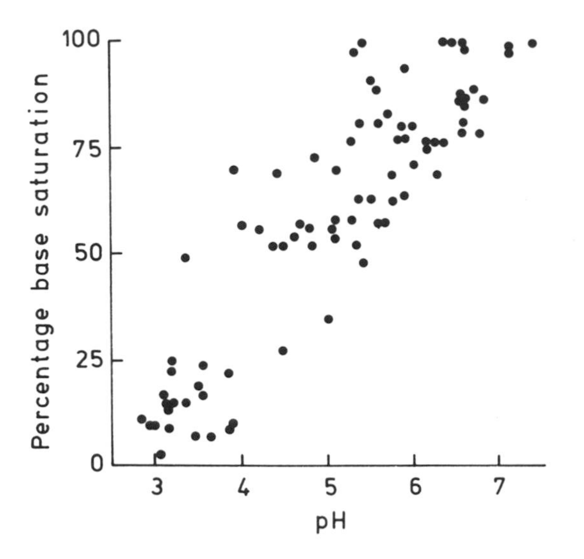

Figure 1.2. The relationship between percentage base saturation and pH in waterlogged habitats in England. (After Gorham 1953:355)

Redundancy may also be necessary in a data collection in order to produce a map showing the distribution of community types or to obtain estimates of the means and variances of species abundances in each community type (Goodall 1978*b*). Redundancy must be summarized, however, in order to describe communities economically or to permit certain multivariate analyses that are not practical for large data sets.

The origin of sample redundancy is simply the collecting of numerous samples in relation to their inherent variability so that replicates exist. Close points in Figure 1.1, representing samples differing in *Festuca* cover by only a few percent and likewise differing little in *Agrostis* cover, exemplify redundant, replicate samples. The origin of species redundancy, however, is the simultaneous effects, upon numerous species distributions, of relatively fewer underlying environmental gradients. For example, the decrease in *Festuca* and the increase in *Agrostis* moving to the right in Figure 1.1 are both due to a single environmental gradient from chalky to clayey soil.

Relationships

A major aim of community ecology is to elucidate relationships among samples, among species, and among samples and species considered jointly. If environmental or historical data are also collected at each sample site or if habitat preferences of the species are known, these additional data may also be related to the community analysis and may be helpful when interpreting community relationships.

Outliers

An outlier is a sample of peculiar species composition that has low similarity to all other samples (Everitt 1978:11–16; Gauch 1980). Similarly, a species is an outlier if it has low similarity to all other species (in its occurrences in samples). For the sake of simplicity, only sample outliers will be discussed here.

Outliers come in degrees because (1) the degree of sample peculiarity is variable and (2) the concept of outliers may be extended, to an arbitrary degree, to groups of more than one sample that are odd in the same way. Community data frequently contain outliers because of disturbed, heterogeneous, or otherwise unusual sites or because of gaps in sampling. Many multivariate methods give unsatisfactory results if outliers are present in a data set, so for these methods it is important to be able to identify outliers and to remove them from the data set prior to analysis.

If a data matrix is composed of two or more blocks of samples, with very few species occurring in more than one block, the matrix is said to be *disjunct.* In a disjunct sample set, samples have mostly nonzero similarities within blocks, but mostly zero (or near-zero) similarities between blocks. *Disjunction* is the common term for large groups of singular samples, whereas *outlier* is the common term for one singular sample (or a few similar but singular samples). The difference in terminology reflects a rather superficial matter, however, namely, the number of samples involved; it does not reflect fundamentally different aspects of the data.

The zero (or near-zero) similarities of outliers to the main body of data imply that the relationships of outliers to other samples are poorly defined by the data. Successful analysis of outliers on the basis of community data does not depend on finding better multivariate methods; rather, it is simply impossible because the data are not informative.

In some cases, outliers may be understood by using corroborative environmental or historical data. For example, in a set of desert vegetation samples, the distinctive flora of an oasis is patently attributable to the plentiful water supply. This interpretation derives, however, from data or insight external to the community data. If outliers are understood, it is on the basis of external information.

Multivariate methods

Multivariate analysis serves two basic roles in community ecology (Clymo 1980). (1) It helps ecologists to discover structure in the data. Discovery in community ecology tends to progress by successive refinement, with a large number of small steps (Poore 1962); multivariate analysis is a major tool in this process. (2) Multivariate analysis provides relatively objective, easy summarization of the data, which both facilitates the ecologist's comprehension of the data and provides a means for effective communication of results.

In contrast to multivariate analysis, many statistical methods are univariate or bivariate. Referring to ordination and classification as "pattern analysis," Williams & Gillard (1971:245–6) contrast multivariate with classical statistical methods:

> First, statistics is concerned with the testing of hypotheses, most commonly in the form of a probability that a null hypothesis is true. Pattern analyses, on the other hand, begin with no specific hypotheses; their function is to elicit, from a quantity of data, some internal structure from which hypotheses can be *generated.* Such hypotheses may, of course, later be tested by statistical methods. Secondly, statistics is at its most powerful when dealing with one, or very few, variates of approximately known distribution; the multivariate analogues of the standard methods tend to be weaker, and are computationally intractable if the system is over-defined or non-orthogonal. In contrast, pattern analysis methods are weakest where statistics is strongest, in the one- or two-variate case; they are at their most powerful when the number of attributes is large, and they can handle without difficulty mixtures of different types of attributes.

Hypothesis generation and testing are both necessary in scientific research, but multivariate analyses contribute principally to hypothesis generation. Interesting research strategies using both multivariate analyses and statistics may be found, however, in the work of Williams & Gillard (1971), Williams & Edye (1974), Edye (1976*b*), Williams (1976: 130–6), and Nishisato (1980).

General or specific interests may prevail in a given community study

(Sneath & Sokal 1973:109–13). General interests involve the relationships of species, communities, and environmental factors, with no singling out of some particular species, for instance, for special attention. General interests constitute a single research perspective because equal emphasis on all variables provides no potential for variation. Specific interests, however, exist without number because variables can be weighted differentially in countless variations. Examples are the distribution and habitat requirements of a particular rare wildflower, the disturbance effects of a newly introduced species, and the effects of soil calcium level on certain plant species. Because there exists one general interest, it is possible to develop methodologies for solving the general problem that are widely applicable and transportable. General methods operate by giving approximately equal emphasis to all the data, and most multivariate methods are essentially of this character and are therefore appropriate. Specific interests cannot be treated with maximum power, however, by standard, general methods because the diffusion of interest impairs the focus on one particular problem. Nevertheless, even for specific problems, multivariate methods are often useful at many steps; in particular, there is often merit in beginning with a general study of the community. It can be argued that extensive and general studies should precede or at least accompany intensive and specific ones (Poore 1956, 1962; Cain & Castro 1959:104; Foin & Jain 1977; also see Green 1979:135–6, 198–9), but the reverse is also done for various reasons.

Multivariate analysis of community data is challenging for several reasons.

(1) Community data are complex, involving the four aspects of data just discussed (noise, redundancy, relationships, and outliers).

(2) Community data are bulky. For example, a modest study with 300 samples and 100 species has a data matrix with 30 000 entries.

(3) The investigator may have a particular interest or question in mind upon which community data bear informatively but indirectly. Community data consist of lists of species abundances in samples (the usual samples-by-species data matrix), but the investigator's interest may be quite different in nature than species lists. For example, the investigator's purpose may be to assign each sample site to one of five community types for producing a local vegetation map. As another example, the investigator may want a single binary bit of information for each site, such as a decision to fertilize or not, or to burn or not. Often the best method for obtaining the desired information is not to

seek it directly but rather to begin with community sampling and to derive the desired information by means of multivariate analysis. This alternative is motivated by the advantages of community samples, including (a) objectivity, (b) speed and low cost, (c) effective correlation with other factors (perhaps environmental or historical) that are more expensive or elusive to observe directly, (d) relevance to a variety of present or potential research interests, and (e) compatability and comparability with related studies in the literature based on community samples.

(4) The variety of data sets, research purposes, and required formats for presentation of results demands both a variety of multivariate methods and guidelines for choosing an appropriate method or methods. To facilitate communication among ecologists, however, there is need for selection of only a modest number of preferred methods for common usage (Shimwell 1971:282; Pielou 1977:331).

To meet these challenges, ecologists have applied, developed, and tested a number of multivariate methods. Many methods are only decades old, and many of the best methods are very recent. In Chapters 3–5, the three basic multivariate strategies, direct gradient analysis, ordination, and classification, will be treated. Here these strategies will just be introduced.

Direct gradient analysis

Direct gradient analysis is used to study the distribution of species along recognized, easily measured, environmental gradients (Whittaker 1967, 1978*b*). For example, community samples may be taken along a depth gradient in a lake or along gradients of elevation and topographic moisture status in a mountain. Direct gradient analysis provides (1) the primary observational basis for ecological models of community structure and (2) relatively well understood field data sets, which are then appropriate for testing multivariate methods because the expected results are known to a fair degree.

The importance of direct gradient analysis in providing the foundation for community ecology cannot be overestimated. Direct gradient analysis must be complemented, however, with additional approaches for several reasons. (1) In the more difficult cases, the important environmental factors may not be evident, making it impossible to apply direct gradient analysis. (2) The investigator's preliminary choice of environmental gradients may be justifiable and adequate for some purposes, but studies demanding a high level of objectivity may require other approaches. (3) An interesting perspective involves sample and species relationships as defined by the community data alone, apart from any environmental or other data. For example, Goodall (1954*a*) emphasizes the analysis of plant communities as a means of studying the environment as scaled and integrated by plants.

Direct gradient analysis is quite different from ordination and classification in implementation in that direct gradient analysis involves simple graphing procedures, whereas ordination and classification usually require sophisticated mathematics and computers. The general function, or purpose, of direct gradient analysis, however, is the same as that of ordination and classification (namely, to summarize and reveal the structure in multivariate data). Despite its simplicity, direct gradient analysis has a unique and important role in the methodological triad of multivariate analysis.

Ordination

Ordination primarily endeavors to represent sample and species relationships as faithfully as possible in a low-dimensional space (Anderson 1971: Orlóci 1974*a*, 1978*a*:103; Everitt 1978; Noy-Meir 1979). The end product is a graph, usually two-dimensional, in which similar samples or species or both are near each other and dissimilar entities are far apart. Frequently an environmental interpretation of the sample (or species) arrangement is offered; this interpretation arises from work external to the ordination itself but is often quite defensible and interesting. The advantage of low-dimensionality is workability for contemplation and communication; the disadvantage is that some degree of fidelity to the data structure must be frequently sacrificed in the projection into only one to a few dimensions.

Ordination pertains to the four aspects of the data as follows: (1) Ordination is effective for showing relationships. (2) Ordination is effective for reducing noise (Gauch 1981). (3) Some ordination techniques merely give outliers an undistinguished location either in the center of the ordination or around the periphery. Depending on the ordination positions of valid samples and on the investigator's experience with ordination techniques, this may cause confusion. Other ordination techniques treat outliers as representing very long, important gradients, thereby compressing the main body of samples unsatisfactorily into a very small space. In any event, the lack of effective analysis of outliers is not to be faulted because community data are uninformative about outliers anyway, but confusion of results and compression of ordination space are undesirable. In many cases it is best to remove outliers from the data set prior to ordinating (or after an initial ordination). Ordination can help identify outliers and disjunctions (Gauch, Whittaker, & Wentworth 1977), but classification may be more efficient (Gauch 1980). (4) Ordination summarizes data redundancy by placing similar entities in proximity and by producing an economical understanding of the data in terms of a few gradients in community composition (which may be interpretable environmentally). This summarization of redundancy is rather different in nature from that of classification wherein similar entities are put into groups.

Ecologists use quite an array of ordination techniques (Pielou 1977; Orlóci 1978*a*:102–85; Whittaker 1978*c*). Popular techniques include weighted averages, polar (Bray–Curtis) ordination, principal components analysis, reciprocal averaging, detrended correspondence analysis, and nonmetric multidimensional scaling. Less common techniques include factor analysis, principal coordinates analysis, canonical correlation analysis, and Gaussian ordination.

Classification

Classification basically involves grouping similar entities together in clusters. It is a fundamental activity in science, as it is in any process of thinking or communicating (Poore 1962; Shimwell 1971:42; Sokal 1974, 1977; Blashfield & Aldenderfer 1978).

Community ecologists use three kinds of classification. (1) *Table arrangement* by the Braun-Blanquet approach seeks to order the samples-by-species data matrix by placing samples and species into the order that best reveals the intrinsic structure of the data (Braun-Blanquet 1932; Poore 1955; Moore, Fitzsimons, Lambe, & White 1970; Mueller-Dombois & Ellenberg 1974:177–210; Westhoff & Maarel 1978). Compositionally similar samples are brought close together, as are distributionally similar species. The nonzero data matrix entries are thereby concentrated into blocks, and lines may be drawn in the matrix to mark off sample and species clusters. Although the original procedure was applied by hand, desire for increased objectivity and for applicability to large data sets has motivated development of computerized approximations to this method (Maarel, Janssen, & Louppen 1978; Hill 1979*b*). (2) *Nonhierarchical classification* merely puts similar samples (or species) into clusters (Gauch 1980). The researcher may control the number of clusters formed. (3) *Hierarchical classification* puts similar samples (or species) into groups and, additionally, arranges the groups into a hierarchical, treelike structure called a *dendrogram,* which indicates relationships among the groups (Sneath & Sokal 1973:260; Everitt 1978:42–64; Gauch & Whittaker 1981; also see Simon 1962).

Classification is relevant to all four aspects of the data noted earlier. (1) Noise can be reduced by combining the samples of a cluster into a single average or composite sample (Gauch 1980). (2) Classification is obviously effective for summarizing redundancy in the data. For example, 500 community samples might be placed into 15 community types that effectively represent the variability present in the data and give a workable number of entities for contemplation and communication. (3) Table arrangements express relationships among samples and species; hierarchical classifications express relationships among samples or species; nonhierarchical classifications indicate clusters of similar entities but do not indicate the larger picture of relationships among the clusters. (4) Outliers can be detected simply by noting samples that fail to cluster with other samples at a given, fairly low, level of similarity (Gauch 1980).

Examples

Direct gradient analysis, ordination, and classification methods will be treated in detail in later chapters. Here examples will be given to introduce the nature of the results obtainable by these methods.

Table 1.1 gives the number of tree stems per hectare using a 10-step soil moisture gradient (from 1 for mesic ravines to 10 for xeric, open, southwest-facing slopes) for 23 tree species in pine–oak woodlands at about 2000-m elevation in the Santa Catalina Mountains, southern Arizona (adapted from Whittaker 1978*b*). The 10 samples are arranged in order of increasingly xeric conditions and the species are easily arranged by corresponding habitat preferences to make the structure of these data evident. This progression in soil moisture status and corresponding dominant species is readily appreciated even by mere reconaissance of the Santa Catalina Mountains, but community samples are necessary to quantify the pattern (as in Table 1.1).

A simple method of direct gradient analysis is the graphing of species abundances along the recognized environmental gradient. As

Table 1.1. *Tree stems per hectare using a 10-step moisture gradient*

	Sample (mesic)									(xeric)
Species[a]	1	2	3	4	5	6	7	8	9	10
Salix lasiolepis	110									
Alnus oblongifolia	106									
Acer glabrum var. *neomexicanum*	12									
Abies concolor	114	2								
Quercus gambelii	64	20								
Pinus strobiformis	78	24	4							
Juglans major	14	12	2							
Robinia neomexicana	16	10	4	2						
Pseudotsuga menziesii	88	130	54	2						
Rhamnus californica ssp. *ursina*	6	26	8	2						
Quercus rugosa	252	186	116	72	44	24	32	32	22	10
Prunus virens	12	76	42	12	2					
Pinus ponderosa	36	644	472	212	48	6	2			
Arbutus arizonica		4	16	54	32	16				
Quercus hypoleucoides	4	446	670	1052	1072	724	248	106	124	22
Pinus chihuahuana		2	8	20	64	72	14			
Quercus arizonica		18	36	94	144	196	200	164	50	38
Arctostaphylos pungens				2	20	34	28	10	24	14
Juniperus deppeana		2	8	8	26	92	102	78	58	50
Arctostaphylos pringlei				12	42	268	108	308	308	68
Garrya wrightii			2	8	18	10	26	136	114	20
Pinus cembroides			2	6	26	58	186	322	228	170
Quercus emoryi						10	16	36	28	152

[a]The species were obtained from a pine–oak woodland at ~2000-m elevation in the Santa Catalina Mountains, Arizona.

Stems over 1-cm diameter at breast height were counted, and each sample is based on five 0.1-ha subsamples.

Source: Adapted from Whittaker (1978*b*:15).

an example, this is done in Figure 1.3 for several of the more common species of Table 1.1. Note that each species is individualistic in its distribution, and community composition varies continuously rather than changing abruptly (Whittaker 1978*b*). The other species from Table 1.1 not shown in Figure 1.3 also follow this pattern. This direct gradient analysis effectively shows the responses of these species to the moisture gradient. Table 1.2 shows a second data set with 25 samples of meadow communities from the Danube Valley, south of Ulm, Germany (adapted from Mueller-Dombois & Ellenberg 1974). A number

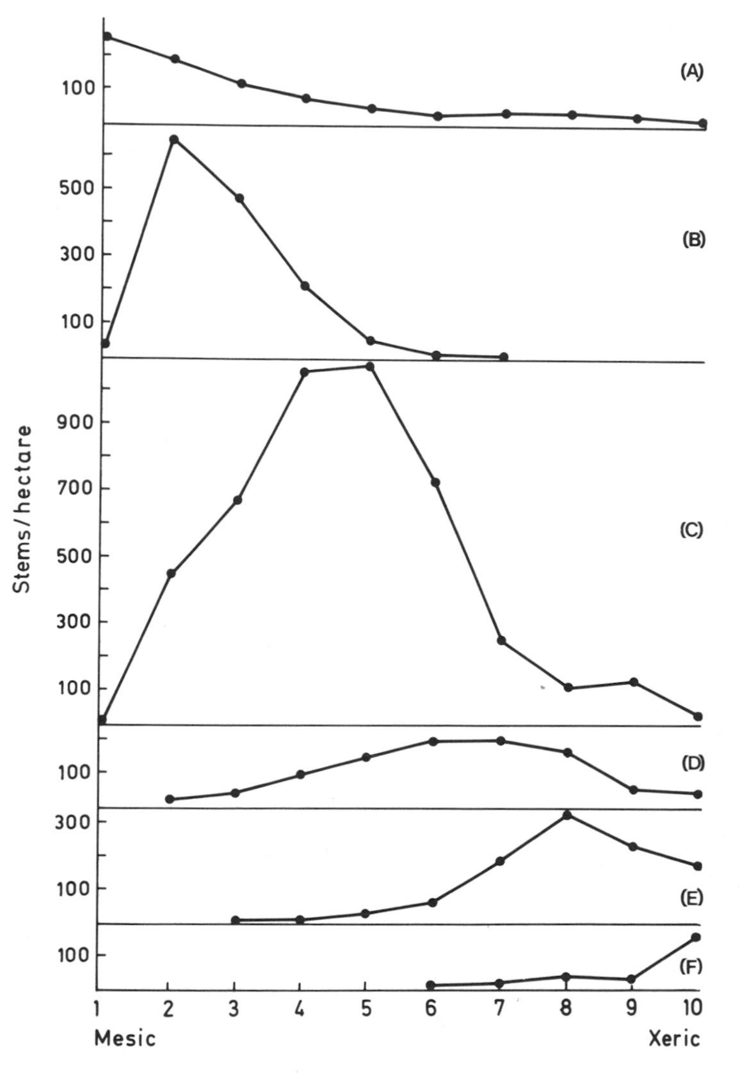

Figure 1.3. Tree stems per hectare using a ten-step moisture gradient for several species in the Santa Catalina Mountains, Arizona. Species shown are: (*A*) *Quercus rugosa;* (*B*) *Pinus ponderosa;* (*C*) *Quercus hypoleucoides;* (*D*) *Quercus arizonica;* (*E*) *Pinus cembroides;* (*F*) *Quercus emoryi.* (Data tabulated in Table 1.1)

of minor species have been deleted for the sake of clarity and the original percentage coverage values have been transformed to a one-digit format by the octave scale of Table 1.2. In the table, the 25 samples are listed in an arbitrary order, beginning with sample 8 (with

Table 1.2. *German meadow samples with species arranged alphabetically and samples arbitrarily*

	Sample[a]																								
	C	B	G	C	G	C	B	G	C	B	C	C	B	O	B	C	G	C	B	C	B	C	G	B	G
Species	8	3	20	14	16	13	1	22	25	15	11	23	24	19	9	6	18	21	10	7	2	17	12	4	5
Achillea millefolium		4	5	6	5	4	5	4	1	3		1	4		3		6	1	5	2	1	1	2	2	3
Alopecurus pratensis	6					4		6	2		5					5	3	6							3
Angelica sylvestris				2		1			1			2				2				3					
Anthriscus silvestris						1			1			1	1												3
Arrhenatherum elatius	6	3	7	6	6	6	1	7	8	6	6	7	7	7	5	4	7	7	5	4	2	6	6	3	4
Bellis perennis	1	1			1	1									1	1	1		2			1	2		1
Briza media					3		2								3				2		2				
Bromus erectus		8					8			8			6		8				7					9	
Campanula glomerata		2			1		1						1		2								2	2	
Cardamine pratensis					1						1														
Carex acutiformis									3		4			3				2		6					
Carex gracilis									2					4											
Carex panicea					3																				
Centaurea jacea		5	3	4	3		1			1		1	1			3		1		4	2	3	2		
Chenopodium album														1											
Chrysanthemum levcanthemum	2	4	4	3	2	2	2	2		2	1	2	5	1	3	4		1	5	3	1	2	4	1	1
Cirsium oleraceum	7		1	7		1		1	4		4	2		1		6		3		7		3			1
Crepis biennis	5	1	2	1				2			1	5	1			2	2	1			1	1	2		3
Deschampsia caespitosa	3			2							7			3				5		6		6			
Euphrasia odontites												1					1								
Festuca ovina							3								2				3		2				
Festuca rubra	1		3			2	6			5	3	1	3	1	4				4			3	3	3	1
Filipendula ulmaria	3			1		1										4									
Geum rivale	5		1		2	1		1	1		4			2		3	1	2		2		1	1		
Glechoma hederacea	1							1			1									1			1		1
Glyceria fluitans														7											
Helictotrichon pubescens			7		6	4	2	3					2		4	5	4	1	1	4	2	4	2		7
Holcus lanatus	3			3		2			6		3	1		1		2		3		2		3			2
Koeleria pyramidata							4								4				4					3	
Lamium album														1											
Leontodon hispidus		1			2															4	2				
Linum catharticum		1			1		1								1									1	
Lychnis flos-cuculi	1				2				1		1			1				1				1			
Lysimachia nummularia	2		1						1		1					1		1				1	1		
Melandrium diurnum				2	2	1		1	4					1		3		1		2		1	1		
Pastinaca sativa				1		2											4								
Phalaris arundinacea														7											
Phleum pratense				2																			1		
Pimpinella magna						2												2		1		1			
Poa pratensis	4	6	5	5	6	3	4	6	6	6	5	6	7	2	6	3	6	2	5	5	9	5	7	5	4
Polygonum bistorta									4					2						2			1		

Table 1.2 (*cont.*)

Species	Sample[a] C	B	G	C	G	C	B	G	C	B	C	C	B	O	B	C	G	C	B	C	B	C	G	B	G
	8	3	20	14	16	13	1	22	25	15	11	23	24	19	9	6	18	21	10	7	2	17	12	4	5
Polygonum convolvatus														1											
Salvia pratensis		4								2			4		3				5						
Sanguisorba officinale				1																					
Scabiosa columbaria							1			1									1		3			2	
Silene inflata										1						1					3				
Taraxacum officinale	2	1	1	1		3	1	1	1	1		4	1			2	2		1	4		1	1	1	4
Thymus serpyllum		1					2												3		1			1	
Tragopogon pratensis					2	2				1						2				1			1		3
Viola hirta		1								1			1		3		1		4						

[a]Octave scale: 1 for rare; 2 for 1% coverage; 3 for 2%; 4 for 3 to 4%; 5 for 5 to 8%; 6 for 9 to 16%; 7 for 17 to 32%; 8 for 33 to 64%; 9 for 65 to 100%.
Braun-Blanquet code: B for the *Bromus–Arrhenatherum* community type, G for *Geum–Arrhenatherum,* C for *Cirsium–Arrhenatherum,* and O for the outlier (sample 19).
Source: Adapted from Mueller-Dombois & Ellenberg (1974:182–3).

accompanying letter codes), and the species in alphabetic order (which is essentially an arbitrary order from an ecological viewpoint). Arbitrary sample and species orders are common in field data prior to analysis, so this table is typical of raw field data. No structure is discernible from perusal of Table 1.2 (other than the trivial matter that species vary in commonness).

Table 1.3 presents the German meadow data as classified and arranged by the Braun-Blanquet approach (adapted from Mueller-Dombois & Ellenberg 1974; this data set is also discussed by Maarel, Janssen, & Louppen 1978; Hill 1979*b*; Maarel 1979*a,b;* and Hill & Gauch 1980). The data are identical to Table 1.2, but the table has been arranged in order to display its inherent structure. The samples are grouped by the Braun-Blanquet analysis into three community types: *Bromus–Arrhenatherum, Geum–Arrhenatherum,* and *Cirsium–Arrhenatherum,* of which the first is relatively distinct and the latter two intergrade. Sample 19 in the *Cirsium–Arrhenatherum* type is considered an outlier because its species composition is rather peculiar, including the presence of several species not found in any other sample. Mueller-Dombois & Ellenberg (1974:189) attribute the out-

Table 1.3. *German meadow samples arranged by the Braun-Blanquet method*

Species	Sample[a] B	B	B	B	B	B	B	B	G	G	G	G	G	G	C	O	C	C	C	C	C	C	C	C	C
	4	10	1	9	15	3	24	2	16	18	12	20	5	22	23	19	13	17	6	11	8	21	25	14	7
Ia																									
Bromus erectus	9	7	8	8	8	8	6																		
Scabiosa columbaria	2	1	1		1			3																	
Thymus serpyllum	1	3	2			1		1																	
Salvia pratensis		5		3	2	4	4																		
Koelria pyramidata	3	4	4	4																					
Festuca ovina		3	3	2				2																	
Ib																									
Campanula glomerata	2		1	2		2	1		1		2														
Viola hirta		4		3	1	1	1			1															
Briza media		2	2	3				2	3																
Linum catharticum	1		1	1		1			1																
IIb																									
Geum rivale									2	1	1	1		1		2	1	1	3	4	5	2	1		2
Holcus lanatus													2		1	1	2	3	2	3	3	3	6	3	2
Melandrium diurnum									2		1			1		1	1	1	3			1	4	2	2
Alopecurus pratensis										3			3	6			4		5	5	6	6	2		
Lysimachia nummularia											1	1						1	1	1	2	1	1		
Lychnis flos-cuculi									2							1		1		1	1	1	1		
Glechoma hederacea											1		1	1						1	1				1
IIa																									
Cirsium oleraceum												1	1	1	2	1	1	3	6	4	7	3	4	7	7
Deschampsia caespitosa																3		6		7	3	5		2	6
Angelica sylvestris															2		1		2				1	2	3
Carex acutiformis																3				4		2	3		6
Filipendula ulmaria																	1		4		3			1	
Pimpinella magna																	2	1				2			1
Polygonum bistorta											1					2							4		2
Arrhenatherum elatius	3	5	1	5	6	3	7	2	6	7	6	7	4	7	7	7	6	6	4	6	6	7	8	6	4
Poa pratensis	5	5	4	6	6	6	7	9	6	6	7	5	4	6	6	2	3	5	2	5	4	2	6	5	5
Chrysanthemum levcanthemum	1	5	2	3	2	4	5	1	2		4	4	1	2	2	1	2	2	4	1	2	1		3	3
Achillea millefolium	2	5	5	3	3	4	4	1	5	6	2	5	3	4	1		4	1				1	1	6	2
Taraxacum officinale	1	1	1		1	1	1			2	1	1	4	1	4		3	1	2		2		1	1	4
Helictotrichon pubescens		1	2	4			2	2	6	4	2	7	7	3			4	4	5			1			4
Crepis biennis						1	1	1		2	2	2	3	2	5			1	2	1	5	1		1	
Centaurea jacea			1		1	5	1	2	3		2	3			1			3	3			1		4	4
Festuca rubra	3	4	6	4	5		3				3	3	1		1	1	2	3		3	1				
Bellis perennis		2		1		1			1	1	2		1				1	1	1		1				

Table 1.3 (*cont.*)

	Sample[a]																								
	B	B	B	B	B	B	B	B	G	G	G	G	G	G	C	O	C	C	C	C	C	C	C	C	C
Species	4	10	1	9	15	3	24	2	16	18	12	20	5	22	23	19	13	17	6	11	8	21	25	14	7
Tragopogon pratensis					1				2		1		3				2		2						1
Anthriscus silvestris							1						3		1		1						1		
Leontodon hispidus						1		2	2																4
Silene inflata					1			3											1						
Pastinaca sativa										4							2							1	
Carex gracilis																4							2		
Cardamine pratensis									1											1					
Euphrasia odontites										1					1										
Phleum pratense											1													2	
Glyceria fluitans																7									
Phalaris arundinacea																7									
Carex panicea									3																
Sanguisorba officinale																								1	
Lamium album																1									
Polygonum convolvatus																1									
Chenopodium album																1									

[a]See Table 1.2 footnote for coding key.
Source: Adapted from Mueller-Dombois & Ellenberg (1974:190–1).

lier's peculiarity to its unusual site near a small tributary of the Danube that is periodically flooded and subjected to some sedimentation. They recommend conducting a larger survey, which would include periodically flooded sites, and transferring this sample to a new type with similar samples. Samples are numbered sequentially as taken in the meadow and are coded according to the Braun-Blanquet analysis noted in Table 1.2. The species in Table 1.3 are listed in five groups: group Ia is narrowly restricted to *Bromus–Arrhenatherum,* Ib mainly restricted to *Bromus–Arrhenatherum,* IIb restricted to *Geum–Arrhenatherum* and *Cirsium–Arrhenatherum,* IIa mainly restricted to *Cirsium–Arrhenatherum,* and the last group to species too common or too rare to be characteristic of any particular community type. Species in the first four groups have informative distributions and are termed *differential* species; the last group contains *nondifferential* species. Comparison of Tables 1.2 and 1.3 shows the gain in clarity obtained by ordering the table in a meaningful manner, even though these two tables contain identical data.

Table 1.4. *German meadow samples arranged by ranked detrended correspondence analysis ordination scores*

Species	Sample[a] B1	B4	B10	B9	B3	B15	B2	B24	G16	G20	G12	G18	G5	C23	G22	C13	C14	C17	C6	C7	C8	C11	C21	C25	O19
Koeleria pyramidata	4	3	4	4																					
Thymus serpyllum	2	1	3		1		1																		
Festuca ovina	3		3	2			2																		
Bromus erectus	8	9	7	8	8	8		6																	
Scabiosa columbaria	1	2	1			1	3																		
Viola hirta			4	3	1	1		1				1													
Salvia pratensis			5	3	4	2		4																	
Linum catharticum	1	1		1	1				1																
Briza media	2		2	3			2		3																
Campanula glomerata	1	2		2	2			1	1		2														
Silene inflata						1	3												1						
Festuca rubra	6	3	4	4		5		3		3	3		1	1		2		3			1	3			1
Achillea millefolium	5	2	5	3	4	3	1	4	5	5	2	6	3	1	4	4	6	1		2			1	1	
Bellis perennis			2	1	1				1		2	1	1			1		1	1		1				
Chrysanthemum levcanthemum	2	1	5	3	4	2	1	5	2	4	4		1	2	2	2	3	2	4	3	2	1	1		1
Carex panicea									3																
Poa pratensis	4	5	5	6	6	6	9	7	6	5	7	6	4	6	6	3	5	5	3	5	4	5	2	6	2
Centaurea jacea	1				5	1	2	1	3	3	2			1			4	3	3	4			1		
Helictotrichon pubescens	2		1	4			2	2	6	7	2	4	7		3	4		4	5	4			1		
Leontodon hispidus					1		2		2											4					
Euphrasia odontites												1		1											
Taraxacum officinale	1	1	1		1	1		1		1	1	2	4	4	1	3	1	1	2	4	2			1	
Pastinaca sativa												4				2	1								
Tragopogon pratensis						1			2		1		3			2			2	1					
Anthriscus silvestris								1					3	1		1								1	
Arrhenatherum elatius	1	3	5	5	3	6	2	7	6	7	6	7	4	7	7	6	6	6	4	4	6	6	7	8	7
Crepis biennis					1		1	1		2	2	2	3	5	2		1	1	2		5	1	1		
Phleum pratense											1						2								
Sanguisorba officinale																	1								
Cardamine pratensis									1													1			
Glechoma hederacea											1		1		1					1	1	1			
Angelica sylvestris														2		1	2		2	3				1	
Filipendula ulmaria																1	1		4		3				
Alopecurus pratensis												3	3		6	4			5		6	5	6	2	
Pimpinella magna																2		1		1			2		
Lysimachia nummularia										1	1							1	1		2	1	1	1	
Cirsium oleraceum										1			1	2	1	1	7	3	6	7	7	4	3	4	1
Melandrium diurnum									2		1				1	1	2	1	3	2			1	4	1
Geum rivale									2	1	1	1			1	1		1	3	2	5	4	2	1	2
Holcus lanatus													2	1		2	3	3	2	2	3	3	3	6	1
Lychnis flos-cuculi									2									1			1	1	1	1	1
Deschampsia caespitosa																	2	6		6	3	7	5		3

Table 1.4 (*cont.*)

	Sample[a]																								
	B	B	B	B	B	B	B	B	G	G	G	G	G	C	G	C	C	C	C	C	C	C	C	C	O
			1			1		2	1	2	1	1		2	2	1	1	1				1	2	2	1
Species	1	4	0	9	3	5	2	4	6	0	2	8	5	3	2	3	4	7	6	7	8	1	1	5	9
Carex acutiformis																				6		4	2	3	3
Polygonum bistorta											1									2				4	2
Carex gracilis																								2	4
Chenopodium album																									1
Polygonum convolvatus																									1
Lamium album																									1
Phalaris arundinacea																									7
Glyceria fluitans																									7

[a]See Table 1.2 footnote for coding key.

Table 1.4 presents the German meadow data arranged by placing the samples (and likewise the species) in rank order by their first-axis ordination scores using detrended correspondence analysis (DCA). (Reciprocal averaging ordination is identical to DCA for ranked positions on the first axis, so it would give identical results.) Comparison of Table 1.4 with Table 1.2 shows again the gain in clarity resulting from meaningful arrangement of the community table. The sample sequence at the top of Table 1.4 cleanly segregates the community types recognized by the Braun-Blanquet analysis, with the minor imperfection that samples 22 and 23 need to be reversed in order to obtain pure blocks of G and C samples. This imperfection is quite unimportant since the *Geum–Arrhenatherum* and *Cirsium–Arrhenatherum* community types intergrade, and the Braun-Blanquet method involves subjective judgments so that different investigators may obtain slightly different results.

The Braun-Blanquet and DCA data matrix arrangements (Tables 1.3 and 1.4) are both satisfactory. They differ in numerous small details, which are of little interest, but there are two notable differences. (1) Detrended correspondence analysis arranges the species in a single sequence, with the species' abundance peaks moving across the table as one reads down the list. The Braun-Blanquet arrangement similarly concentrates matrix values in blocks (in this case mainly along the matrix diagonal) for the differential species, but the nondifferential species are listed in a separate, final group with no indication of their

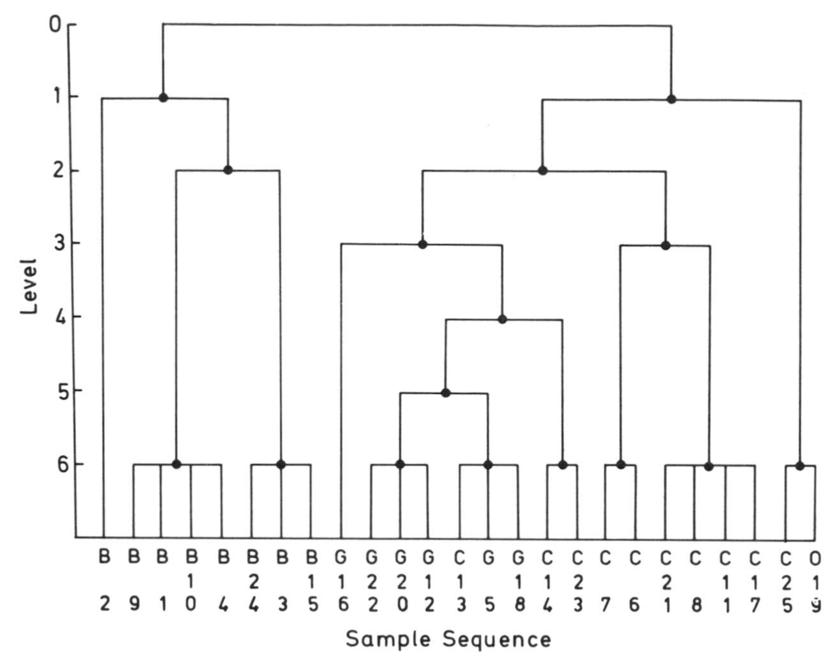

Figure 1.4. Dendrogram of two-way indicator species analysis (TWINSPAN) of German meadow samples. (Data and coding from Table 1.2)

places in the species sequence. (2) Braun-Blanquet analysis proceeds by a fairly complex process requiring considerable training, and different investigators may obtain somewhat different results. On the other hand, the DCA arrangement requires computer processing but minimal training, and all investigators obtain identical results.

Hierarchical classification of the German meadow data using two-way indicator species analysis (TWINSPAN, Hill 1979*b*; Gauch & Whittaker 1981) is shown in Figure 1.4 using a dendrogram (a branching arrangement illustrating the relative similarity of samples). The sample sequence for the 25 samples is indicated along the abscissa. The levels from 0 to 6 on the ordinate indicate progressively finer divisions of the samples into increasingly homogeneous groups. Samples joining each other at the bottom of the figure (level 6) are most similar, whereas samples joining at the top (level 0) are most dissimilar. The first division clearly separates all *Bromus–Arrhenatherum* samples from the others. Among the *Bromus–Arrhenatherum* samples, sample 2 is most distinctive and splits off first; then samples 9, 1, 10, and 4 are placed in one group and samples 24, 3, and 15 in the other. In the right-hand side of the first division, the outlier (sample 19) and sample

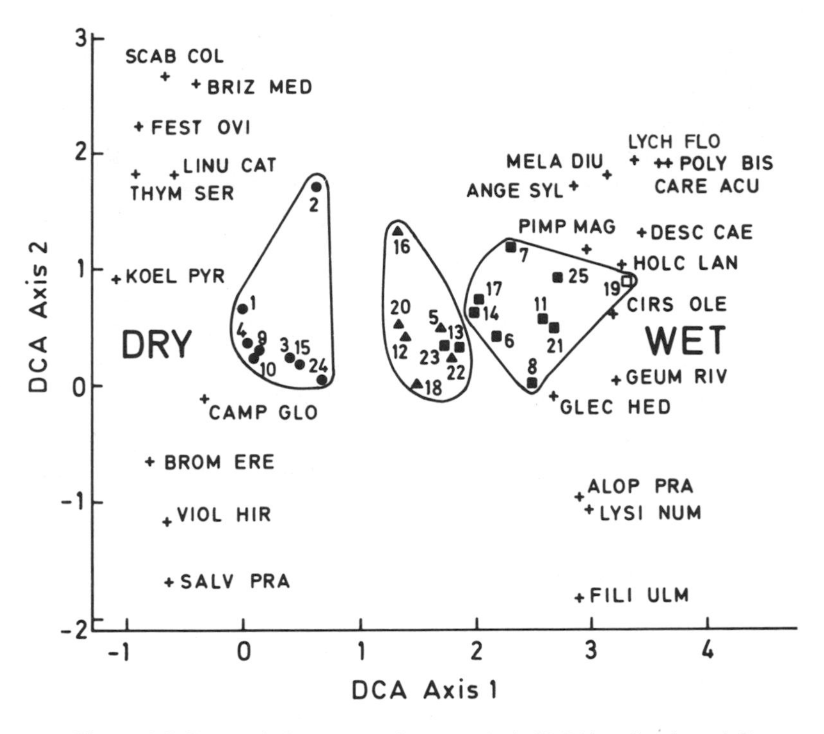

Figure 1.5. Detrended correspondence analysis (DCA) ordination of German meadow samples and species. The samples were previously classified by Braun-Blanquet analysis into three community types: *Bromus–Arrhenatherum* (●), *Geum–Arrhenatherum* (▲), and *Cirsium–Arrhenatherum* (■) with one outlier (□). An environmental interpretation of the first DCA axis is offered: a dry-to-wet soil moisture gradient. Also indicated are three sample clusters resulting from composite clustering, a nonhierarchical clustering technique. (Data tabulated in Table 1.2; also see Table 1.4)

25, which is similar to the outlier, split off first; then the *Geum–Arrhenatherum* samples with a few *Cirsium–Arrhenatherum* samples are separated from most of the *Cirsium–Arrhenatherum* samples, and so on through progressively finer divisions. Note the trend across the sample sequence (from B to G to C to O), which faithfully reflects the moisture gradient (from dry to wet).

Nonhierarchical classification of these meadow samples using composite clustering (Gauch 1979, 1980) is shown in Figure 1.5. Here the samples are put into three clusters: eight *Bromus–Arrhenatherum* samples, six *Geum–Arrhenatherum* samples with two *Cirsium–Arrhenatherum* samples, and the remaining eight *Cirsium–Arrhenatherum* samples and the outlier. Nonhierarchical classification merely identifies these three groups as indicated by the three clusters of points in Figure

1.5; it does not derive any relationship among the clusters. The spatial information in the figure is derived from an ordination of these data, which will be discussed next.

Detrended correspondence analysis (DCA) ordination of the meadow data is also shown in Figure 1.5. Ordination places similar entities in close proximity in a low-dimensional space. *Bromus–Arrhenatherum* samples fall together to the left, with sample 2 being somewhat peculiar; *Geum–Arrhenatherum* samples occupy a middle position, with sample 16 somewhat peculiar; and *Cirsium–Arrhenatherum* samples ordinate to the right, with sample 25 and the outlier (sample 19) furthest to the right. Simultaneously, DCA ordinates the species; ordination positions are shown for the differential species. (The nondifferential species mainly occupy positions near the center of the ordination field because most are not indicative of any particular habitat; they are not shown in this figure in order to avoid cluttering.) *Bromus erectus* falls to the left on the first ordination axis and dominates the corresponding samples (as may be seen in Table 1.4); likewise, the right-hand side features *Deschampsia caespitosa, Cirsium oleraceum, Geum rivale,* and *Holcus lanatus.* This difference in dominant species indicates a dry-to-wet moisture gradient (Hill & Gauch 1980), as indicated in Figure 1.5. At the extreme wet end of the gradient, the outlier from a periodically flooded site contains the aquatic grasses *Glyceria fluitans* and *Phalaris arundinacea.* The second DCA axis spreads the samples less than the first axis. The second axis may show noise mainly; no environmental interpretation is offered here.

At this introductory point, only two general comments are offered on the utility of multivariate analysis of community data. (1) Any of several multivariate methods is useful to reveal the basic, general structure of community data. This structure may be quite obscure in the raw, unorganized data set. The advantage of meaningful organization of the data increases with data set size and complexity. (2) A combination of ordination and classification methods applied to the same data set is especially useful. Figure 1.5 combines an ordination and a nonhierarchical classification into a single figure. Similarly, the DCA ordination positions (Figure 1.5) are helpful for understanding the divisions imposed by the TWINSPAN hierarchical classification (Figure 1.4). The spatial, graphical output of ordination, the cluster assignments of classification (hierarchical or nonhierarchical), and arranged data matrices are complementary for elucidating and communicating patterns in community data.

Evaluation of multivariate methods

Multivariate methods are so numerous that comparisons and evaluations are needed. Evaluations are based on several criteria and performance is tested with a variety of data sets. Before going into detail, however, two basic points should be made. (1) Moore, Fitzsimons, Lambe, & White (1970:1) state the fundamental criterion for evaluating multivariate methods with great precision: "In a phytosociological context, the efficiency of a method may be judged by its power to return a maximum understanding of the structural complexity of vegetation and of its relationship to environmental factors for a minimum of time input." Likewise, Greig-Smith (1971:150) says, "The crucial test of a technique is essentially an empirical one. Does it prove practical in use? Are the results capable of readier, or more exact, or more certain interpretation than those obtained by other means?" (Also see Crovello 1970; Shimwell 1971:278; and Greig-Smith 1980.) Note that understanding communities is the goal and that multivariate methods are a tool. (2) Progress in understanding communities must be viewed in a context of successive refinement. The requirement is that, given the present degree of understanding of communities (or of some particular community), multivariate methods facilitate taking the next step.

Criteria for evaluating multivariate methods will be considered in detail in later chapters, but a general introduction is appropriate here. Multivariate methods serve as a tool in the ongoing effort to understand communities better, and in this role several performance criteria are important (Williams & Lance 1968; Sneath 1969; Moore, Fitzsimons, Lambe, & White 1970; Greig-Smith 1971; Rand 1971; Sneath & Sokal 1973:63; Szőcs 1973; Dale 1975; Moral & Watson 1978; Orlóci 1978*a*:181, 239; Whittaker & Gauch 1978; Maarel 1979*a*; Moral 1980; Wilson 1981). Multivariate methods should be:

(1) accurate in representing data structure (involving appropriate treatment of noise, redundancy, relationships, and outliers);

(2) effective in relating community gradients to environmental, historical, and other data and in hypothesis generation about such relationships;

(3) robust, giving consistently useful results despite variation in the properties of diverse data sets and despite noisy data;

(4) objective, unless considerable subjectivity is allowed or required for some reason;

(5) lucid, both for the investigator and for readers of the published work;

(6) effectively integrated within an overall research program; and

(7) economical in terms of required data collection, demand upon the investigator's time to produce a given analysis, demand upon the investigator's time to first learn how to apply and interpret the multivariate method, and required computer time and memory.

Multivariate methods are tested using a variety of field data and simulated data (Gauch, Whittaker, & Wentworth 1977; Gauch & Whittaker 1981). Simulated data offer the advantages of both precisely known, expected results and ease and economy in varying data set properties. Field data offer realism, although the expected results defined by direct gradient analysis are necessarily of limited precision. As the advantages of simulated and field data are different and complementary, it is best to use both for evaluating multivariate methods. Furthermore, several data set properties (including number and length of gradients, noise level, numbers of samples and species, degree of natural clustering, and presence of outliers or disjunction) are known to affect the performance of multivariate methods, so it is necessary to use a number of data sets (on the order of one to a few dozen) in order to obtain a balanced evaluation.

The human factor

Community ecology is a product of interaction between communities and ecologists through observations and analysis; like human knowledge in general, it is a joint product of the observer and the observed (Whittaker 1956, 1962; also see Williams 1976:125, 128 and Wimsatt 1980). Methodologies, accomplishments, and limitations of community ecology are consequent as much as anything on the human factor (Moore, Fitzsimons, Lambe, & White 1970; Greig-Smith 1971). Any characterization of community ecology without explicit recognition of the human element is necessarily incomplete and superficial.

Communities are difficult objects for us to study. We face two challenges. (1) The number and individuality of communities and community components are staggering in comparison to the number of individual items we can consider. A single community is composed of a vast profusion of living organisms, with each species different from others, each individual of a species unique, and all individuals and the environment varying in time (Poore 1956; Pielou 1977:1). The community is too rich to be observed or described entirely, and in its full detail, each community is unique (Greig-Smith 1971). (2) Communities are integrated, whereas our minds must approach communities by a succession of individual thoughts (Simon 1962; Daubenmire 1968:270; Macfadyen 1975). Much insight and creativity are required to decompose or abstract the community into effectively considered subunits (Simon 1962; Solomon 1979). As Cormack (1979:191) observes, "Nature does not usually conduct her experimentation with a view to revealing her own secrets most efficiently." Essentially everything in the physical world is conceivably relevant to the life of community organisms. As an extreme but realistic example, the origin of a high-energy gamma ray in a distant galaxy and its trajectory through space as affected by the gravitation of billions of stars could affect whether a zygote survives, mutates, or is destroyed by chromosomal damage, and the resulting fate of the individual could affect a community drastically. Necessarily, the finitude of human efforts requires that our studies be both partial and piecemeal. Communities have countless complex interactions and causal connections, some of which defy observation and analysis (Elsasser 1969). Cormack (1979:178) notes the usual dilemma: "Biological realism varies inversely with analytical tractability."

The human response to complexity has been characterized by Simon (1962, 1981). Regarding communities, the response has several characteristic features.

Objectivity and repeatability are important methodological considerations because the overall task involves a large number of persons (Sneath & Sokal 1973:11; Daubenmire 1968:270–1). Objectivity of a study considered as a whole (from observation to final reporting of results), not objectivity of individual steps in the procedure, is the crucial point because subjectivity can be banished from one step only to reappear elsewhere in a worse, more subtle, or more complex form.

Efficiency is important because community ecology is at best time-consuming (Moore, Fitzsimons, Lambe, & White 1970). As with objectivity, the essential matter is overall efficiency, rather than efficiency of a single step.

Sampling of communities is required because exhaustive enumeration is impossible (Daubenmire 1968:79; Mueller-Dombois & Ellenberg 1974:31; Pielou 1974:98; Orlóci 1978*a*:1). Choices must be made as to the number, size, and placement of samples, as well as the number of observations to record at each. The resulting data have properties arising from the communities studied and additional properties arising from the process and statistical nature of sampling.

Display options must be human engineered to enable us to effectively visualize and communicate large volumes of information (Simon 1962; Crovello 1970; Gauch 1977:34; Everitt 1978; Orlóci 1978*a*:170; Noy-Meir 1979). Arranged data matrices (Moore, Fitzsimons, Lambe, & White 1970) and ordination graphs (Noy-Meir 1979) are effective. Although ordinarily considered in the context of presentation of results, display options are also important for what may properly be considered the observational stage of research in that important features of the data cannot be perceived apart from these tools (Moore, Fitzsimons, Lambe, & White 1970; Solomon 1979).

Computers are a helpful and cost-effective tool for reducing certain human limitations. Rapid access to large data bases and relatively objective and inexpensive data reduction have had an enormous impact on procedures in community ecology. The human-with-a-computer is in a different position from the unaided human (Simon 1981). Computers cannot supply a creative element and they cannot be supplied all the information perceived by a trained ecologist, but they can reduce the time required for routine analyses and, hence, enable ecologists to have more time for creative thinking (Lambert & Williams 1962; Williams 1976:128; Greig-Smith 1980; Pignatti 1980). By saving labor and time, computers can speed up the ecologist's process of successive refinement. Williams (1962), in an interesting paper, "Computers as Botanists," found that, in two out of three community studies, the patterns emphasized by computer analysis were preferable to those from ecologists' initial assessments, and he attributed the computer's success to its freedom from preconceptions. Obviously this is a small sample size, but it is entirely fair to say that community ecologists have found multivariate analyses useful routinely (Maarel, Orlóci, & Pignatti 1976; Orlóci 1978*a*). Dahl (in Moore & O'Sullivan 1970:30) rightly concludes that "computers are good servants but bad masters."

Nonexperimental methods characterize much community ecology because the time scale of the life of many species and, consequently, of changes in community populations is long compared to the human life span. Likewise, many communities cover areas of hectares to thousands of hectares. Consequently, the temporal and spatial scales of many communities make experimental manipulation impractical or impossible. For communities with small, short-lived species, some interesting experimentation has been done, but such experimentation is rare and the results are complicated by the fact that any manipulation (burning, exclosure by a fence, addition of mineral nutrients or water, and so on) has a multitude of effects and side effects (Poore 1962; Daubenmire 1968:269–70). Often ecologists must content themselves with studying those combinations of factors provided by nature. This concession has the consequence that statistical tests available from appropriately designed experiments in many branches of science are not available, and desired information may have to be extracted laboriously from data that are informative, although not in a direct or simple manner. On the other hand, it can be argued, by reference to the design of factorial experiments, that simultaneous study of several factors can be efficient and comprehensive (Snedecor & Cochran 1967:339–80; also see Poore 1956, 1962; Greig-Smith 1971). Furthermore, those conditions that actually exist in nature are frequently those of interest anyway.

Considerable subjectivity remains unavoidable in community ecology (Crovello 1970). Characteristically, numerous procedures exist for the same general ends, none of which are perfect, and for which there are conflicting claims of relative merit (Shimwell 1971:278). On balance, ordinarily, if several community ecologists studied the same area with similar resources and time, their results would show the same major features despite differences in detail (Moore, Fitzsimons, Lambe, & White 1970; Gauch 1977:18).

Several schools of community ecology exist because of limited exposure to views of distant workers (especially those writing in other languages), somewhat different properties of plant and animal communities in different parts of the world, and the force of habit and reluctance to change old procedures and theories (Shimwell 1971:42–62; Whittaker 1962, 1978*a*). This affects what is observed and sampled in the first place, as well as data analysis procedures and the conceptual view of communities from which questions and hypotheses are raised and into which results are framed. Exchanging and comparing methods and results are important in identifying the most effective methods and approaches using groups of complementary methods.

Community ecology must remain partially an art (Crovello 1970; Moore, Fitzsimons, Lambe, & White 1970; Greig-Smith 1971; Solomon 1979). Certain aspects, such as the application of a given sampling procedure or the computation of a given ordination or classification, are mechanical, but choice of the sampling method or analysis in the first place is an art. Successive refinement, by its very nature, is complex and defies complete description. From fieldwork and scientific training in general, the ecologist gathers much valuable information that cannot be processed by a mechanical procedure nor made available to a computer. Because ecology is partly an art, the quality of results is influenced by the investigator's experience (Simon 1962; Williams 1976:125). Creativity must involve trial and error (Simon 1962; Solomon 1979), and exceptionally creative workers must, with fortitude, accept the risks of many errors and frequent frustration.

Human resources in the study of communities may be considered on three levels: the scientific community as a whole, an individual expert, and an individual novice. (Here an "individual" is construed loosely to mean either an individual person, literally, or a small team of several persons, perhaps an ecologist, a few field technicians, and a computer consultant.) An expert is a person spending much of his or her time for at least several years on community ecology and having good exposure to developments in community ecology worldwide. A novice is a person having limited experience with community ecology, frequently having primary expertise in related fields, and for whom community ecology methods may be merely tools serving within more general research purposes. Fruitful communication is needed among these three levels. Community ecology experts must scan the results of the scientific community for relevant material (Simon 1962), develop and test new methods, and communicate with one another in a detailed, technical manner. Experts must also appreciate the working environment of the novice or allied professional who has a minimum of time and experience but who needs reasonably good access to community ecology (Greig-Smith 1971; Shimwell 1971:282). This does not mean that the novice is devoid of a conceptual, theoretical framework. Indeed, the purposes and procedures of community ecology are more easily understood and remembered when their bases are understood. What is required is that the novice be presented with the best methods and the most common recommendations, rather than a compendium of everything that experts have ever tried or considered (Williams 1976:28). Thus equipped, the novice should have reasonable success in most cases but occasionally may need to go to the literature or to an expert for further information.

Redundancy and underlying relationships are requisite features of complex systems in order for a system to be available to human observation and understanding (Simon 1962). Complex systems without these features largely escape our notice, being beyond our capacities of memory or computation. It is no mere coincidence that redundancy and relationships are aspects of both community data and human comprehensibility–community ecology is a joint product of the observer and the observed.

Successive refinement

Successive refinement has been emphasized as the working mode of community ecologists (Poore 1956, 1962; Moore, Fitzsimons, Lambe, & White 1970; Hill 1973; Macfadyen 1975; Noy-Meir 1979; Solomon 1979; Aris & Penn 1980; Southwood 1980). For example, in a list of 10 advantages of the Braun-Blanquet approach, Poore (1956:46) cites the opportunity for successive refinement as "the most important characteristic of this method." Successive refinement bears upon all stages of a community study; consequently, this topic appears frequently throughout this text. Successive refinement and purposes are two topics within community ecology that cannot be extracted neatly from other topics and treated thoroughly in a single place (because, to use the terminology of Simon (1962), their complexity is not nearly decomposable). At this point, some of the more general features of successive refinement will be treated.

Successive refinement essentially involves repeated cycles of knowledge, questions, and observations (compare Solomon 1979). One begins with a point of knowledge (arising from observation, conversation, the literature, one's conceptual framework, conjecture, or whatever). According to an old adage, every answer raises ten questions. This adage is quite true, although the "ten" may be replaced with a much larger number. Little creativity is needed to raise these ten new questions. Practicality usually limits further investigation to, say, one of these ten questions. Much creativity is then required to make reasonably frequent selection of the most interesting and telling question. Effective selection requires insightful comparison with one's research purposes, purposes stated at the outset of a research program and purposes developed in the light of recent results. The question one selects then necessitates the choice of further observations, and subsequent analysis leads to new knowledge. The new knowledge is then the beginning point for the next cycle of successive refinement. Much insight is needed to decide when adequate results are achieved and, thus, when it is time to move on to the next cycle; there are penalties for cycling too rashly or too slowly (Poore 1962). Creativity is enhanced by experience and by interaction with other scientists (both in community ecology and in other fields, these two groups enhancing distinctly different components of creativity). A complex interplay among observation, analysis, and synthesis requires that progress involve a large number of small steps and implies that research steps can be specified at the outset only in part.

For example, one may notice that around Ithaca, New York, eastern red cedar (*Juniperus virginiana* L.) is restricted to sites along a prominent lake (Cayuga) with limestone outcrops. This point of knowledge easily sparks a multitude of questions: Is the increase in cedar due to calcium, another element, or several elements or to temperature effects modified by the lake? Are other species showing the same distribution pattern? Is cedar a reliable indicator of limestone? Is association with limestone a consistent pattern for eastern red cedar throughout its range? Were one's research purpose a regional survey of old fields with good agricultural potential, the question of the reliability of cedar as an indicator of limestone might be worthwhile. Selection of this question then leads to choices of pertinent further observations (of a rather obvious nature in this case). Once this question is answered, further questions could be asked. Depending on whether cedar's indicator value for limestone is strong or weak, however, relevant subsequent questions differ. One cannot plan too many steps ahead because the choices eventually bifurcate into an infinitude of possibilities, and many choices cannot be anticipated at the outset.

Successive refinement involves a large number of small steps, and, consequently, frequent recourse to overall research purposes is imperative. By analogy to walking, if one takes N steps of unit length in random directions (without purpose), the expected distance from the origin is only the square root of N and the expected distance in any particular direction is zero. Research purposes are required to direct the steps in a consistent, effective direction, otherwise the same amount of work and expense may yield little knowledge. A researcher spending less than 10% of his or her time considering and reconsidering research purposes is likely to get bogged down. Clearly, purposes originate at the outset of a research program, but it is valuable also to incorporate additional or refined purposes that originate during the course of a research project. Alternation between detailed particulars and overall purposes is essential and must be frequent (Crovello 1970). This alternation should be encouraged by the basic design and routine of a research program. It takes time to design and implement a research project in a manner that encourages successive refinement and emphasizes purposes. No alternative is more efficient in the long run.

Purposes of multivariate analysis

Like all facets of community ecology, purpose is complex and subject to successive refinement (Daubenmire 1968:32–5, 92–5; Crovello 1970; Mueller-Dombois & Ellenberg 1974:10–21; Dale 1975). Some basic purposes of multivariate analysis should be evident already and further treatment of purpose will follow. At this point, an introductory statement of purpose is fitting and three general purposes may be noted.

(1) Multivariate analyses of community data are used to summarize redundancy, reduce noise, elucidate relationships, and identify outliers. Analysis of community data apart from other data may be termed *internal* analysis [or intrinsic analysis (William & Lance 1968) or taxometric analysis (Whittaker & Gauch 1978)]. Multivariate analyses provide relatively automatic, objective means for shifting the level of abstraction from raw data conveniently collected in terms of species abundances in samples to community-level properties such as community types and low-dimensional representations of community gradients (Shimwell 1971:1; Whittaker 1978*c*:3).

(2) Multivariate analyses serve to relate communities to other kinds of data, including environmental and historical data. This may be termed *external* analysis [or extrinsic analysis (William & Lance 1968) or ecological analysis (Whittaker & Gauch 1978); also see Williams 1962; Groenewoud 1976; and Lepart & Debussche 1980].

(3) Results from multivariate analyses serve to improve our understanding of communities, making possible better models of communities. Improved community models then provide the basis for designing more appropriate and effective multivariate methods. Improved methods then allow sharper analyses of communities, with these reciprocating advances leading to successive refinement of both community models and multivariate analyses.

Sometimes an ecologist's goal is merely to characterize communities (internal analysis), but more frequently the goal is both to characterize communities and to interpret these results environmentally (internal and external analysis). It must be emphasized that these two tasks, internal and external analysis, are quite different, as is the power of multivariate analysis to accomplish these tasks. Definition and arrangement of community units (internal analysis) routinely progress well by multivariate analysis of the samples-by-species data matrix because this matrix contains sufficient information for this purpose (Lambert & Williams 1962). In contrast, environmental interpretation (external analysis) involves searching for correspondences between community variation and environmental variation (Williams 1962). The samples-by-species data matrix being analyzed, however, contains no environmental data. Consequently, environmental interpretation is entirely outside the domain of multivariate analysis of community data itself. Environmental interpretation will be discussed further in later chapters, but here the point is emphasized that additional data and additional analyses are required. Multivariate analysis is usually a partial or weak tool for environmental interpretation of community variation; rather, interpretation "requires the services of a trained ecologist" (Lambert & Williams 1962:799). The task of external analysis is aided greatly, however, by the community summarization afforded by internal analysis; consequently, multivariate analysis contributes indirectly to external analysis. Community ecology employs an extensive methodology in which multivariate analysis is not the whole but rather a vital part.

The following list presents representative, specific purposes of community studies (Greig-Smith 1971; Pielou 1977:1–5; Orlóci 1978*a*:1–2, 102–3, 189–91; Whittaker 1978*c*:3–6), reflecting the tendency to progress from description to understanding to prediction and management (Cain & Castro 1959:104; Crovello 1970):

- Description
  - Description of a given community
  - Delimiting and naming of communities
  - Mapping of communities within a region
  - Identification of recurring species groups
  - Assignment of new community samples to previously defined community types
- Understanding
  - Structure of communities
  - Regulation and maintenance of communities
  - Distribution of species and communities along environmental gradients
  - Competitive interactions of species
  - Species niches and habitats
- Prediction and management

Prediction of community from environment or environment from community
Prediction of course of succession or response to disturbance
Land use recommendations
Management of grazing, forest, and recreational areas
Relating community data to other data bases (fire, harvesting, weather)
Data reduction for inventory

These purposes are important because plant and animal communities are interesting, appealing, and crucial to mankind. Communities are important, both in their natural state and as modified by past or potential human disturbances.

A mathematical perspective on the purpose of multivariate analysis is complementary to the preceding ecological perspective and helps to clarify the essential functions and limitations of multivariate analysis. Multivariate analysis is basically a means for summarizing multivariate data. The input is a data matrix, typically containing many thousands of numbers, and the output is for each sample (or species or both) a classification assignment or, alternatively, one to a few ordination scores specifying location in a low-dimensional space. The degree of summarization (generalization) involved may be estimated by using information theory (Shannon & Weaver 1949; Gauch 1977:38). For example, assume that a community data matrix has 100 species and 300 samples, with species abundances measured to one significant digit resulting in ten states of the measurement. The information content of each abundance measurement is then $\log_2 10 \approx 3.3$ bits, but assuming that zeros are common and that the frequencies of the positive states are unequal, this estimate may be reduced somewhat to perhaps 2.5 bits. As there are 30 000 matrix entries, the information content of the data matrix is then 75 000 bits. Now, suppose that multivariate analysis is used to classify the 300 samples, assigning each sample to 1 of 32 classes. Assuming that the classes are of roughly equal sizes, the information content of each assignment is $\log_2 32 = 5$ bits, and the information content of the entire classification is then 1500 bits. Likewise, suppose that multivariate analysis is used to ordinate both samples and species in a two-dimensional ordination and that the ordination space has about 40 significantly different areas populated roughly equally (the 40 areas arising perhaps from eight distinctive zones along the first axis and five along the second axis). Each ordination location involves

$\log_2 40 \approx 5.3$ bits of information, and the total information content of the ordination is then 2120 bits. The ratio between the input information content in the data and the output information content in these multivariate analyses (classification and ordination) is then about 2 or 3%. Were the objective of multivariate analysis to assign these 300 sample sites to just two classes representing decisions to burn or not to burn, the output information content of 300 bits in relation to the input 75 000 bits would yield an even smaller ratio of only 0.4%.

These small percentages should not be taken to imply that better multivariate methods could extract tens to hundreds of times as much information from the data. Indeed, the whole point of multivariate analysis is to obtain a briefer description of the data having less information, different in kind, than the raw data (Orlóci 1974*a*). The critical matters are the kind of information retained and the directness of its relationship to environmental gradients and management concerns, rather than the quantity of information (other than the general constraint that it be decidedly smaller than the input information but not vanishingly small). Ideally, multivariate analysis selectively retains information on relationships in the data while summarizing redundancy and selectively eliminating noise.

Raw data in community ecology are usually too bulky and too complex for direct assimilation, given the capacities of the human mind. Results from multivariate analysis, however, contain practical quantities of information for contemplation and for communication of results. The basic purpose of multivariate analysis in community ecology is thus to form a bridge between the properties of communities and the properties of human mental processes.

An overview

Community ecology develops in a complex interplay of (1) communities and community samples, (2) ecologists and their conceptual frameworks, and (3) multivariate and other methods of analysis. Because of this complexity, community ecology has a diversity of schools and a considerable subjective element, in addition to which the working environment is usually multifactorial and nonexperimental. Progress must be sought in a mode of successive refinement, because crucial questions may not be evident until some work has already been done. On the other hand, some routine purposes are ordinarily served by standard methods having reasonable data base requirements.

Community ecology is also relevant to ecologists whose primary interest is biogeography, autecology, or applied management (Poore 1962; Bakker 1979; Dobben 1979). Poore (1962:35) emphasizes that all disciplines bearing on the biota are "closely associated." Even if an ecologist's primary interest is the distribution of a rare species or the management of a particular area, much may be gained by prefacing such studies with a general, community-level study. "The very detailed examination of one area or situation, though frequently very valuable, may easily prove to be time partially wasted, because the situation investigated is unrepresentative; or it may give negative results because, through inexperience, the wrong features of the environment have been observed" (Poore 1962:39). Poore argues for working from the general to the specific; for beginning ecological studies on the community level (also see Cain & Castro 1959:104). This recommendation accords with fundamental considerations on the architecture and description of complexity (Simon 1962).

An overview of scientific methodology is needed in order to appreciate the distinctive strengths and weaknesses of the multivariate analysis research approach and to appreciate the complementary roles of other approaches. (1) Deductive derivation of results from first principles is a major research approach in the mathematical and physical sciences. In ecology, construction of models and deduction of their implications serve to summarize and simplify theory and to show the consistent implications of a given stance. The theoretical approach has limitations, however, in that, for many ecological disciplines, there is a paucity of first principles, an indefinite goal or question to pursue, and a general impediment from the sheer individuality of the entities under study. As emphasized already, the result of these difficulties is a research approach typified by successive refinement (Poore 1956; McIntosh 1980; Simberloff 1980; Strong 1980). (2) Experimental manipulation offers the most direct test of hypotheses and is invaluable when applicable as noted earlier; however, the time and space scales in community ecology frequently put experimentation outside the bounds of human resources (Mertz & McCauley 1980). (3) Univariate statistics, analysis of variance, and plots of two variables at a time are valuable tools for examining a few variables but are not effective for multivariate data (Everitt 1978:5). (4) Collection and presentation of field data are essential in ecology. Important features of the data may be perceptible, however, only after an appropriate analysis. Likewise, the significance of a particular measurement may be missed when separated from the theory showing its broader implications. (5) Multivariate analysis summarizes the main features in multivariate data and helps with generating hypotheses. (6) Several conclusions may be drawn concerning the above points: (a) Each scientific research approach has distinctive strengths. (b) The wide range of ecological endeavor implies that all these approaches have important roles. (c) To say that one research approach is better, in general, than another is of equivalent mentality to saying that a pH meter is more powerful than a microscope. Specific ecological research projects may vary, nevertheless, in the frequency with which various research aproaches are used. For example, community ecology makes heavy use of multivariate analysis, whereas physiological problems emphasize experimental methods. (d) These several research approaches are not necessarily exclusive. For example, multivariate methods may be used to analyze experimental results and these findings may be assimilated into a theoretical framework. This text emphasizes multivariate analysis but encompasses something of the breadth of community ecology by also treating field sampling methods and theoretical models of community structure.

Direct gradient analysis, ordination, and classification are complementary methods for multivariate analysis of community data. Direct gradient analysis is effective with known, evident environmental factors and provides test data for evaluating more indirect methods. Classification may be more suitable for extremely diverse data sets, and ordination for less diverse data sets (Greig-Smith 1971). Ordination may be better for elucidating environmental correlations, but classification may be better for applied managment (Greig-Smith 1971). A generally applicable strategy involves nonhierarchical classification, first, for initial summary of redundancy and identification of outliers, followed by ordination and hierarchical classification to reveal relationships; both steps reduce noise.

The reality that community ecologists study has been characterized by Whittaker (1952:31) as "loosely ordered, complexly patterned, multiply determined." An ecologist need not apologize for "the difficulty of his field, the necessary limitations of his data, statistical involvements of his work, partial indeterminacy of his results, and the slowness and laboriousness of progress." Computer analyses using multivariate techniques are important tools for extricating useful results from complex community data.

2

Sampling methods

Community sampling is the initial, observational phase of community studies. The inherent strength or weakness of a study and the range of potential data analyses that will be subsequently appropriate are determined and fixed to a great degree at this first step, data collection (Cain & Castro 1959:2; Poore 1962; Greig-Smith 1964:20; Mueller-Dombois & Ellenberg 1974:32). An investigator planning a sampling procedure faces three questions, which will be considered in detail in the following three sections. (1) What general considerations and tradeoffs affect the practicality and effectiveness of sampling procedures? (2) What community sampling procedure is best for a particular study? (3) What corroborative environmental and historical data should be gathered at each sample site?

The first question involves general principles, which can be treated reasonably well in this chapter. They will be discussed primarily in the context of terrestrial vascular plants, but other kinds of communities will be discussed later in this chapter. The second and third questions involve so many permutations of research purpose, level of accuracy, scope of study, kind or kinds of communities, intended subsequent analysis, and so on, that thorough treatment here is impractical. Thus, for these questions, only a sketchy response will be offered, with the burden carried by references to the literature. This is fair enough because an investigator planning a project on, say, phytoplankton communities in small lakes should begin with a survey of relevant literature, which will reveal customary sampling procedures, which may be adopted or else adapted in view of the investigator's particular needs.

Unfortunately, the term sample is used inconsistently in the literature. Especially with authors having a statistical orientation, sample refers to the entire data collection, and the individual members are termed sampling units. On the other hand, ecologists often refer to the individual members as samples and to the entire data collection as a sample set. Here the latter convention is adopted.

As will become apparent, designing a sampling procedure involves numerous subjective decisions (Crovello 1970). The quantity and quality of data obtained may be less than ideal, even with a good sampling procedure. Nevertheless, very imperfect data suffice for many quite interesting research purposes (Whittaker 1952; Green 1977).

General considerations

Requirements and tradeoffs

Regardless of the particular plant or animal community of interest, many considerations in designing a sampling procedure are much the same. These general considerations include desirable sampling requirements and their implied tradeoffs, provision for successive refinement, sampling accuracy, incorporation of research purposes, and the size, shape, number, and placement of samples.

The desirable qualities of community samples are easily listed, although this is worth doing explicitly. Unfortunately, many of these qualities have strong, unavoidable, negative interactions with other qualities, calling for difficult tradeoff decisions. Several sampling requirements follow:

(1) Appropriate. Among the many kinds of information that may be gathered, some are more appropriate than others to the character of the community (Whittaker 1978*b*:9), the investigator's research purposes (Cain & Castro 1959:2–3), and the requirements imposed by plans for subsequent data analysis [particularly if statistical analysis is anticipated (Brown 1954:8; Snedecor & Cochran 1967; Kershaw 1973:21)]. Sampling must be appropriate in intensity and breadth to support research conclusions, as is often emphasized, but likewise, sampling should not occupy large amounts of time and resources beyond those required for the research objectives. Poore (1962:51 and 65) notes that "the proper province of plant sociological studies should be to describe vegetation and to discover and define problems for solution by more exact methods" and that overly quantitative work cannot provide "the wide coverage necessary for the initial assessment of critical problems," and may even stifle progress.

Data may be appropriate even if superficially tangential to the main research purposes. Multivariate analysis of a samples-by-species data matrix produces (1) a summary of sample and species relationships and (2) community-level descriptors, such as community types from classification or arrangement of the samples and species in a low-dimensional space from ordination. The second point, the emergence of community-level descriptors, deserves careful attention. The quantity of information in the results of a multivariate analysis is smaller than that in the original data matrix by an order or orders of magnitude, as was noted in the introduction. This shift to a briefer, more economical expression is effectively a shift from the level of samples and species to the level of communities (Greig-Smith 1971; Orlóci 1974*a*; Gauch 1977:38). An important consequence of this shift in level of abstraction is that if one's primary interest is on the level of communities, collection of data on the level of species in samples may be appropriate (provided that multivariate analysis is employed subsequently). Likewise, if one's primary interest concerns environmental factors, analysis of the vegetation may offer the most effective approach because the vegetation itself scales and integrates the environment from the viewpoint of the plants (Goodall 1954*a*). Effects on animal communities may also be interesting for the same reason.

(2) Homogeneous. Community samples should be homogeneous in structure and composition if the research purpose is to represent community types by samples (Poore 1955, 1962; Daubenmire 1968:79–80; Mueller-Dombois & Ellenberg 1974:46; Green 1979:35–8) or to relate vegetation to environment (Poore 1962). Samples should be of uniform environment also (without apparent differences of soil type, moisture status, and disturbance) in order to allow comparison of vegetation and environment and to avoid differences in the vegetation that environmental nonuniformity ordinarily imposes (Daubenmire 1968:80). An additional advantage of placing samples within uniform areas, away from marked discontinuities, is that the exact location of a sample then makes little difference (Daubenmire 1968:80). On the other hand, if the research's purpose is to know only the volume of timber in a forest or the amount of forage in a pasture, it is sensible to ignore communities and to sample a bounded area as a unit without requiring sample homogeneity. Such data, however, serve only one purpose (Daubenmire 1968:80).

Attempts at taking homogeneous samples encounter three problems. (1) Foremost is the problem of scale. Plants and animals are rarely distributed at random. Rather they are patterned on several size scales, and likewise, the environment is patterned on several scales (Poore 1962; Greig-Smith 1964:54–93; Goodall 1970; Pielou 1974:131–201, 1977:111–266). Consequently, any sample size will be appropriate for some species and environmental factors, too small to be representative for others, and too large to be homogeneous for yet others. The result is that two species may show negative, positive, or zero correlation in their distributions depending on the sample size (Greig-Smith 1964:106). The community units of community ecology are not naturally delimited (Goodall 1954*b*; Poore 1962). Ecologists must "draw artificial lines separating portions of vegetation from contiguous more or less similar vegetation" (Poore 1962:47). If the purpose of a study is to describe all the vegetation within a given region, this problem is not serious. If a study concentrates on a certain community type, however, such as a beech–maple forest, criteria for including or rejecting sites are difficult to formulate because of the inherent continuity in community variation. (3) Sites vary in their degree of homogeneity and in the type and severity of disturbances, so ecologists often choose the more uniform, or undisturbed, sites after reconnaissance of the study region (Poore 1962). Samples judged to be homogeneous will often show an internal pattern in the form of scattered patches of individual species (clustering or contagion). A homogeneous sample should not, however, show trends of changing species composition from one side of the sample to the other. Criteria for selecting suitable sites are often complex and rather subjective, and proper application of such criteria may presuppose reconnaissance of the study region and general experience with community ecology methods.

Absolute homogeneity in community samples is unattainable. One might think the best alternative would then be to devise statistical measures for assessing relative homogeneity (such as the three possibilities given by Goodall 1954*b*). Statistically significant tests of homogeneity are made rarely, however, for three main reasons: "that the use of statistical techniques on every stand would excessively slow down the accumulation of results; that statistical techniques can hardly be applied in many of the situations where surveys should be made; and that the stands of vegetation are often far too small" (Poore 1955:235). Statistics should not be used to produce an air of precision that is only tenuously related to real properties of communities.

The problem of obtaining reasonable sample homogeneity may be resolved along two lines. (1) Although rigorous procedures for attaining sample homogeneity are ordinarily impractical or elusive, experience shows that the customary subjective procedures for selecting sample sites and for evaluating a "general visual impression of homogeneity" (Goodall 1954*b*:173; Poore 1962:49) are adequate: "its justification lies in its proven usefulness and the fact that the judgements of experienced ecologists tend to coincide" (Poore 1962:47; also see Mueller-Dombois & Ellenberg 1974:46–7). Community ecology is necessarily both an art and a science. One implication of the art is the value of the cumulative experience derived from years of research. Later in this chapter, sample sizes and various procedures that have withstood the test of time will be described. (2) The critical matter is homogeneity within sample sites relative to overall variation in the data set (Poore 1962). When the data set includes diverse communities, overall community relationships can emerge quite clearly despite considerable imperfections in the individual samples because differences in between-community variation exceed within-community variation (Whittaker 1952; Green 1979:27–8). If, however, the overall range in community variations is small in a given study, greater sample homogeneity is required in order to keep interesting community differences dominant. Although absolute sample homogeneity is impossible, the required relative homogeneity is ordinarily attained by standard methods without unreasonable effort.

(3) Objective and standardized. The possibilities for sampling procedures are uncountable. Consequently, selection of a sampling procedure has a considerable subjective element (Crovello 1970). Once selected, however, the sampling procedure should be, to as great a degree as is required and practical, applicable in an objective, standardized way, while also being unambiguous and operational (Pielou 1974:98). In this regard, it is desirable to record all species abundances in the same units, as the resulting commensurability simplifies subsequent analysis (Orlóci 1978*a*:8). Sometimes different strata are recorded in different units, however, such as percentage covers for herbs and basal areas for trees.

Standardization of the sampling method is required for data to be comparable, which, after all, is basic to science (Cain & Castro 1959:1, 104–5; Orlóci 1978*a*:2–3). Standardization is especially important for projects involving teams of field-workers (Greig-Smith 1964:2) or involving comparison of results with other studies (Cain & Castro 1959:2; Daubenmire 1968:270).

A good degree of consistency of results is obtained with experienced field-workers (Greig-Smith 1964:1–2). In cases where taxonomic and ecological experience are limited, it is necessary to design sampling methods that require minimal expertise and yet are adequate for the purposes at hand.

(4) Efficient. "Since collection of quantitative data in the field is at best a time-consuming task, it is imperative that the samples taken should be such as give the maximum amount of information in return for the effort and time involved" (Greig-Smith 1964:20; also see Poore 1962, Crovello 1970, and Moore, Fitzsimons, Lambe, & White 1970).

The desirable properties of sampling methods just listed lead inexorably to conflicts requiring difficult tradeoff decisions.

The desire for standardized methods is in conflict with the special requirements of a great diversity of communities and research purposes. For the most part, the conflict must be resolved by avoiding either extreme of excessively few or numerous methods. Adoption of a modest number of standardized sampling methods allows a given study to be carried out by a sampling method that is reasonably appropriate to it and that has also been applied in a number of related studies of interest. A small sacrifice in appropriateness may be amply repaid by gaining comparability with other studies and compatibility with other data bases. One should avoid "enslavement to his or anybody's methods" (Cain & Castro 1959:105), but on the other hand, one should not sacrifice comparability with other studies unless one's preferred method clearly gives substantially different information than standard methods and one's research purposes clearly call for a unique approach.

Another conflict involves the number of samples collected and their quality. The greater the number of items recorded at each site and the greater the precision of each measurement, the greater will be the cost and time involved per sample and, consequently, the fewer the samples that will be obtained. This tradeoff will be discussed later in this chapter.

In general, the requirement that sampling be rapid runs contrary to most of the desired properties of community samples. In the final analysis, "the exigencies of the situation must prevail over the ideal" (Cain & Castro 1959:105). Selection of the most useful field methods in view of the exigencies is an art. In general terms, the conflict must be handled by skillfully designing projects of intermediate ambition – challenging enough to constitute interesting and worthwhile advances but realistic in their claims and implementation.

Successive refinement

Data collection occurs at the beginning of a study, so community sampling decisions must be made beforehand, bearing in mind that the sampling procedure has strong implications for all subsequent steps clear through to the conclusion of a study. Often, at the conclusion, one sees ways in which the sampling procedure could have been improved, and one would do things differently if the study were to be done again.

Not all community studies are equally challenging. Many are, in fact, routine, such as when a standard, reliable approach is simply applied to another study region. Other studies are more original. Desirable sampling procedures may then be far from obvious, and the discrepancy may be great between how one sees things at the beginning and at the end. As the difficulty increases, provision for successive refinement becomes increasingly important.

Successive refinement should be allowed for unless a study is of a routine nature. The primary requirement for successive refinement is a three-step approach: reconnaissance, pilot study, and main study (Cain & Castro 1959:104–7; Gauch 1977:8–9). First, reconnaissance helps to indicate the variety of communities present and the probable environmental factors of importance. Second, a pilot study can evaluate and sharpen these impressions prior to expenditure of the bulk of available time and resources. Also, it can reduce the discrepancy at the end between what one did and what one wishes had been done. The pilot study may well include some test samples from which one may judge the interaction of time required and kinds of desired data obtained for the kinds of communities studied. Finally, in the main study, most of the data are collected and analyzed.

Reconnaissance is preliminary examination of a territory (Brown 1954:192; Cain & Castro 1959:3–4, 104–5; Daubenmire 1968:267; Gauch 1977:8). "In reconnaissance the country is traversed rapidly by the most convenient means, and the eye searches for the most general and obvious features" (Cain & Castro 1959:105). Several questions arise. (1) What are the major community types? What are their relative areas and what size and shape areas do they cover? (2) What species are present and which are common? (3) What community sampling procedures might be appropriate? (4) What are the apparent correlations with environmental factors such as climate, soils, and topography? How could these factors be measured or estimated? (5) Are there important historical or disturbance factors? How can these factors be measured? Should disturbed areas be sampled, and if not, what unambiguous and operational criteria can be used to exclude disturbed sites? (6) How accessible are various regions? What other practical considerations are relevant, such as availability of maps, taxonomic authorities, background studies, and local experts? (7) What special requirements arise from the purposes of the study?

Reconnaissance, as preliminary examination of a territory, should also be taken to include exposure to relevant literature. Science is a cumulative effort and others' experiences may spare one from beginning successive refinement at an elementary or inefficient point. Occasionally, effective work has been done at the level of mere reconnaissance but, invariably, this involves persons of exceptional and long experience (Cain & Castro 1959:3–4).

A separate pilot study ideally follows reconnaissance, if time permits (Cain & Castro 1959:106–7; Gauch 1977:8–9; Green 1979:31–2). In the *pilot study,* a small amount of data is collected prior to the main sampling effort. It can test the sampling methods, and the resulting small sample set can be analyzed by the multivariate techniques to be used on the main study in the hope of eliminating certain problems that may not arise until the data are analyzed. Such preliminary analyses can sometimes supplement the reconnaissance by suggesting important relations of communities to environments, which the samples should adequately represent. The investigator can judge whether adequate (and not too time-consuming) data on species composition are being obtained and whether environmental measurements are appropriate. (The decision may be reached to omit some measures considered redundant or invariant or to intensify others, such as the decision to seek additional soils data.) A pilot study should represent the variety of communities in order to test the sampling method under various circumstances, but it should also include replicate samples in order to see how noisy the data are. Practical aspects should be considered also, such as improvements on the field data sheet, modification of any procedure that requires unreasonable time, and clarification of decisions or measurements that appear to be applied inconsistently. During the pilot study, field data sheets may be designed for efficient keypunching and computer processing. If there are not too many species (at least common ones), they may be listed on the data sheet in order to save time in the field. A standardized data sheet is especially important when there are several groups of field-workers and when someone else does the keypunching.

At the conclusion of the reconnaissance and pilot study, one should have general familiarity with the landscape, a suitable and efficient sampling method, reasonable taxonomic ability, and an idea of the kinds and approximate numbers of samples to be taken in the main study. If numerous changes in the sampling method are made, an additional pilot study may be advisable.

In the *main study,* the bulk of the data are collected and analyzed. Later in this chapter, standard sampling procedures will be discussed. If reconnaissance and the pilot study are done well, successive refinement should have progressed sufficiently to allow for an effective main study, which will not conclude with serious regrets about procedures and results (Green 1979:31–2).

Scope, accuracy, and purposes

The scope of a study refers to the area covered and the variety of communities encountered. An extensive study may cover a state or a country and use a large number of relatively rapid samples; an intensive study may look at a single community type within a limited area and use a smaller number of more exacting samples. Intensive studies may (but will not necessarily) seek greater completeness in species composition and greater accuracy in measuring environmental factors and species abundances. Accuracy in estimates of species abundances may be considered on three levels: one-digit estimates, exact quantitative measurements, and mere presence or absence.

Field data for terrestrial vegetation are often collected at the intermediate level of accuracy of one-digit estimates (Braun-Blanquet 1932:26–36; Becking 1957; Cain & Castro 1959:135–44, 179, 202–6, 235, 246, 249; Bannister 1966; Mueller-Dombois & Ellenberg 1974:58–63; Gauch 1977:49; Maarel 1979*a*; also see Hamer & Soulsby 1980). Estimation is made by quick visual inspection, and most scales have between five and ten values. The Braun-Blanquet and Domin scales have been most used, and a comparison of them is made by Westhoff & Maarel (1978). Gauch (1977:49) presents an octave scale that is essentially logarithmic to the base 2, having ten values 0–9 (convenient for computer processing) and a level of accuracy appropriate for visual estimates of species abundances. Input data for the octave scale should be in the range 0 to 100 (such as percentages), and data values are converted according to Table 2.1.

A common alternative to one-digit estimates is direct estimation of species cover in percents (by 1% units up to 5%; 2 or 5% units up to 20%; and 10% units up to 100%). Such estimates can give somewhat more information on the community with little increase in time needed

Table 2.1. *Conversion table from data values to octave scale*

Input	Output
0	0
$0 < x < 0.5$	1
$0.5 \leq x < 1$	2
$1 \leq x < 2$	3
$2 \leq x < 4$	4
$4 \leq x < 8$	5
$8 \leq x < 16$	6
$16 \leq x < 32$	7
$32 \leq x < 64$	8
$64 \leq x \leq 100$	9

Source: From Gauch (1977:49).

to record the data; for some persons it may be easier and quicker to think in terms of percentage cover rather than to remember and use the intervals of some scale. For later analysis, these percentages are easily converted to one of the traditional scales if desired.

Intermediate accuracy has much in its favor, compared with extremes in either direction. For extensive surveys with very diverse communities, it may be argued that the bulk of the information lies in qualitative differences, that is, in species presences and absences (Greig-Smith 1971). Furthermore, some multivariate analyses, especially among classification techniques, can accept only presence/absence data. The additional effort required for a visual estimate of abundance instead of a record of mere presence or absence is small, however, especially if one considers the unavoidable overhead time for travel during sampling, choosing and marking out sample sites, and so on. This small extra effort produces more informative data, suitable for a greater variety of purposes.

On the other hand, exacting measurements require so much time that the number of samples obtainable drops drastically (Poore 1962; Wikum & Shanholtzer 1978). Given that species abundances of replicate samples typically have considerable scatter (from ±15% to 100% of the mean abundances), it is pointless to make individual measurements to an accuracy of 1%, or even 5%. Movement of a sample site only a fraction of its length would usually be ecologically irrelevant yet it would alter species abundances considerably. Visual estimation involves a degree of error that can be problematic if systematic biases are present and much accuracy is required (Greig-Smith 1964:2–4; Orlóci 1978*a*:21). Usually, however, the inherent variability, or noise, of community samples is larger than errors from visual estimation (Orlóci 1978*a*:21). As Brown (1954:17) notes, if the standard errors of species abundances due to sampling and due to inherent variability between samples are denoted by σ_S and σ_I, respectively, the aggregate error σ is given by

$$\sigma = (\sigma_S^2 + \sigma_I^2)^{1/2}$$

Thus if the sampling error is one-third of the inherent variability, the aggregate error increases only 5%. If, for example, the inherent variability is 30%, sampling with 10% variability yields an aggregate error of ~ 31.6%.

The best strategy, in general, is to estimate species quantities with an accuracy equal to or somewhat better than the inherent noise level in replicate community samples. This provides as much information as is meaningful and not more.

Another consideration in favor of intermediate accuracy for estimates of species abundances is its adequacy for purposes of multivariate analysis. The results of ordination are affected little by even severe rounding of the input data, so little difference, or benefit, accrues from more accurate field measurements (Gauch 1981; also see Green 1977). Likewise, classification should be insensitive to small changes in the input data (Hill 1977).

Clear research purposes are essential. "Be able to state concisely to someone else what question you are asking. Your results will be as coherent and as comprehensible as your initial conception of the problem" (Green 1979:25).

Green's description of several stages in formulating purposes have been adapted here (Green 1979:25–7). (1) Begin by stating the objective in general terms (for example, to describe the forest vegetation in Tompkins County, New York). (2) Increase the precision of the purpose by specifying criteria and procedures (to describe those forests of Tompkins County that are natural or at least 50 years old, using a total of about 200 tenth-hectare samples for vascular plants, with samples placed, in light of reconnaissance, to represent as fully as possible the variety of forests in the county). Produce a detailed experimental design. (3) Show the project design to another ecologist for comment. If a statistical element is present, consult a statistician before the project begins (Crovello 1970). If certain environmental factors are important (such as soils) or certain taxonomic groups present special difficulties, further consult suitable specialists. Revise the project design, as necessary, to make it appropriate, efficient, and practical. In so far as is reasonable, try to anticipate and provide for the various possibilities and interests that may develop during the project. (4) Reevaluate the project design in terms of the originally stated purpose (Crovello 1970).

It may be advisable to go through these four steps twice – first, for the research project as a whole and second, for the design of sampling procedures in particular – in order to assure that the sampling methods serve the research purposes properly.

Size, shape, and number of samples

Sample size may be adapted to the characteristics of the community being sampled, especially to the size of individual plants or clones (Cain & Castro 1959:131; Green 1979:38–43). Green (1979:38) recommends, as a rule of thumb, that samples for nonmobile organisms be no smaller than 20 times the size of the organisms (also see Williams 1971*b*); for mobile organisms, the size of avoidance movement replaces the size of the organism in this rule. Ideally, all the samples of a sample set should be of the same size, but when a variety of communities are encountered, variable sample sizes may increase sample efficiency.

Table 2.2 shows the suggested sample sizes for vegetation samples listed by Westhoff & Maarel (1978). Mueller-Dombois & Ellenberg (1974:48) give a similar, although briefer, list. Cain & Castro (1959: 146) also give a list, with recommended sizes typically a few times smaller. Mueller-Dombois & Ellenberg (1974:47) note a tendency for North American ecologists to use larger sample areas in forests than do Europeans.

Two relatively objective methods for selecting a sample size have been developed. (1) Most important is the species–area method for determining a minimal area (Braun-Blanquet 1932:52–5; Cain & Castro 1959:108–21, 165–77; Greig-Smith 1964:153–6; Daubenmire 1968:89; Shimwell 1971:14–17; Kershaw 1973:172–5; Mueller-Dombois & Ellenberg 1974:47–54; Orlóci 1978*a*:24). Nested plots, from 8 to 10 in number, each twice as large the next smaller plot, are inspected for species and a species–area graph is made with the number of species on the ordinate and the sample area on the abscissa. Typically, the number of species encountered rises rapidly at first and then increases more slowly.

Table 2.2. *Suggested size for samples of various kinds of vegetation*

Type of vegetation	Size (m^2)
Epiphytic communities	0.1–0.4
Terrestrial moss communities	1–4
Hygrophilous pioneer communities	1–4
Dune grasslands	1–10
Salt marshes	2–10
Pastures	5–10
Mobile coastal dune communities	10–20
Hay meadows	10–25
Heathlands	10–50
Alpine meadow	10–50
Dwarf shrub	10–50
Calcareous grasslands	10–50
Chaparral	10–100
Temperate sclerophyll shrubland	10–100
Weed communities	25–100
Shrub communities	25–100
Steppe communities	50–100
Temperate deciduous forest	100–500
Mixed deciduous forest (North America)	200–800
Tropical secondary rain forest	200–1000
Tropical swamp forest	2000–4000

Source: Adapted from Westhoff & Maarel (1978:307).

Various criteria can be used to determine the cutoff point at which the curve becomes rather flat, and the corresponding sample area becomes the recommended minimal area. In practice, the curve may fail to show a clear flattening. It must be conceded that various criteria have been suggested for determining the minimal area; the procedure is only partially objective. Raunkiaer's (1934) law of frequencies (Cain & Castro 1959:159–65; Greig-Smith 1964:15–19; Shimwell 1971:10–12; but see McIntosh 1962), and Preston's (1948) work on the commonness and rarity of species (Cain & Castro 1959:159–65; Greig-Smith 1964:17) can provide the reader with an understanding of species–area curves. (2) The second method is the computation of standard errors of the mean for species abundances measured in a sequence of sample areas (Greig-Smith 1964:20–43; Daubenmire 1968:89–92; Mueller-Dombois & Ellenberg 1974:76–80), followed by a decision on sample area based on a desired confidence level.

Objective methods for determining sample size have been useful and require the investigator to try a variety of sample sizes initially and to compare results carefully. It would be unrealistic, however, not to mention two limitations. (1) As discussed earlier regarding sample homogeneity, plant distributions (and environmental factors) involve many size scales, so any sample size may be appropriate for some species but, unfortunately, too large or too small for others. Likewise, plant species vary in their spatial pattern; distributions are sometimes random, occasionally regular, and most frequently clumped (Brown 1954:32; Greig-Smith 1964:54–93; Pielou 1974:131–201; Green 1979: 39). The spatial pattern has complex effects on desirable sample sizes (Fekete & Szőcs 1974). At best, the concept of an ideal sample size actually refers to an overall optimum because no sample size is ideal for each and all of the species. (2) Accuracy can be improved by increasing the size of the samples, the number of samples, or both (unless a certain sample size is required because of the size scale of the community or environmental pattern). If sufficient data are available relating sample variability (error) and cost for samples of different sizes, statistical methods can be used to derive the optimal sample size (Brown 1954:16; Green 1979:38), but reliable data on costs are rare. Useful principles can be stated, but in the end, the ecologist is usually left with personal judgment (Crovello 1970).

In view of the complexities and considerable subjectivity of deciding upon a sample size, one might ask how critical this decision is. As one example, Goff & Mitchell (1975) show that sample size affects ordination results. Nevertheless, in many cases sample size is "much less critical than the experimenter imagines" (Brown 1954:16). Williams (1971*b*) recommends erring, if at all, on the generous side. If plants are distributed at random, the accuracy of abundance estimates is affected only by the total area sampled, not by the size of individual samples (Greig-Smith 1964:29; Mueller-Dombois & Ellenberg 1974:77). Plant distributions are most frequently clumped, however, and consequently, a larger number of smaller samples is more accurate (Brown 1954:12–16; Greig-Smith 1964:28; Daubenmire 1968:91; Green 1979:39). On the other hand, excessively small sample size is not recommended because the work involved in selecting, demarcating, and observing the resulting large number of samples may become prohibitive, particularly if this also involves additional traveling time (Brown 1954:16; Green 1979:39). Also, as sample area becomes smaller, it becomes increasingly difficult to define its boundaries, and errors due to edge effects increase (Brown 1954:16).

Sample plots are often square, but rectangular and circular plots are also common (Brown 1954:19; Greig-Smith 1964:29; Daubenmire 1968:87–8; Kershaw 1973:32). Because most plant distributions are clumped, a rectangle can best go through patches of different species, making rectangular plots efficient and representative. Special attention should be placed on carefully aligning the long axis of a rectangular plot to keep the plot within homogeneous vegetation. As the ratio of length to width increases, the amount of border relative to area increases, which increases errors from edge effects (including, generally, a bias toward counting too many individuals near the border). Very long plots also may be more cumbersome and more difficult to see from one spot. Consequently, rectangular plots are customarily about two to four times as long as they are wide. Circular plots have the greatest area relative to border (although only 11% better than a square) but are least efficient in sampling clumped distributions (although only slightly worse than a square). On the whole, a rectangle that is two to four times as long as it is wide is ordinarily most accurate, but in some situations the modest gain in accuracy compared to a square may not be justified if a square is more convenient or leads to data more comparable with those of existent studies of interest.

The number of samples may be determined at the outset by the available time. Likewise, if the purpose is to map a given region at a given level of resolution, the required number of samples is predetermined; otherwise the number of samples desired is affected by the accuracy of individual samples, the required accuracy of results, and the degree of community variation within the study region.

Replicate samples for all (or at least most) community types are recommended, in general, particularly when research involves the following: (1) Differences among communities may be of interest but can only be demonstrated by comparison to differences within communities (Green 1979:27–8); the latter necessitates replicate sampling. (2) Some measures for the distance (dissimilarity) between samples used in ordination and classification algorithms are based on equations that require an estimate for the similarity among replicate samples (including the percentage similarity measure, Bray & Curtis 1957). Replicate sampling provides the direct, most accurate estimate.

As few as three replicate samples of each distinctive community type may suffice for ordinary descriptive purposes (Green 1979:40). If the samples are rather small and noisy or if statistical comparisons among slightly differing communities are anticipated, more replicates may be needed. If the needed number of replicates is uncertain, preliminary sampling and data analysis are advisable to provide guidelines for the main study.

Statistical considerations regarding the number of samples are discussed by Greig-Smith (1964:20–49), Mueller-Dombois & Ellenberg (1974:76–80), and Green (1979:38–43, 126–36). Unfortunately, most of this reasoning presumes that species occurrences, sample locations, or both are distributed at random and this is generally unrealistic. The general results are that more accuracy is gained by increasing sample number than sample area, and that accuracy rises rapidly for the first 10–50 samples of each community type and thereafter returns diminish (Greig-Smith 1964:33; Mueller-Dombois & Ellenberg 1974:78; Green 1979:39–41).

Whittaker (1978*c*:13–14) gives three recommendations for selecting the number of samples in a sample set. (1) Along a well-defined gradient of vegetation and environment, samples are taken at fixed intervals (such as elevation up a mountain). Relatively few samples, on the order of 5 to 20, may suffice. (2) In a substantially disturbed landscape, samples may be taken from all or many sites of sufficiently large, undisturbed, homogeneous vegetation. Typically, 50 to 100 samples are taken. (3) In an area of complex environmental variation, samples are taken at frequent but unspecified intervals as the investigator encounters new combinations of community composition and environment. For example, for each 300-m elevation belt on a given parent material in mountains, 50 to 60 samples are recommended.

A simple, workable, decision-making process applicable to most situations is as follows (see, for instance, Green 1979:27–8). After reconnaissance of the study region and selection of a sample area and shape, selection of the number of samples requires two estimates. First, how many community types are distinctive enough to merit consideration, given one's study purposes? Second, how many samples are needed to describe each community type adequately? Multiplication of the two resulting numbers gives the desired estimate of the number of samples required.

Placement of samples

Samples may be placed in the study region by four methods: (1) random location, (2) regular placement in a grid or transect, (3) perferential selection of sites considered especially typical, homogeneous, representative, or undisturbed, and (4) stratified sampling, in which the study region is subdivided into compartments by some criterion and each compartment is then sampled using random siting (Cain & Castro 1959:132–3; Greig-Smith 1964:21; Mueller-Dombois & Ellenberg 1974:39–42; Goldsmith & Harrison 1976). These sampling possibilities have different merits and practicality. The appropriateness of these possibilities varies with the scope of a study, be it within-community, small-scale pattern (such as contagion), or community pattern over a large landscape.

The first sampling method, random placement, has unique advantages from a statistical viewpoint for accurate within-community species abundance estimates (Cain & Castro 1959:122–3; Grieg-Smith 1964:21, 24; Kershaw 1973:28–30; Mueller-Dombois & Ellenberg 1974:39; Pielou 1974:104–7; Goldsmith & Harrison 1976; Orlóci 1978*a*:22). Random placement is possible only if all potential sample sites can be given an equal probability of being picked. Random selection from a complete list of potential sample sites can be done using a random number, and for a territory marked on a map, one can use a pair of random numbers for coordinates and locate the site by surveying. Less formal approaches to randomization are likely to be inadequate (Greig-Smith 1964:24). If any sample site is selected at random more than once, the duplicate picks are discarded so that all samples are in different places.

Random placement gives each place an equal probability of being picked. This equality may be desired for accurate species abundance estimates within a stand. It is not, however, necessarily appropriate for the study of a landscape and works poorly for many aspects of spatial pattern. Equal probability allows computation for the samples of an estimate of the mean abundance of each species and the standard error of the mean (Pielou 1974:104–7). If the species abundances are counts of individuals, multiplication by the number of potential sample sites gives an estimate of the total population size and its standard error. The *t*-test can be used to determine a confidence interval of, say, 95%. If a certain predetermined confidence interval is desired, the number of samples required can be judged roughly in advance, provided that some preliminary sampling supplies an approximate value for the variance of the samples (Pielou 1974:107). For rarer species, a greater number of samples is required for a given level of accuracy, so a compromise may be struck with a practical number of samples intended to be accurate for moderately common species. Furthermore, two sets of samples can be compared by a *t*-test, using their means and standard errors to determine whether species abundances differ significantly (Greig-Smith 1964:22).

Random sampling also has disadvantages. (1) Random sampling may be less accurate than systematic or preferential sampling (Greig-Smith 1964:22; Moore, Fitzsimons, Lambe, & White 1970). Unfortunately, the accuracy of nonrandom samples cannot be computed statistically, even though nonrandom samples may, in fact, be more accurate. (2) Unless the sampling intensity is extremely high, coverage of the study region will be uneven. By chance, some places will be oversampled and other places undersampled (Pielou 1974:103; Goldsmith & Harrison 1976; Orlóci 1978*a*:22). (3) Enumerating potential sampling sites or accurately locating sites in a coordinate system is tedious and requires a lot of time (Greig-Smith 1964:22). (4) When a study region contains environmental irregularities or disturbances to be avoided, a sufficiently objective and operational exclusion rule may be unattainable, making proper random sampling impossible.

Moore, Fitzsimons, Lambe, & White (1970) conclude that for random sampling the advantages are outweighed by the disadvantages. Indeed, random sampling is attempted rather infrequently, and among these attempts, not all claims to randomness are confirmed under scrutiny. Especially for study of a variety of communities over a landscape, random sampling is rarely done and rarely effective.

The second sampling method, regular placement, employs a transect, or grid, of samples (Cain & Castro 1959:121, 124–5; Greig-Smith 1964:22; Williams 1971*b;* Kershaw 1973:32–9; Goldsmith & Harrison 1976; Whittaker 1978*c*:13). Regular sampling can cover the ground evenly and give accurate results (unless the distance between samples corresponds in scale to a periodicity in the communities, as in dunes with periodic ridges and slacks or arctic vegetation with polygonal patterns). When interest centers on variability in communities or spatial pattern, regular sampling is effective (Greig-Smith 1964:23, 54–93; Kershaw 1973:32). Likewise, when interest concerns community response to apparent environmental gradients, samples placed regularly along these gradients are effective (Kershaw 1973:32). Another advantage is that regular placement of samples in the field is convenient and rapid (Greig-Smith 1964:22).

Strip transects are sequences of samples taken next to one another or at a regular distance along a line (Cain & Castro 1959:124–5; Kershaw 1973:32–7). Transect lines usually run across, rather than with, topographic gradients so that the lines equitably encounter the variety of communities and environments present. If replicate samples are desired within one uniform community type, however, the transect line should stay within this community type. In some cases, transect width and spacing are set to result in a sampling intensity of a given percentage of the study area.

Regular grids of samples provide straightforward, two-dimensional coverage of a study region (Cain & Castro 1959:121; Greig-Smith 1964:54–93; Kershaw 1973:32–9). Studies of spatial patterns of species or communities often employ a regular sample grid, but nested quadrats and strip transects are also common. A grid may cover a region entirely or only certain portions, if interest involves only certain community types or habitats (in terms of environmental parameters or degree of disturbance).

The third sampling method, preferential sampling, is used most frequently by plant ecologists, especially by Europeans (Poore 1962; Moore, Fitzsimons, Lambe, & White 1970; Mueller-Dombois & Ellenberg 1974:32–3; Goldsmith & Harrison 1976; Orlóci 1978*a*:22–4). The ecologist subjectively (1) selects sample sites that appear to be homogeneous and (2) distributes samples among the community types equitably (Moore, Fitzsimons, Lambe, & White 1970). An experienced ecologist may produce satisfactory results by preferential sampling using considerably fewer samples than needed by any other approach. The disadvantage of preferential sampling is that the ecologist's biased preconceptions may introduce appreciable errors, and in any case, statistical tests are invalid. Such data are commonly employed for multivariate analysis, however, with the understanding that results are for descriptive purposes only, which may be entirely adequate for the purposes of a study (Crovello 1970).

The final method, stratified sampling, has two steps: first, the study area is divided into a number of compartments, and second, randomly sited samples are taken in each compartment (Brown 1954:12–13; Greig-Smith 1964:23; Kershaw 1973:34; Mueller-Dombois & Ellenberg 1974:39; Pielou 1974:107–12; Goldsmith & Harrison 1976; Orlóci 1978*a*:22–3; Tomlinson 1981). The basic intent of stratified sampling is to combine the advantages of systematic and random sampling (namely, accuracy and statistical validity, respectively). This hybrid approach is particularly worthwhile in the commonest case of a heterogeneous study area. "If the area proves to be uniform after all then the data can all be lumped together, but on the other hand if the area is variable then information is available which is relative to each sub-division of the plot and is accordingly of considerable interest" (Kershaw 1973:34).

The number of compartments required is a matter of judgment. The community variability, the extent of the study region, and the purpose of the study are major considerations, together with the desired resolution of results and the practicality of the implied number of samples.

There are two basic kinds of criteria for defining compartment boundaries. One is simple geometric dissection of the study area, the other is some criterion based on community types or environmental factors, using prior knowledge of the study region, for example, shrubby and nonshrubby patches in a partially shrubby grassland or elevation belts in a mountain. In either case, the actual definitions of boundaries in the field should be as clear and objective as possible. Also, criteria should be chosen to avoid subsequent circular reasoning. "Whatever criteria are used, it must be kept in mind that the compartments cannot be validly compared on the basis of the criteria by which they are delimited since such a comparison would involve a circularity of argument. An example of such an inadmissable argument is the use of climatic compartments to prove the existence of natural zones in vegetation, knowing that climatologists like to draw climatic boundaries between groups of stations to coincide with apparent boundaries between vegetation zones" (Orlóci 1978*a*:23).

Compartments may be of equal areas (not necessarily of the same shape). If compartments also have equal numbers of random samples, each point has the same probability of being sampled. Provided that the compartments are not comparable in size to the variability within the area, statistical tests are valid and there is no loss of accuracy, in comparison to random sampling (Greig-Smith 1964:23). Equal sample numbers in each compartment are best if the variances within compartments are equal or unknown. If some compartments have greater variances than others, however, it is desirable to sample more variable compartments more intensively (Pielou 1974:112). Alternatively, compartments of unequal areas may be used; Pielou (1974:108–12) describes the statistical calculations appropriate to this case. Sample numbers may be chosen proportionally to compartment areas or more variable compartments may be sampled more intensively, depending on the requirements of subsequent statistical tests planned by the investigator.

A disadvantage of stratified sampling may be that it requires more work in the field than does either of its parent methods (Goldsmith &

Harrison 1976; Orlóci 1978*a*:23), although Brown (1954:13) does not ordinarily expect increased labor. It does require numerous decisions, and these decisions can be made properly only after reconnaissance and perhaps a pilot study.

Samples from extremes of the environmental gradients encountered are particularly important in helping the researcher to interpret community gradients in terms of environmental gradients. Samples from environmental extremes also improve ordination results (Mohler 1981). Sites having environmental extremes may be rare, however, making their representation poor if random placement of sample sites is used. Either preferential or stratified sampling can, however, improve the representation of samples from environmental extremes.

Standard community sampling procedures

From the preceding discussion, it is clear that the selection of effective sampling procedures requires numerous considerations, including (1) the kinds of communities sampled, (2) the kinds of environmental and historical corroborative data needed to aid interpretation of results, (3) the scope, accuracy, and purposes of the study, (4) the requirements imposed for comparability with other studies, (5) the requirements imposed for valid application of anticipated data analysis methods, (6) the practical limitations, and (7) the personal preferences arising, for example, from the nationality or school of the investigator. Because choice of sampling methods is a complex art, it is worthwhile to consider standard methods that have proven to be effective.

Standard sampling methods are treated here by providing references to the literature because detailed treatment of such numerous possibilities is not within the scope of this book. For economy, mainly references to texts are provided; Schultz, Eberhardt, Thomas, & Cochran (1976) include a bibliography on sampling methods.

Typically, the taxonomic unit employed in sampling is the species, and sample sites are visited once. The resulting community data are a two-way samples-by-species matrix of abundance values. Species abundance measures include mere presence or absence, percentage cover, density (number of individuals), frequency (percentage of quadrats or points having a species present), biomass, or some weighted average of two or more such quantities. The ordinary case of samples-by-species two-way matrices will be discussed first; at the end of this section, other possibilities will be discussed.

Terrestrial vegetation

Terrestrial plant communities have probably received more attention from ecologists than all other communities combined. Because terrestrial plants are fixed in their locations and certain assemblages of species (community types) are encountered repeatedly, a community-level viewpoint is intrinsically appropriate. Several decades of research worldwide have provided numerous sampling procedures, some of which found widespread favor and eventually became standard methods.

Several texts include substantial treatment of sampling methods for a variety of vegetation types (Braun-Blanquet 1932; Cain & Castro 1959; Greig-Smith 1964; Daubenmire 1968; Shimwell 1971; also see Barbour, Burk, & Pitts 1980), and Brown (1954) treats grasslands extensively. Pielou (1974) discusses statistical considerations. The broadest and most thorough presentation is by Mueller-Dombois & Ellenberg (1974) including relevé, quantitative plot, line-intercept, and plotless methods, with explicit details and numerous examples from the literature for a wide variety of vegetation. Smartt, Meacock, & Lambert (1974) investigate the properties of various measures of species abundance.

Although sampling techniques are diverse, much research in community patterns over landscapes has used either relevés or tenth-hectare (or other) quadrats. Ecologists usually place samples by (1) moving through the study area taking each homogeneous sample site encountered that differs (in composition, topographic moisture conditions, soil, and so on) from the sample preceding it, while (2) obtaining replicate samples but not, beyond some point, duplicating the kinds of samples already taken, while also (3) intentionally seeking out scarce types of samples (such as ravines) to ensure that these are represented.

Much American work uses tenth-hectare (20×50 m) standard samples (Whittaker 1978*b*). Trees are tallied by species and diameter at breast height (permitting calculation of basal areas) and shrubs by diameter at the base. For the herbs and low shrubs, cover estimates are recorded in 25 quadrats of 1 m^2 placed along the sample midline or randomly in its area. Rare plants outside these quadrats are listed. Cover of trees and shrubs may be measured by recording the beginning and end of foliage of each species along the 50-m tape (line-intercept technique) or estimated visually. Such samples (which mostly take 1 to 3 hours for two people) represent an intermediate level of sample effort, compared to rapid relevés and to more demanding point-centered samples of the Wisconsin school (Cottam & Curtis 1956; Whittaker 1978*b*; Mueller-Dombois & Ellenberg 1974:110–12). American ecologists usually know and sample only vascular plants; European phytosociologists usually also sample bryophytes and lichens.

Vegetation samples based on a list of species with visual estimation of their coverages are termed *relevés*. Such samples and the supplementary data that may be taken are described by Braun-Blanquet (1932), Mueller-Dombois & Ellenberg (1974:45–66), and Westhoff & Maarel (1978). Relevés have no standard area, but as noted earlier, these authors provide suggested areas for various vegetation types. Relevés are the most frequently used kind of vegetation sample, with over 200 000 relevés collected in Europe alone (Pignatti 1980).

Tenth-hectare sampling gives greater accuracy, whereas relevé sampling gives more samples. The scope and purposes of a study may imply a clear preference between these two options. In other cases, the precision of a study will be determined largely by the number of days of sampling available but not significantly by the choice between these options.

Terrestrial animals

The mobility of animals and their much greater number of species make the community viewpoint diffuser and more difficult than for plants. For example, a bird species needing nesting sites, food, and water may occupy a territory of considerable size and may even migrate thousands of kilometers. Its mobility in time and space makes many of its interactions or associations with other species weak. Furthermore, the difficulty in estimating population sizes for moving animals is much greater than for plants; consequently, a researcher may not have adequate resources to study the numerous species present in a community. Insects, or invertebrates, with small ranges may interact more substantially, making a community approach more viable; however, the efficiency of any particular collection method is likely to vary drastically with the various species. In any event, the fraction of animal ecology done from a community perspective is relatively small, and of this work, much actually involves only a guild, which is a taxonomic or functional subset of the animals present, such as insect herbivores on cabbage or fruit-eating birds at some site.

Andrewartha (1961), Kendeigh (1961), Caughley (1977), and Tanner (1978) provide methods for estimating the density of animal populations. Pielou (1974) treats censusing, two-phase sampling, estimation by the regression method, capture–recapture, and the removal method. Southwood (1978) emphasizes insect populations. Schultz, Eberhardt, Thomas, & Cochran (1976) provide a bibliography on population estimation methods.

Aquatic communities

Sampling methods for lakes and streams have been summarized by Wetzel & Likens (1979), giving methods for phytoplankton, zooplankton, fish and other animals, littoral plants and animals, and benthic fauna of lakes and streams. (Measurement of activities of organisms is also treated, including primary productivity, feeding, and decomposition.) Corroborative data collection is discussed for environmental, disturbance, and historical factors. It is stressed that spatial and temporal fluctuations in populations on many scales necessitate the careful design of sampling procedures. Furthermore, the existence of up to a dozen or more distinct stages in the life history of some species is an obvious problem for population estimation. This text also discusses some basic statistical considerations for sampling. Wood (1975) treats aquatic plant community sampling, and Lind (1979) discusses common methods in limnology.

Marine communities

Marine environments encompass diverse habitats: beaches and dunes, the intertidal zone and the tide pools, estuaries and peripheral areas, coral reefs, ocean waters, and the ocean bottom (Zottoli 1978). Besides the obvious physical challenges for sampling, the spatial and temporal patterns in the distribution of marine organisms can be problematic (Steele 1977). Schlieper (1972) presents sampling methods for marine plants, animals, and microbes and for physical and chemical parameters. Holme & McIntyre (1971) discuss methods for marine benthos. Chardy, Glemarec, & Laurec (1976) ordinate marine benthic data, and Gladfelter, Ogden, & Gladfelter (1980) classify coral reef fish communities.

Microbial communities

The small size of microbes causes a high surface-area-to-volume ratio, resulting in intimate contact with their environment (R. Campbell

1977:1). Furthermore, many heterotrophs have extracellular enzymes that act outside the controlled chemical environment inside the cell. Consequently, microbes are sensitive to environmental parameters such as temperature, light, pH, organic and inorganic nutrients, carbon dioxide, oxygen, and water. Microbes are generally so well dispersed that any capable of growing in a particular environment will probably be there; conversely, if absent, there is probably a reason why the organism cannot grow there (R. Campbell 1977:2). In short, small size tends to result in wide dispersal and geographic distribution and in a tight relationship between active microbial populations and the environment.

Small size also implies an enormous number of distinctive microhabitats (in addition to familiar habitat differences). "You must adjust the scale of your thoughts and imagination to encompass both the large habitat variations and also the very small dimensions over which variations in microbial communities can occur" (R. Campbell 1977:2). For example, individual soil crumbs differing in mineral and organic matter content offer distinctive microhabitats.

The numbers, biomass, and biological activity of various microbial species may be poorly related because of differences in the size, condition, growth rate, and metabolic activity of various species and individuals (Brock 1966:163–80; R. Campbell 1977:2–6). An ecologist must discern which of these quantities is most relevant to research interests. There are numerous techniques for estimating these species parameters, but technical problems are considerable, often resulting in tedious and expensive assays and in rather limited accuracy. Unlike a higher plant or animal ecologist, the microbial ecologist cannot merely go out in the field to see and count individuals.

Cairns (1977) and Droop & Jannasch (1977, 1980) consider aquatic microbial communities, Stevenson & Colwell (1973) estuarine microbes, and Colwell & Morita (1974) marine microbial communities. Griffin (1972) and Krupa & Dommergues (1979) treat soil microbiology; Brown (1978), all soil organisms from a community ecology perspective; and Thayer (1975), microbial interaction with the physical environment.

Paleoecology

Like ecology, paleoecology considers the relationships between organisms and their environment, but in the past (Birks 1973:3–6; Janssen 1979). Reconstruction of past communities has methodological challenges because the relationship between the abundance of an organism and the quantity of its fossil remains can vary over several orders of magnitude, and some fossils are transported and mixed with others. Reconstruction of past climate is even more difficult (Nairn 1964). The rewards of careful research are great, however, because paleoecology provides insight on the background of present communities and on community response to various climatic or environmental perturbations, which, in many cases, for practical reasons, cannot be studied by experimental methods.

Braun-Blanquet (1932:336–40) describes early pollen analyses and results. Distinctively ecological treatments of fossil studies include those by Schäfer (1972), Imbrie & Newell (1964), Krasilov (1975), and McKerrow (1978). Birks (1973) presents an extensive study of both present and past vegetation on the Isle of Skye (off the northwest coast of Scotland, with an area of 1720 km^2), with thorough treatment of methods and with community analysis similar to the Braun-Blanquet method. Application of recent multivariate methods to fossil communities has been effective and contributes methodological advances in analyzing paleoecological data.

Special problems

Community ecologists usually study a samples-by-species data matrix, but there are other possibilities.

Instead of species, the basic taxonomic unit may be the genus, family, or order (Maarel 1972; Dale & Clifford 1976; Moral & Denton 1977; Maarel 1979*a*); conversely, finer divisions into subspecies or phenotypes may be used. For many species, individuals change dramatically in their ecological requirements as they age or grow, so abundances may be recorded separately for several age or size classes. For example, forest trees increase in size several orders of magnitude while growing from seedling to mature tree, changing the space they occupy (hence the environment they encounter) and their physiology and ecological requirements (Goff 1968; Goff & Zedler 1972; Zedler & Goff 1973; Peet & Loucks 1977). In forests, the tree and undergrowth strata may be studied separately and compared (Bratton 1975*a*; Moral & Watson 1978; Onyekwelu & Okafor 1979); likewise, see Jonasson (1981) for analyses of various taxonomic groups in a heath. Occasionally the units for recording plant or animal abundances are not taxonomic at all but rather are based on general features of the organisms such as growth form, flowering time, and guild membership (Knight & Loucks 1969; Bridgewater 1978). From the viewpoint of multivariate analysis, use of descriptors other than species poses no problem.

Sometimes the most natural way to view a data set is as a three-way (or many-way) matrix, instead of the usual two-way matrix. As noted in the introduction, many-way data must be reorganized, summarized, or separated into one or more two-way matrices if ordination and classification are to be used.

Successional communities can be sampled by two strategies: sampling a permanent site (or sites) many times or sampling many sites at a single time and attempting to reconstruct the time sequence (Daubenmire 1968:100–13; Goff 1968; Kershaw 1973:40–64; Maarel & Werger 1978; Maarel 1979*a*; Veblen & Ashton 1979; Austin 1977, 1980*a*; Birks 1980; Glenn-Lewin 1980; Huschle & Hironaka 1980; Maarel 1980*b*; Noble & Slatyer 1980; Onans & Parsons 1980; Peet & Christensen 1980; Persson 1980; Regnéll 1980). Because of the long lifetimes of species in some communities, only the latter may be possible. If the lifetimes of the organisms are brief, experimental manipulation may help elucidate mechanisms and interactions, and enough replicates to evaluate stochastic factors may be feasible (Sousa 1979). Except for simple cases, succession studies can be taxing because (1) the driving disturbances may be complex (like forest fires, which vary in temperature and duration in several scales in three dimensions), (2) a complex spatial pattern may occur with nearby patches in various stages of succession, (3) successional pathways may branch and loop, rather than follow a simple sequence, (4) stochastic factors may take large roles, (5) the inherent complexity may necessitate a large data base, and (6) the application of results to a new territory seems to require great care.

Ecological investigations may involve collection of several kinds of data, resulting in several two-way data matrices. For example, a study might produce six two-way data matrices: sites-by-plants, sites-by-soils, sites-by-birds, birds-by-morphology, birds-by-foods, and foods-by-nutrients. Analysis of each two-way matrix is a standard task, but, naturally, some integrated perspective may be desired in the end. A traditional technique for such problems is canonical correlation analysis, but tests with ecological data have been unpromising because of nonlinearity in the data (Gauch & Wentworth 1976). More effective approaches are presented by Gauch & Stone (1979) and will be discussed in Chapter 6.

Environmental and historical data

Plant, animal, and microbial communities on land and in lakes and oceans are affected by a great diversity of environmental and historical factors (with historical factors here including what might be called disturbance factors). Ordinarily, community data for each sample site are accompanied by corroborative environmental and historical measurements to aid interpretation of the community patterns. As Williams (1962:310) notes, "Interpretation is always of one set of information in the light of another set of information; it consists simply of finding the joint pattern in *two* sets of data." In community ecology, species and community patterns are interpreted principally in terms of environmental gradients. Because ordination and classification are ordinarily based on community data alone (exclusive of environmental data), environmental interpretation is a separate, subsequent step. Often, interpretation is by informal comparison of community and environmental patterns, but statistical approaches are also used. In the references cited earlier in this chapter, comments are made on the environmental and historical factors ordinarily most important in the various communities and on methods for measuring such factors.

Environmental and historical data collection presents four general problems for community ecology studies. (1) The more important environmental and historical factors need to be identified and practical methods for measurement devised. Sometimes important factors are far from obvious. Factors that are important only in a few samples can be especially problematic. (2) As McIntosh (1970:264) observes, "In nature, no factor ever varies alone." Environmental factors tend to form complexes, that is, a number of factors varying together. For example, altitude may correlate with temperature, humidity, rainfall, and soil properties; soil pH may correlate with several element concentrations and, perhaps, moisture status. These correlations make multivariate analysis of environmental data worthwhile for producing simplified descriptions of the environment, but also they may obscure the interpretation of results beyond the level of rather general, basic environmental factors. (3) Environmental and historical factors vary on many scales in time and space. This makes any sampling scheme effective for some factors but not for others. Ordinarily environmental and historical factors are assessed using the same size scale as is used for gathering the community data. (4) Because environmental and historical data are usually multivariate and the community data are also multivariate, at the end, when results are interpreted, the difficult task often arises of comparing two sets of data, both multivariate.

In addition to the references already supplied, a few general references on collecting environmental data may be useful. Munn (1970) and G. S. Campbell (1977) treat measurement of environment from a distinctively biological perspective, while the National Academy of Sciences (1971) considers animals and Etherington (1975) and Chapman (1976) plants. Geiger (1965), Shaw (1967), and Seeman, Chirkov, Lomas, & Primault (1979) treat ground-level climatology. Lee (1978) treats forest microclimatology. Kinne (1979) considers marine environmental factors. Fritschen & Gay (1979) discuss environmental instrumentation and data acquisition. Green (1979) explains statistical considerations for applications in environmental biology.

3

Direct gradient analysis

Direct gradient analysis is used to display the distribution of organisms along gradients of important environmental factors. By contrast, ordination and classification techniques generally start with the analysis of community data alone and, only later, use environmental data for interpretation. Direct gradient analysis is a major research approach in community ecology, forming a methodological triad with ordination and classification. Ramensky (1930) and Gause (1930) originated direct gradient analysis, but active research began around 1950 (Whittaker 1948, 1967, 1978*b*). The results of numerous direct gradient analyses have become the foundation for the Gaussian model of community structure, and this model has served an important role in testing and designing multivariate methods.

Basic purposes and example

Before presenting the methods and results of direct gradient analysis in detail, a typical example will be considered to introduce basic purposes and methods.

Figure 3.1 depicts the native vegetation and topography of Nelson County, North Dakota, at a hypothetical site. Prairie, meadow, and marsh occur on a rolling plain of low relief with soil drainage appearing to be the major environmental factor (Dix & Smeins 1967). As one might gather from Figure 3.1, mere reconnaissance of Nelson County would suffice to ascertain the primary importance of drainage. Likewise, it would be easy to determine the characteristic species of prairie, meadow, and marsh. Casual inspection, however, resolves only the easiest questions; a quantitative approach is required for more difficult questions.

In order to study this vegetation in an accurate, quantitative manner, Dix & Smeins took 100 community samples to represent the range of vegetation present in the county. Homogeneous stands of 0.1 ha were sampled, recording frequency in 30 quadrats of 0.5×0.5 m for com-

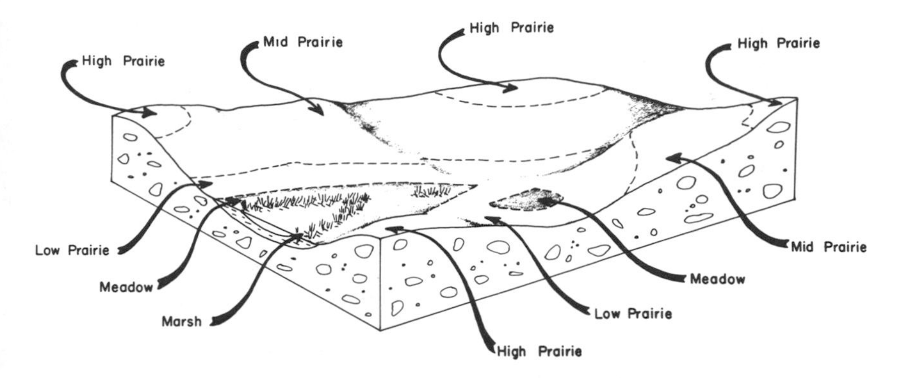

Figure 3.1. A hypothetical diagram of the topography and plant communities of a site in Nelson County, North Dakota. (From Dix & Smeins 1967:43)

mon species and a presence list for all species found in the stand. Environmental measurements included slope, exposure, drainage and soil profile, texture, carbonate, pH, conductance, sulfate, chloride, and water-retaining capacity. These data permitted a quantitative approach to the vegetation of Nelson County and its relation to environment. Figure 3.2 shows a direct gradient analysis of the distribution of 16 important species along a drainage regime gradient.

Dix & Smeins comment on this figure as follows: (1) The species are individualistic in their responses to the environmental gradient, having scattered modes and variously wide or narrow distributions (in contrast to the alternative of forming groups of species with similar distributions). (2) Consequently, the plant communities form a vegetational continuum (in contrast to forming distinctive associations or types without intermediates). The terms prairie, meadow, and marsh for community types are useful for descriptive purposes, but it is recognized that some stands are borderline in the continuum between these types and, therefore, their assignment to a given type is partly arbitrary. It may also be noted that most species have one optimum along the environmental gradient and decline to either side; most response curves are approximately bell shaped or Gaussian.

The following are four representative questions about the distribution of species and communities in relation to environmental gradients requiring a quantitative approach. The purpose of direct gradient analysis is to gather and organize community and environmental data to answer questions such as these.

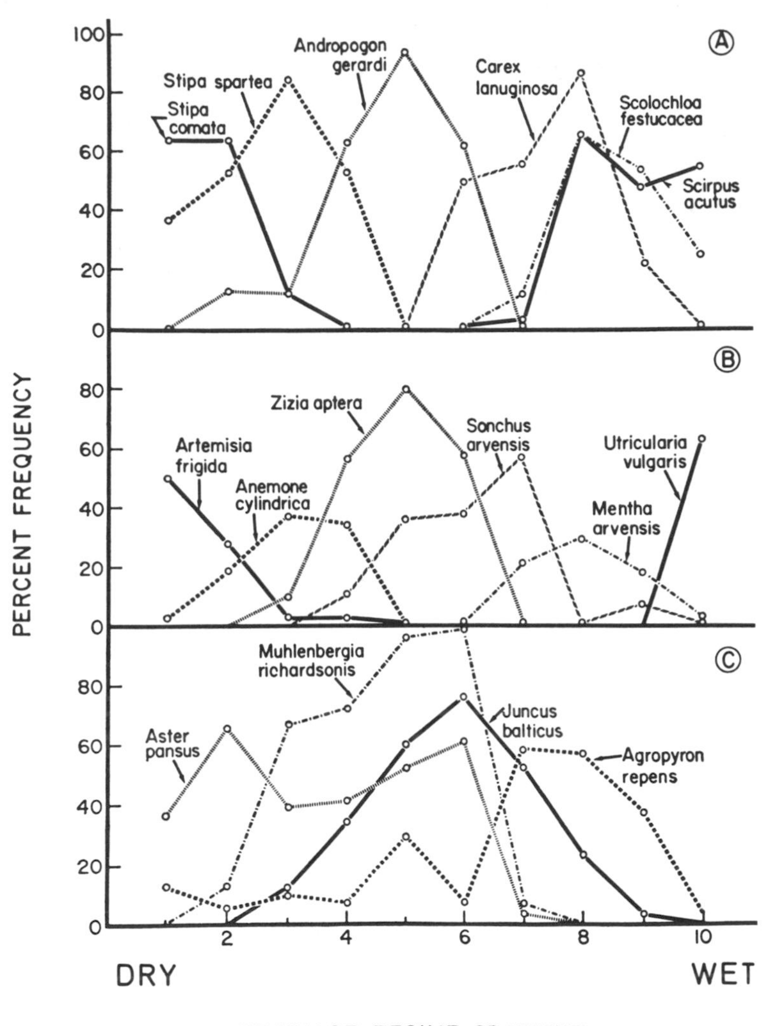

Figure 3.2. Distribution of 16 plant species along a drainage regime gradient in Nelson County, North Dakota. Panel *A* shows graminoids; *B*, forbs; and *C*, graminoids and forbs of ubiquitous distribution. (From Dix & Smeins 1967:33)

(1) Precisely which environmental factor in a complex of factors principally affects the distribution of organisms and communities? This question is difficult because many factors vary together (McIntosh 1970). This complicates identification of the most important environmental factors. In the example of Figure 3.2, Dix & Smeins (1967:29)

correctly (and conservatively) conclude that the vegetation reacts strongly to "the drainage regime or to some factor or factor complex related to it." For some purposes such a result may be adequate, but other purposes may require a more precise result. Dix & Smeins investigated soil moisture in greater detail and found that elevation correlates partially with a factor complex including drainage, soil moisture, soil type (from regosols on high land to humic-gley soils at the bottoms of depressions), soil texture, soil conductivity, salinity, and sulfate concentration. Of this factor complex, they found the gradient in soil moisture to correlate most strongly and consistently with the gradient in community composition. Even this result is still complex because (a) soil moisture itself is complex, having components of amount, depth, and duration of water (and, furthermore, means, variances, maxima, and minima for these components) and (b) the effects on the plants of low and high soil moisture are different (low moisture limiting growth by desiccation and high moisture limiting growth by reduction of oxygen availability to roots). In contrast to this primary environmental factor complex of soil moisture, other factors were found not to vary substantially in the study region (including soil carbonates, chloride, pH, and probably organic matter) or not to affect substantially the distribution of plants (including the water-retaining capacity of the soils). Although direct gradient analysis is useful in identifying ecologically important environmental factors, additional research approaches, such as experimental manipulation, may be needed. There are several exceptional papers using multivariate methods for discerning important environmental factors by Loucks (1962), Grigal & Arneman (1970), Walker & Wehrhahn (1971), Forsythe & Loucks (1972), Jeglum (1974), Vitt & Slack (1975), Jonescu (1979), Lepart & Debussche (1980), Westman (1980, 1981), and Wentworth (1981).

(2) How can environmental factors best be measured or estimated? Often a given environmental factor such as soil moisture can be measured or estimated by a variety of methods, which vary in accuracy, expense, directness, objectivity, time, equipment, and training required. For example, Dix & Smeins begin with a drainage regime estimate for each stand based on topography, using a scale with six categories. Subsequently, a phytosociological scale is developed using the plants themselves as indicators of drainage conditions. The 120 common species are tabulated according to their abundances in the six categories of the original drainage estimate, and those species with an average frequency in a category at least 10% greater than in any other

category are selected as indicator species. In this case, 48 species are selected and given indicator values from 1, for the driest category, to 6, for the wettest category. A new phytosociological stand index is then calculated for each stand as a weighted average (the sum of species relative frequencies is multiplied by the indicator values and the result is divided by the sum of species relative frequencies; finally, this result is multiplied by 100 to obtain whole numbers). This stand index ranges from 100 to 600; a stand with indicator species only in category 1 (dry) would have a stand index of 100. If a stand contains relative frequencies of 20 for *Stipa comata,* 15 for *Achillea millefolium,* 5 for *Lithospermum canescens,* and 60 for species other than indicator species and if these three indicator species are in categories 1, 2, and 3 respectively, the stand index is

$$\{[(20 \times 1) + (15 \times 2) + (5 \times 3)] / (20 + 15 + 5)\} \times 100 \approx 163$$

indicating that this stand is rather dry but not quite at the dry extreme. Dix & Smeins found the phytosociological scale more accurate than that based on the original visual assessment of topography and drainage (also see Shimwell 1971:240 and Persson 1981). The drainage regime gradient in the abscissa of Figure 3.2 uses the phytosociological scale, breaking its interval of 100 to 600 into ten equal sections numbered 1 (dry extreme) to 10 (wet extreme).

(3) What additional, secondary, environmental gradients affect community distributions? Environmental gradients beyond the primary one or ones may also be interesting. The study of secondary gradients is difficult because they are overshadowed by the primary gradient or gradients and they often involve relatively few species and samples of the data set. These difficulties make a careful, quantitative approach especially advantageous for secondary gradients. For example, in the Nelson County study, *Hordeum jubatum* and especially *Scirpus paludosus* responded favorably to a secondary gradient of salinity. Likewise, *Taraxacum officinale* and *Agropyron repens* responded to a secondary gradient of disturbance (even more so than to the soil moisture gradient, which was the primary gradient for most species). Nelson County has numerous natural and human disturbances that could also be studied by direct gradient analysis.

(4) What general principles emerge from direct gradient analyses to characterize the combining of individual species into communities? One basic question is whether communities are well-defined natural

units with little if any transitional mixtures or whether communities usually intergrade continuously (Whittaker 1967). In this direct gradient analysis of Nelson County vegetation, the latter viewpoint is supported. Likewise, how can communities be modeled and simulated mathematically? The answers to these questions are the bases for models of community structure and have a strong bearing on the anticipated effectiveness of various viewpoints and research methods in community ecology.

Some terminology is useful at this point (Whittaker 1978*b*). A gradient in community composition is termed a *coenocline* (as in Figure 3.2 or a *coenoplane* or *coenocube* in two or three dimensions of variation as illustrated later in this chapter). A gradient in environmental conditions is termed an *environmental complex-gradient* because a complex of factors is always involved (although, for simplicity, this term is sometimes shortened to *environmental gradient,* with the reader understanding that a complex-gradient is always implied). The complex-gradient and coenocline together form a gradient of communities and environments, termed an *ecocline.* An abrupt or relatively rapid change in an environmental complex-gradient is termed an *ecotone;* this term is also used for a rapid change in a coenocline or ecocline. For example, an abrupt change in soil parent materials supporting distinctive floras constitutes an ecotone. The term *community type* has been used already with the informal, though lucid, meaning of a recognizable community characterized by a distinctive species assemblage. At this point a more precise definition is desirable. In the context of the Braun-Blanquet approach (especially in its first two or three decades), groups of species are believed to associate together in natural units, ordinarily without transitions or mixtures, and community types are intended to describe these natural units (Westhoff & Maarel 1978). This context is not implied here. Here a community type is described intensively by its characteristic species assemblage or extensively by listing the subset of samples within a sample set that belongs to the community type. This definition does not assume either that community types are natural or, conversely, that they are humanly imposed categories; neither is it implied that the assignment of a community type to each sample is entirely objective or unambiguous. These qualifications reflect practical limitations; they are not to be taken as criticisms because community types are of universal and unquestionable utility for describing communities.

Methods

The example in Figure 3.2 of direct gradient analysis of the vegetation of Nelson County is typical, having species abundances plotted along a recognized environmental gradient. There are, however, many variations and permutations on this basic theme. Although direct gradient analysis remains the principal ingredient, some variations also involve ordination, classification, and other methods to some degree. Various possible methods will be described here, and some will be illustrated in the following section on results.

The variables measured along an environmental gradient may be other than species abundances: They may be more general, such as families or orders, growth form, geographic origin of species, number of species or individuals, primary productivity, and community type; or they may be more detailed, such as size or age classes of species populations, life-cycle stages, or particular phenotypes, ecotypes, or subspecies. Also, the variables may be abstract or mathematical properties, such as sample similarity measurements.

The gradient used may be other than a simple, recognized, environmental gradient. The possibilities include an abstract mathematical gradient produced by ordination, a synthetic index that integrates a number of environmental measurements, and mere spatial position along a transect of samples. Several gradients may be used, rather than only one, to produce a multidimensional direct gradient analysis.

The purpose of a study may be broader than in customary direct gradient analysis. Besides relating species and community distributions to environmental gradients, these results may be linked to additional data bases concerning, for example, fire management.

Community ecology data are usually rather noisy, so some form of smoothing or averaging is a common aspect of direct gradient analysis methods; only occasionally do researchers publish tables or figures using raw data. More often, species abundance curves along gradients are drawn with hand smoothing or with running averages. Frequently individual samples are combined and averaged to produce composite samples that are less noisy than the original samples. The criterion for combining samples may be based on stratification of an environmental measurement, a classification or ordination of the samples, or various other possibilities.

Most of these variations in direct gradient analysis can be summarized by saying that both the organisms and the environment may be

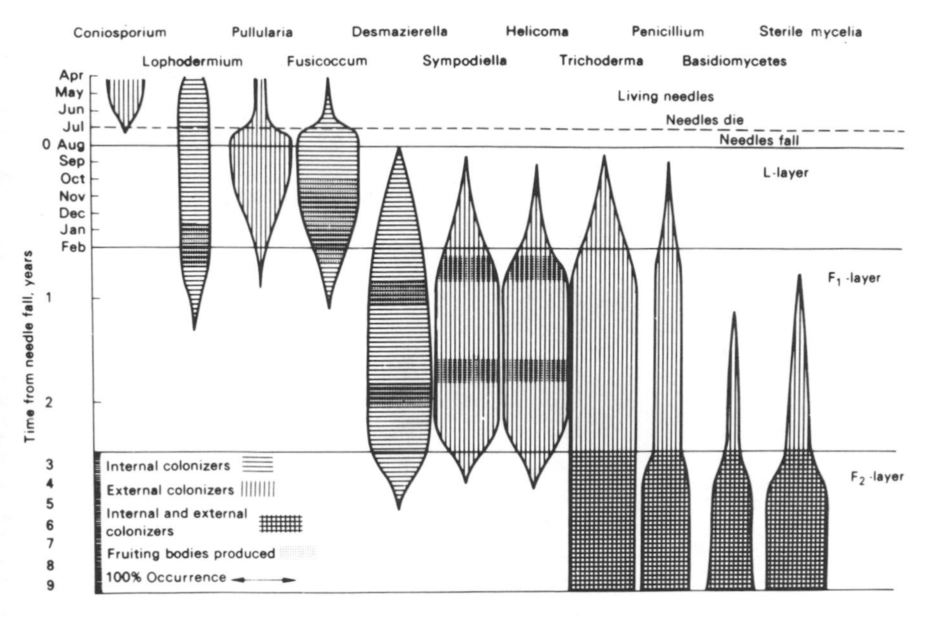

Figure 3.3. Temporal and spatial changes in fungal populations colonizing pine needles in litter layers of a Scots pine (*Pinus sylvestris*) forest in England. (From Richards 1974:79 based on Kendrick & Burges 1962)

approached from a variously analytic or synthetic perspective and on a variously concrete or abstract basis.

Results

Several examples of direct gradient analysis may be considered in order to show results for a diversity of taxa and a diversity of environments and to provide a sequence from simple to complex and abstract applications.

Figure 3.3 shows a coenocline of fungal species colonizing pine needles in pine forest litter along a temporal and litter depth gradient (Richards 1974; also see Visser & Parkinson 1975). Time and depth gradients are related monotonically here because of the continuing deposition of litter and thus constitute a complex-gradient. Populations are shown for *Coniosporium, Lophodermium, Pullularia,* and *Fusicoccum* even before pine needles fall because these fungi infect the needles while still on the trees. Jones (1974) shows similar results for aquatic fungi responding to a salinity gradient.

Figure 3.4 shows the vertical distribution in an African rain forest of

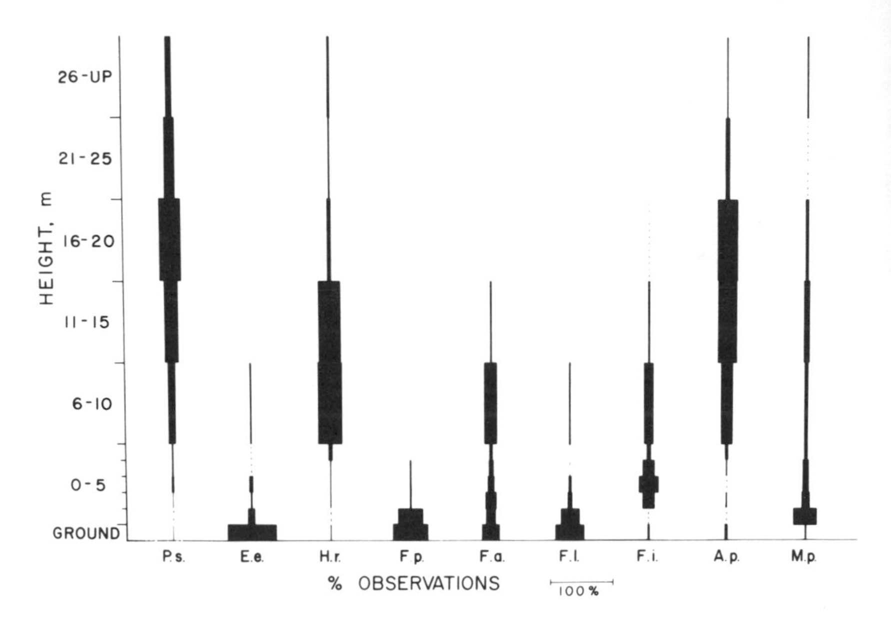

Figure 3.4. Vertical distribution of initial observations of nine squirrel species in an African (Gabon) rain forest. The species from largest (left) to smallest (right) are: *Protoxerus stangeri* (P.s.), *Epixerus ebii* (E.e.), *Heliosciurus rubofrachium* (H.r.), *Funisciurus pyrrhopus* (F.p.), *Funisciurus anerythrus* (F.a.), *Funisciurus lemniscatus* (F.l.), *Funisciurus isabella* (F.i.), *Aethosciurus poensis* (A.p.), and *Myosciurus pumilio* (M.p.). (From Emmons 1980:37. Reprinted by permission of the publisher. Copyright Ecological Society of America)

nine squirrel species (Emmons 1980). The foraging height is a major element in the resource partitioning of these species, along with diet. Adam (1977) presents a similar direct gradient analysis for three rodents in the Ivory Coast. Direct gradient analyses of bird species are presented by Terborgh (1971), Gauthreaux (1978), and Noon (1981).

Figure 3.5 shows an ecocline in chemical factors and animal and plant populations along a complex-gradient of distance downstream from a sewage pollution source (Macan & Worthington 1951). The distance from the source, oxygen, and ammonia nitrogen are all aspects of the pollution complex-gradient; any could have served as a gradient for direct gradient analysis, but distance is used here. Many taxa show typical Gaussian responses; others show bimodal responses because of persistence for a couple of miles after sewage outfall, disappearance for a distance, and then reestablishment after pollution wanes. Decline in pollution is a continuous function of distance, but a classification with four levels of pollution is indicated. Macan & Wor-

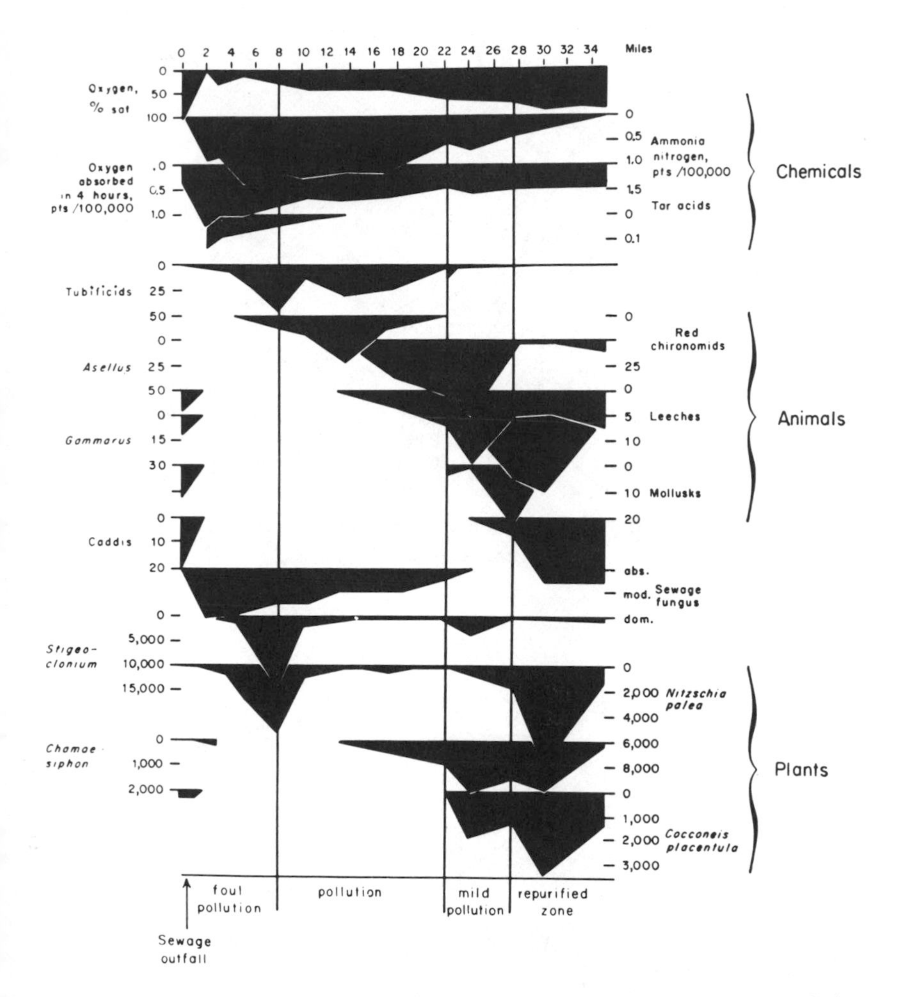

Figure 3.5. Gradients in chemical factors and animal and plant populations caused by sewage pollution in the River Trent, England. (From Macan & Worthington 1951:228)

thington (1951:272) present a similar figure for copper pollution in the River Churnet; also see Balloch, Davies, & Jones (1976); Dor, Schechter, & Shuval (1976); Ozimek (1978); and Arfi, Champalbert, & Patriti (1981).

Figure 3.6 shows tree and insect species populations in the Great Smoky Mountains, Tennessee, along a coenoplane with two complex-gradients, moisture status (from mesic coves to xeric ridges and peaks) and elevation (from 1500 to 6500 ft) (Whittaker 1952, 1956; see Brush,

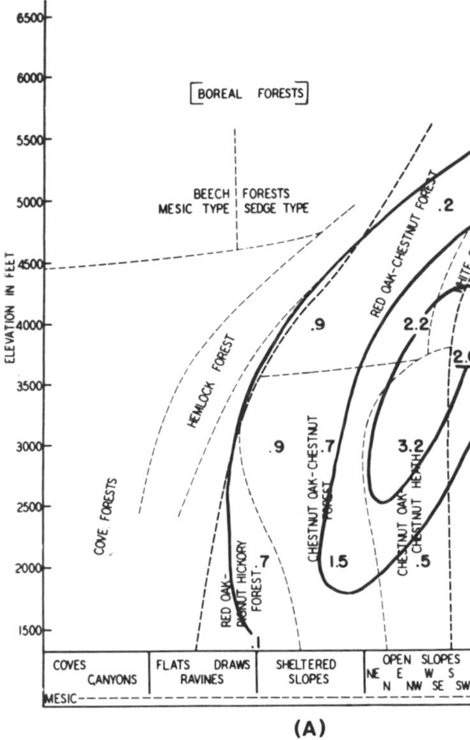
ELEVATION IN FEET
6500
6000
5500
5000
4500
4000
3500
3000
2500
2000
1500
BOREAL FORESTS
HEATH BALD
BEECH FORESTS
MESIC TYPE
SEDGE TYPE
GRASSY BALD
RED OAK-CHESTNUT FOREST
WHITE OAK-CHESTNUT FOREST
TABLE MOUNTAIN PINE HEATH
HEMLOCK FOREST
COVE FORESTS
CHESTNUT OAK-CHESTNUT FOREST
CHESTNUT OAK-CHESTNUT HEATH
PITCH PINE HEATH
RED OAK-PIGNUT HICKORY FOREST
VIRGINIA PINE FOREST
COVES CANYONS
FLATS DRAWS RAVINES
SHELTERED SLOPES
OPEN SLOPES NE E W S N NW SE SW
RIDGES & PEAKS
MESIC
XERIC

(A)

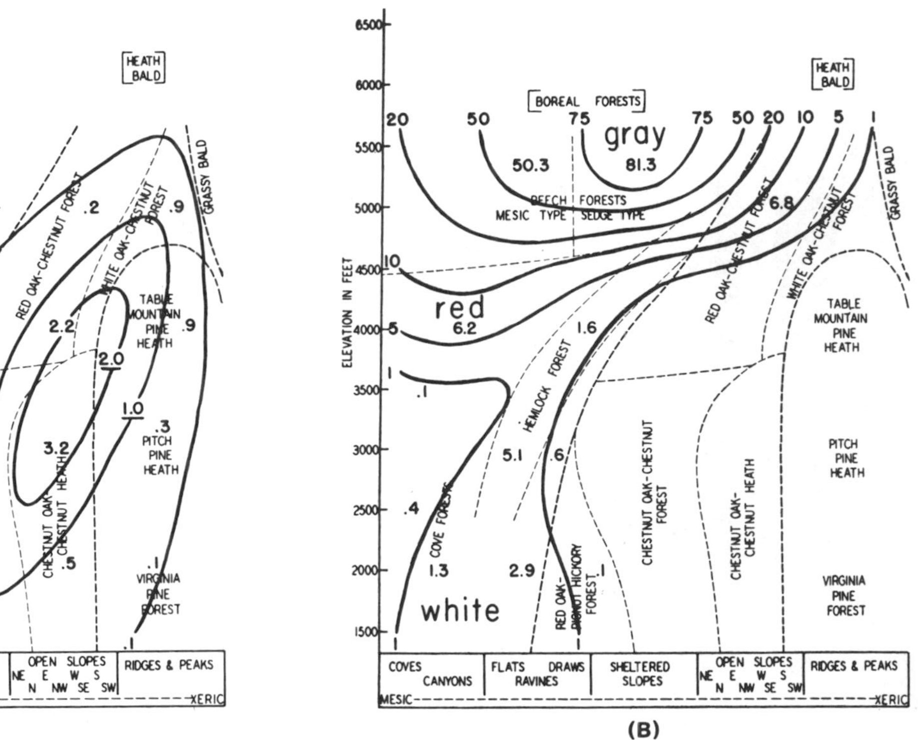
ELEVATION IN FEET
6500
6000
5500
5000
4500
4000
3500
3000
2500
2000
1500
BOREAL FORESTS
HEATH BALD
gray
red
white
BEECH FORESTS
MESIC TYPE
SEDGE TYPE
GRASSY BALD
RED OAK-CHESTNUT FOREST
WHITE OAK-CHESTNUT FOREST
TABLE MOUNTAIN PINE HEATH
HEMLOCK FOREST
COVE FORESTS
CHESTNUT OAK-CHESTNUT FOREST
CHESTNUT OAK-CHESTNUT HEATH
PITCH PINE HEATH
RED OAK-PIGNUT HICKORY FOREST
VIRGINIA PINE FOREST
COVES CANYONS
FLATS DRAWS RAVINES
SHELTERED SLOPES
OPEN SLOPES NE E W S N NW SE SW
RIDGES & PEAKS
MESIC
XERIC

(B)

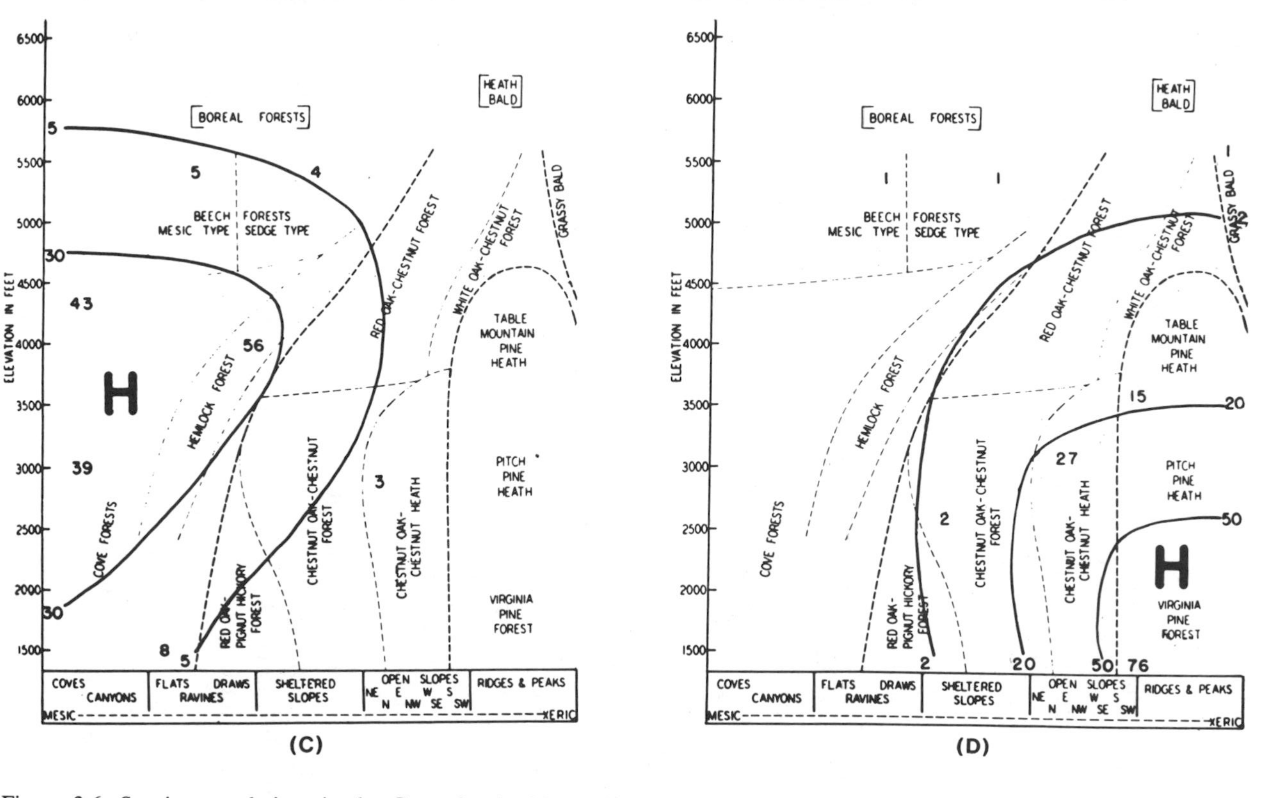

Figure 3.6. Species populations in the Great Smoky Mountains, Tennessee, along gradients of moisture status and elevation. Shown are two tree species, (*A*) *Sassafras albidum* and (*B*) *Fagus grandifolia*, the latter with three ecotypes, and two insect species, (*C*) *Polypsocus corruptus* (Corrodentia) and (*D*) *Kleidocerys resedae* (Hemiptera) (*A–C*, from Whittaker 1952:80; *D*, from Whittaker 1956:74. Reprinted by permission of the publisher. Copyright Ecological Society of America)

Lenk, & Smith 1980 and Roi & Hnatiuk 1980 for similar results). In the background, dashed lines indicate community types dominating each combination of moisture and elevation. *Sassafras albidum* trees grow best in moderately xeric, mid-elevation stands. *Fagus grandifolia* is a complex tree species with three ecotypes (white, red, and gray) showing maximal occurrence in different environmental conditions. Insects favoring mesic habitats (*Polypsocus corruptus*) and xeric habitats (*Kleidocerys resedae*) are also shown. Insect distributions are affected by environmental gradients and by the vegetational gradients (also see Harada 1980 for similar results with mites). This figure suggests that the principles of species individuality and community continuity persist in the case of multiple environmental gradients and are valid for a wide range of taxa, a suggestion confirmed by additional drawings for numerous tree and insect species published in the sources for these figures (Whittaker 1952, 1956). The classic studies by Whittaker precipitated the general acceptance of these principles and presented community types as useful, although arbitrary, groupings rather than as intrinsic, natural features of communities. The supporting quantitative fieldwork was extensive, including counting 25 000 tree stems of 100 species in 300 samples and 30 000 insects of 600 species in 500 sets of sweep samples from 15 sites. Whittaker's (1960) vegetation study of the Siskiyou Mountains, Oregon, extended the direct gradient analysis approach to three gradients (moisture, elevation, and soil parent material), and Kessell's (1979) study of Montana forests used four gradients (moisture, elevation, drainage basin, and time since last fire); also see Itow (1963) and Sobolev & Utekhin (1978).

Figure 3.7 concerns abundances of several fossil marine benthic invertebrate genera in an Ordovician basin near Utica, New York (Cisne & Rabe 1978). The height in the standard stratigraphic section measures age from oldest (0 m) to youngest (38 m). The whole time span, roughly 2 million years, is too short to be accurately measurable by radioactive decay dating methods in these 440 million-year-old rocks. The transect is laid out directly down a slope, parallel to the directions of downslope sediment movement, so that the distance from the zero point, at the shallow water end, measures relative water depth. The horizontal panels depict the distribution of fossils along the transect at times when volcanic ash layers were deposited along the slope. (These ash layers are the time markers in relation to which the ages of strata along the transect have been determined.) The vertical panels depict the temporal distribution of fossils in selected stratigraphic sections.

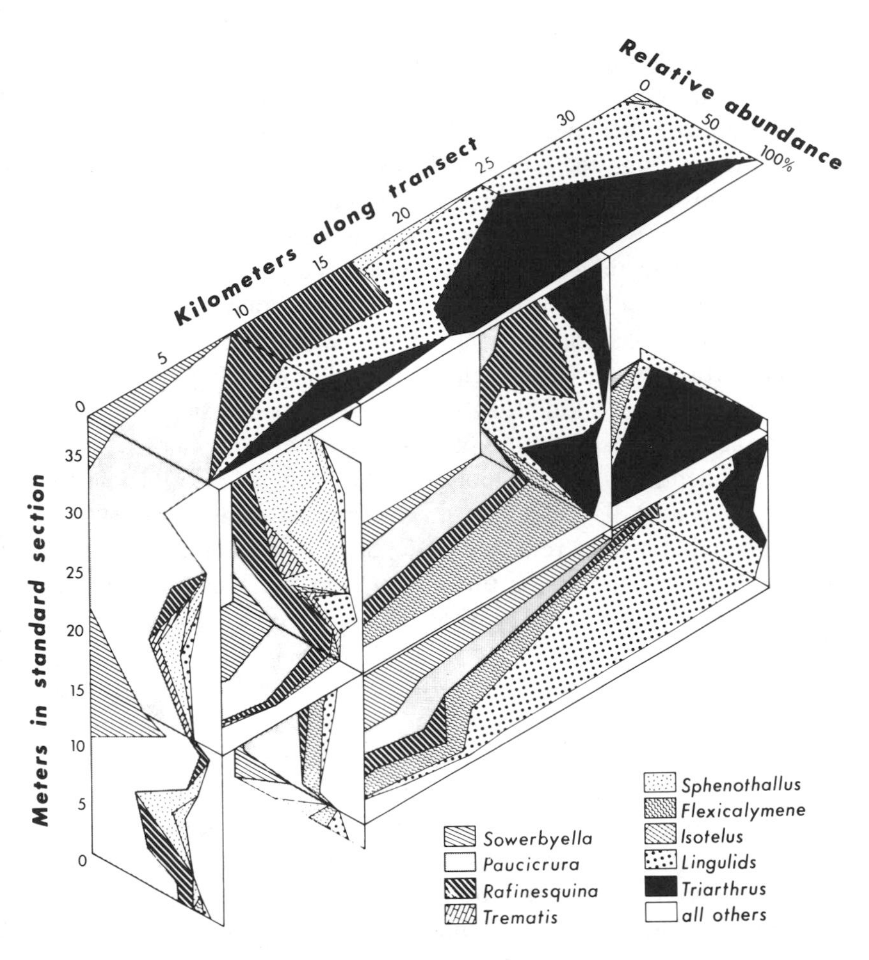

Figure 3.7. Fossil marine benthic invertebrate genera in an Ordovician basin near Utica, New York, with two environmental gradients: depth in standard section (reflecting an estimated time span of 2 million years) and kilometers along transect (reflecting location down a basin slope into deepest waters, with deeper waters corresponding to sites furthest along the transect). (From Cisne & Rabe 1978:348)

The distribution of fossils at any one time reflects water depth. As can be seen in the horizontal panel for the fifteenth meter in the standard section, the brachiopods *Sowerbyella* and *Paucicrura* are the most abundant invertebrates in the shallowest water (at 0 km along the transect), the trilobite *Flexicalymene* is generally the most abundant one at intermediate depths, and the trilobite *Triarthrus* and lingulid brachiopods are the most abundant ones in the deepest waters (at 34 km on the transect). The brachiopod *Trematis* is sometimes fairly common at shallow depths, and the conularid cnidarian *Sphenothallus* and the trilobite *Isotelus* are sometimes abundant at intermediate depths. All three horizontal panels show basically the same patterns of fossil distribution, although the patterns themselves are variously positioned toward the upslope or downslope direction at any one time, reflecting changes in water depth. The vertical panels show more clearly how depth changes with time. They repeat changes associated with shallowing or deepening on the horizontal panel. For instance, in going from 15 to 18 m in the vertical panel at 24 km along the transect, one sees the same sort of change in the relative abundance of fossils as one sees in going from 24 to 34 km along the horizontal panel for the fifteenth meter in the standard section. Note also how the distribution of fossils records the depth fluctuation between the 15-m and 25-m points in vertical panels at 0, 9, and 24 km on the transect. The overall pattern of environmental history reflected by fossils is a deepening of waters in downslope areas as the region tilts about a tectonic hinge located near the 5-km point on the transect. Thus the section on the shallow side of the hinge (at 0 km on the transect) shows relatively little faunal change over time, while sections well downslope (at 24 km) show a dramatic change. Cisne & Rabe found direct gradient analysis and ordination of fossil communities effective tools for studying paleoenvironmental gradients, even for detection of rather subtle features in local paleogeography and tectonics. In turn, these results help with time correlation.

It is important to realize that although published figures usually show only smoothed curves, the original data are ordinarily rather noisy. Figure 3.8, panel (*A*), shows the abundances of black oak (*Quercus veluntina*), white oak (*Q. alba*), red oak (*Q. rubra*), and sugar maple (*Acer saccharum*) in upland forest stands of southern Wisconsin along a succession gradient from early (left) to mature (right), plotting the original data for 95 stands (Curtis & McIntosh 1951). Figure 3.8, panel (*B*), shows the same data after averaging successive groups of five stands and smoothing (using the formula: previous datum plus twice the present datum plus next datum, divided by four). The scatter in these original data is typical for ecological community data, and the contrast between the original data and smoothed curves emphasizes the utility of averaging and smoothing to reduce noise and, likewise, the desirability of obtaining a substantial data base to provide enough samples for accurate averages. Goodall (1954*a*) and Goldsmith & Harrison (1976) show similar figures.

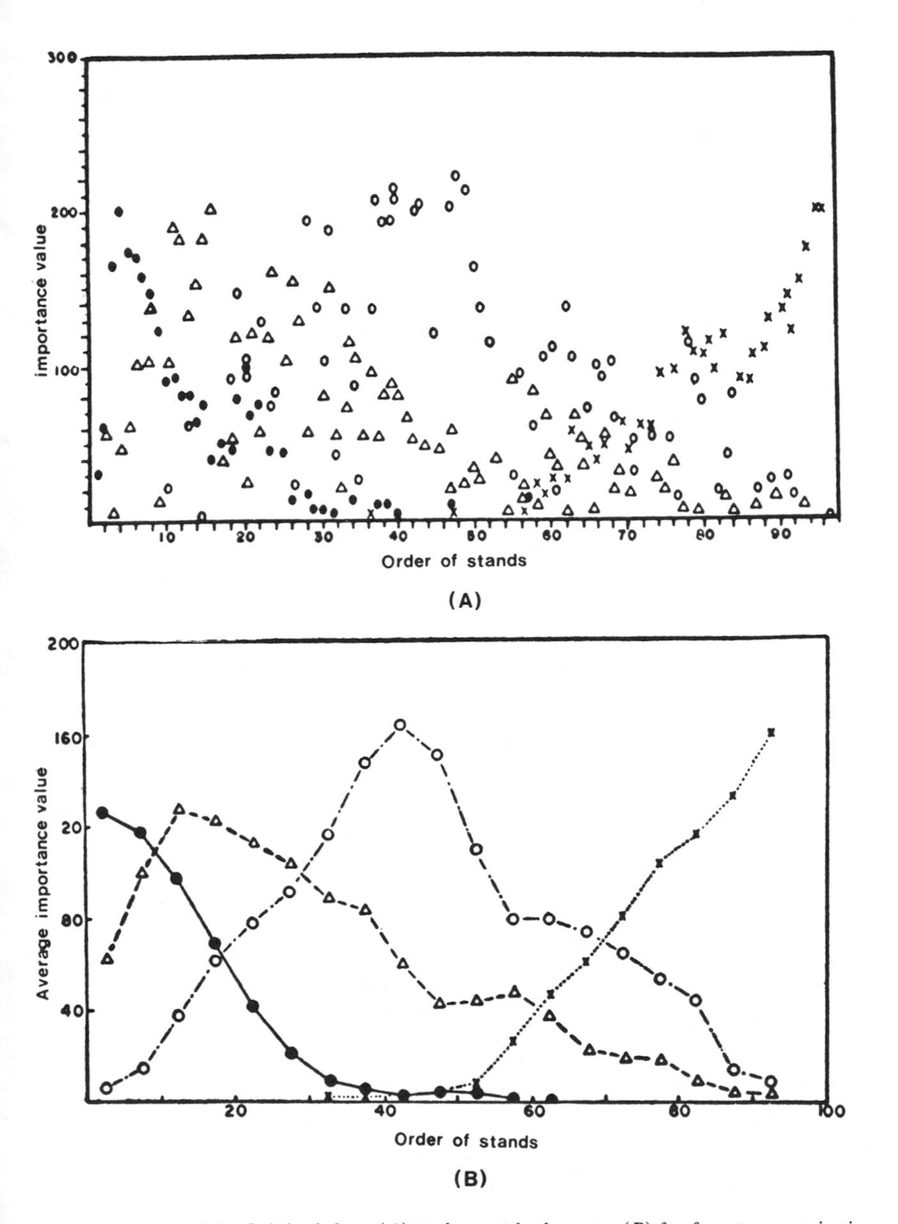

Figure 3.8. Original data (*A*) and smoothed curves (*B*) for four tree species in upland forests of southern Wisconsin along a succession gradient: black oak (●), white oak (△), red oak (○), and sugar maple (X). Species abundances are measured by the importance value, which combines relative (percent) density, dominance, and frequency and has a possible range of 0 to 300. The order of these 95 stands was derived by a careful, but subjective, procedure, which expresses succession from early (left) to mature (right). (From Curtis & McIntosh 1951:487–8)

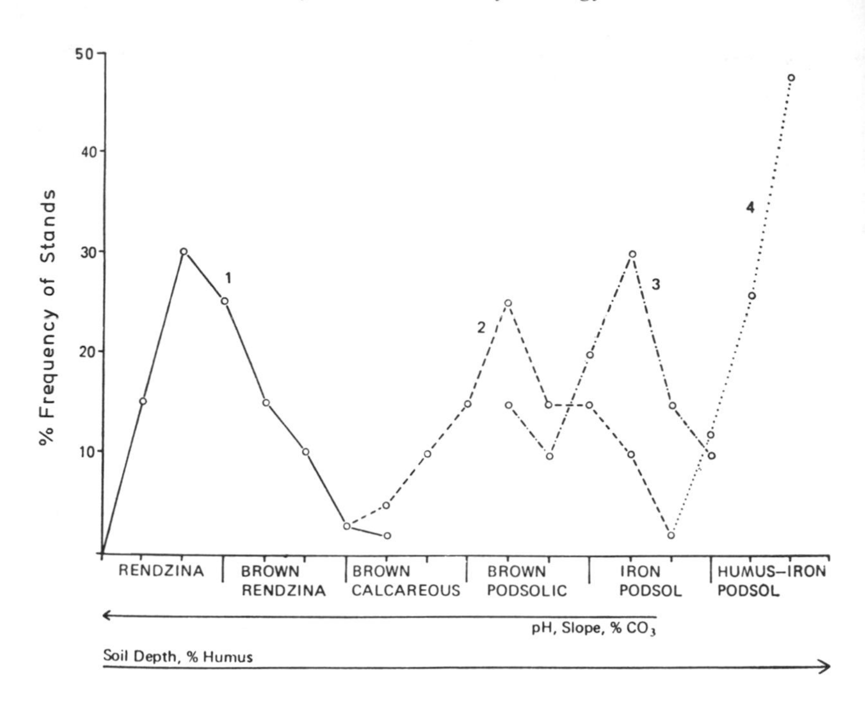

Figure 3.9. Distribution of plant community types along a complex soil gradient in the Peak District of Derbyshire, England. From left to right, the soils have increasing depth and humus, and decreasing pH, slope, and carbonate. The community types are (*1*) *Helictotricho–Caricetum flaccae* limestone grassland, (*2*) *Agrosto–Festucetum* grassland, (*3*) *Nardo–Galietum* grass heath, and (*4*) *Trichophoro–Callunetum* heath. (From Shimwell 1971:241)

The examples shown thus far present species abundances along environmental gradients. Although this is the central approach in direct gradient analysis, other variables or gradients are also used, as will be illustrated in the following examples.

Figure 3.9 depicts a gradient in frequency of plant communities along a soil complex-gradient in the Peak District of Derbyshire, England (Shimwell 1971; also see Borowiec, Kutyna, & Skrzyczyńska 1977; Moral & Watson 1979; Huntley 1979; Brush, Lenk, & Smith 1980; and Feoli & Feoli Chiapella 1980). The soil gradient, from left to right, involves increasing soil depth and humus and decreasing pH, slope, and carbonate. The corresponding plant communities are *Helictotricho–Caricetum flaccae* limestone grassland, *Agrosto–Festucetum* grassland, *Nardo–Galietum* grass heath, and *Trichophoro–Callunetum* heath. Entire assemblages of species as represented by these com-

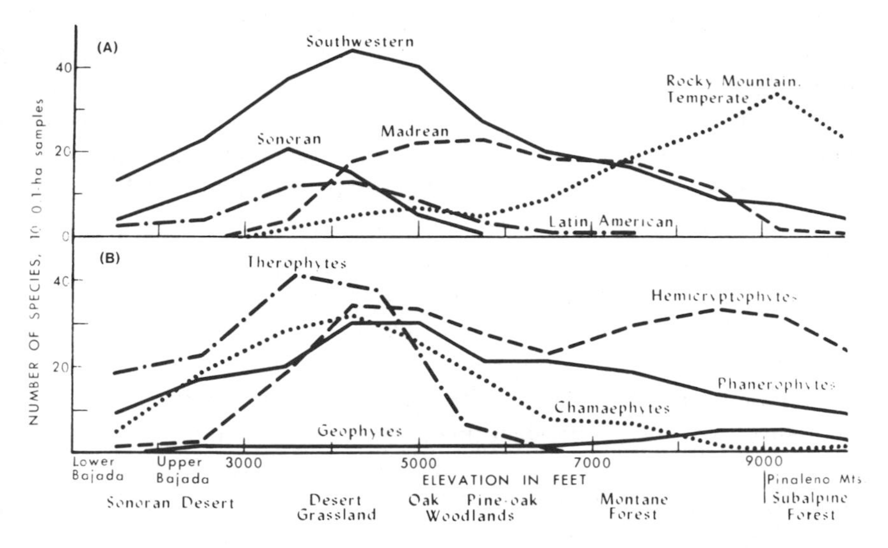

Figure 3.10. Community-level variables for the vegetation along an elevation gradient on the south side of the Santa Catalina Mountains, Arizona, for sites on open, subxeric slopes. Panel (*A*) shows geographic affinity of the species present and panel (*B*) shows life-forms. (From Whittaker & Niering 1964:14)

munity types show approximately Gaussian curves with scattered modes, much like the curves for individual species considered earlier.

Community-level vegetation characteristics are shown in Figure 3.10 along an elevation gradient on the south side of the Santa Catalina Mountains, Arizona, for sites on open, subxeric slopes (Whittaker & Niering 1964; also see Whittaker 1978*b*). Panel (*A*) shows the geographic affinity of the species present: high-elevation forests have mainly northern, Rocky Mountain affinities; at middle elevations, woodlands contain many Madrean species; at low elevations, deserts incorporate mainly southwestern and Sonoran species (these regions are defined in detail by Whittaker & Niering 1964). Panel (*B*) shows the life-forms of the species present. Life-forms, as developed by Raunkiaer (1934), classify species primarily on the basis of location of the shoot meristem or embryonic tissue during the unfavorable season. The life-forms, in order of decreasing height of the meristem or embryonic tissue, are: phanerophytes (trees and shrubs), chamaephytes (shrubs primarily below 25 cm), hemicryptophytes (surface deciduous and evergreen perennial herbs), geophytes (herbs with subterranean meristems), and therophytes (annual herbs persisting by seeds; this classification is described in greater detail by Whittaker & Niering

1964). At high elevations, hemicryptophytes predominate, as is typical for cool-temperate forests. At low elevations, therophytes predominate; annual herbs surviving the dry season as seeds. Direct gradient analysis of community-level characteristics may be extended to two or more dimensions, as in analyses along moisture and elevation gradients in the Great Smoky Mountains of primary productivity (Whittaker 1966), tree species diversity (Whittaker 1956), and insect species diversity (Whittaker 1952); also see Itow (1963), Bakuzis & Hansen (1965:42–7), Peet (1978), and Ellenberg (1979).

Rather than community change along environmental gradients traditionally studied by direct gradient analysis, Figure 3.11 concerns a spatial pattern within communities. Both panels concern vegetation with shrub patches in grassland matrices: (*A*), a mesquite grassland in Texas (Whittaker, Gilbert, & Connell 1979; Whittaker & Naveh 1979); and (*B*), a mallee in New South Wales, Australia (Whittaker & Naveh 1979; Whittaker, Niering, & Crisp 1979); also see Shmida & Whittaker (1981). Community samples were taken in a straight transect of about 100 contiguous 1-m^2 quadrats, intended to traverse several shrubby patches. The gradient on the abscissa is merely physical location along this 100-m transect. The ordinate is a score for each quadrat, which integrates the entire species composition into a single number reflecting grassy (100) to shrubby (0) composition. (This score is obtained by an ordination technique, reciprocal averaging, which will be discussed in the next chapter.) Consequently, this figure shows the nature of the vegetation (from grassy to shrubby) as one moves along a 100-m transect within a complex community.

These direct gradient analyses of within-community spatial patterns differ from the traditional analysis of between-community variation with which this chapter began both by intent and methods. Different methods are used, both for the abscissa and the ordinate: instead of a customary environmental gradient such as moisture status or elevation, the environmental gradient is merely direct physical distance along a transect; instead of individual species abundances, the variable is an ordination score, which integrates the overall species composition into a single number. A traditional analysis with a single grassy-to-shrubby transect could be used to characterize this coenocline. The analysis shown here, however, has the advantages that (1) numerous grassy and shrubby patches are encountered in this long transect, which shows the spatial pattern more fully, thereby (2) spatial pattern can be described and quantified in terms of several parameters as discussed by Whit-

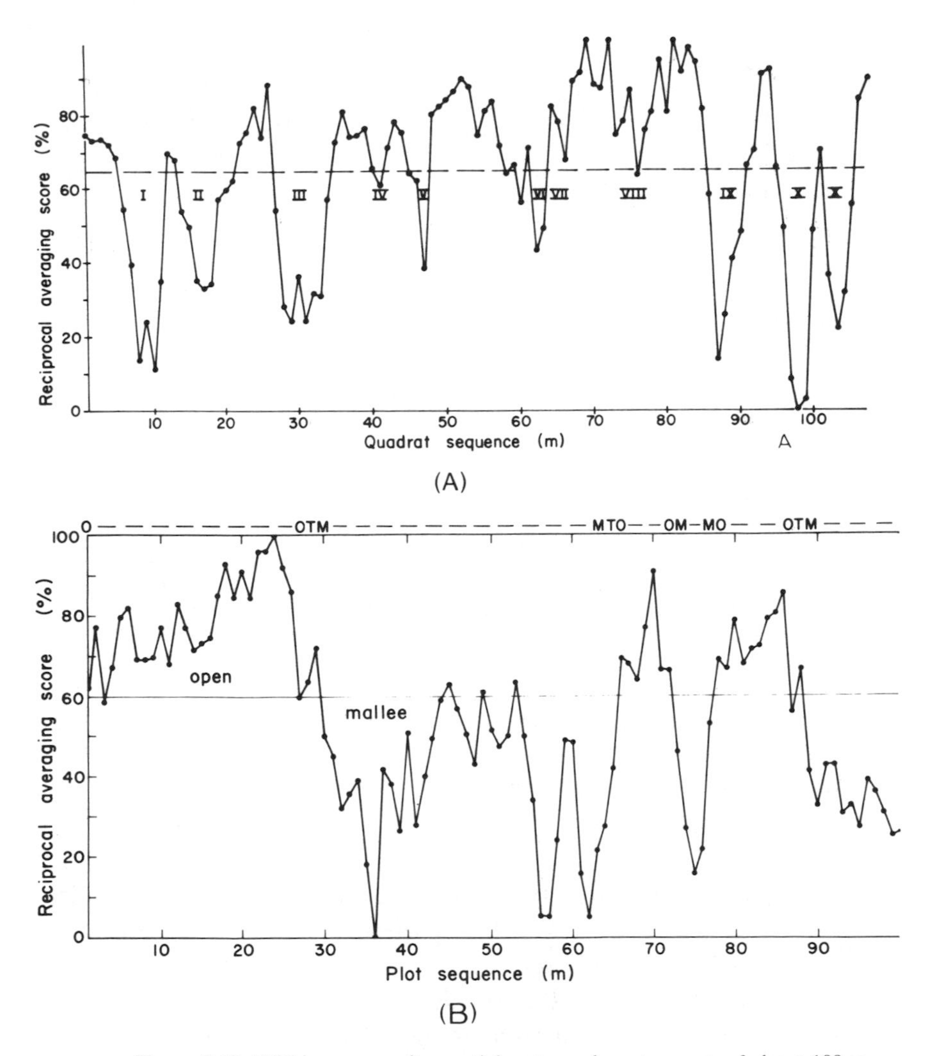

Figure 3.11. Within-community spatial pattern along transects of about 100 m in vegetation with shrub patches in grassland matrices: (*A*), a mesquite grassland in Texas; and (*B*), a mallee in New South Wales, Australia. The reciprocal averaging ordination score integrates the entire species composition of each 1-m^2 quadrat into a single number, which in these cases reflects grassy (100) to shrubby (0) species composition. For example, in (*A*), the transect begins with five open, grassy quadrats, followed by two transitional quadrats, then three shrubby quadrats, and so on, in a loosely repeating pattern. (*A*, from Whittaker, Gilbert, & Connell 1979:139; *B*, from Whittaker, Niering, & Crisp 1979:70)

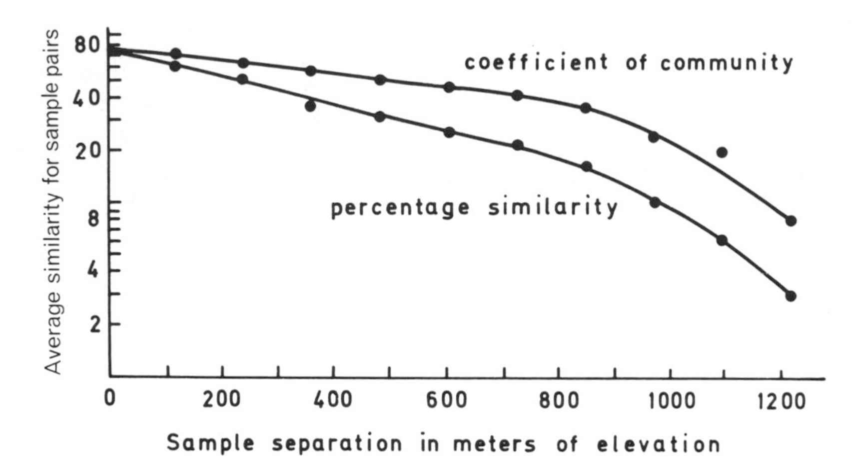

Figure 3.12. Community similarity as a function of sample separation along an elevation gradient for vegetation in the Great Smoky Mountains. The two indices of community similarity shown, percentage similarity and coefficient of community, are defined in the text, as is the ordinate term. (After Whittaker 1960:321. Reproduced by permission of the publisher. Copyright Ecological Society of America)

taker & Naveh (1979). Further discussion of mosaics in natural communities may be found in the work of Bratton (1976), Dierschke (1977), Whittaker & Levin (1977), Cormack (1979), Veblen & Ashton (1979), and Wolff (1980). The approach of Figure 3.11 also differs from traditional analyses of spatial pattern, which concern contagion for one species or species association for two species. Such traditional analyses are appropriate for cases in which there is no strong coordination among many species in their patterns of occurrence. This case, however, is likely to be rare. In contrast, the approach of Whittaker & Naveh (1979:157) is integrative, "dealing with the whole flora at once to reveal major axes of differentiation in the community, axes to which most or all species relate." Its applicability is limited, though, to communities with well-defined patterns and a reasonable number of species (at least about 20). Trial applications to ordinary forest undergrowth were not rewarding (because of failure to meet the first of these two conditions).

Community similarity as a function of sample separation along an elevation gradient in the Great Smoky Mountains is shown in Figure 3.12 (Whittaker 1967; also see Taylor 1977). The abscissa shows sample separation in meters of elevation. The ordinate shows an abstract, mathematical property of the vegetation coenocline of these mountains, namely the average community similarity for pairs of samples separated by increasing differences in elevation. The two community similarity indices used have the range 0 (for entirely unlike community samples) to 100 (for identical samples); the curves do not extrapolate to 100 for replicate samples (on the left) because of sample error and noise. The percentage similarity for a sample pair equals 200 times the sum over species of the smaller of the two samples' abundances, divided by the sum over species of the sum of the two abundances. The coefficient of community is the same except that all abundances are first converted to 1 for presence and 0 for absence. Similar results to those of Figure 3.12 have been found for marine benthic living and fossil communities along a water depth gradient (Cisne & Rabe 1978) and bird communities along an elevation gradient (Terborgh 1971). Gauch (1973*b*) mathematically derives the relationship between sample similarity and ecological separation assuming a Gaussian model for coenoclines, and Swan (1970) derives this relationship by simulation studies (also see Groenewoud 1976). These theoretical results agree well with those shown in Figure 3.12.

A final example, from Glacier National Park, Montana, concerns an applied study for resource and fire management (Kessell 1979). In this study, direct gradient analysis functions within a complex system having several components. (1) Direct gradient analysis is used to estimate plant and animal populations and fuel quantities as functions of four major gradients: elevation, time since last burn, moisture status, and drainage basin. For example, Table 3.1 gives median fuel quantities for forests as a function of time since last burn. The direct gradient analysis uses two additional gradients of local importance (primary succession and alpine wind–snow exposure), and there are various disturbance factors, but these matters need not be discussed here. (2) From aerial photographs for a 40 000-ha area, a high-resolution (10-m) site inventory describes site elevation, topography, vegetation type, stand age, and cultural and hydrological features. This information suffices to define the position of each site along the four major gradients. (3) There is a weather model, and weather data are kept current. (4) A fire behavior model predicts the spread rate, intensity, and flame length for given conditions of weather, vegetation, and fuel quantity. Figure 3.13 shows one aspect of the fire model, the relationship between fire spread and wind speed. In using the fire model, the site inventory gives a site's gradient position; from this information, direct gradient analysis is used to estimate crown, shrub, dead and down, and

Table 3.1. *Median fuel quantities for forest stands in Glacier National Park, Montana, stratified by stand age (time since last burn)*

	Stand age (years)								
Sample	0–10	11–44	45–69	70–89	90–109	110–29	130–49	150–69	Over 169
Size (*N*)	12	7	21	58	15	29	6	18	48
Fuel loads (tons/ha) for:									
Litter	0.67	2.24	2.00	2.08	1.56	1.74	3.97	2.30	2.15
Grass and forbs	0.48	0.45	0.33	0.30	0.34	0.24	0.17	0.36	0.21
Branchwoods									
1 hr	0.42	1.05	1.06	0.72	0.88	0.96	0.86	1.02	1.02
10 hr	2.77	1.18	1.62	1.18	1.24	1.35	1.65	2.19	1.70
100 hr	6.02	3.37	3.38	1.69	3.51	3.43	5.98	1.86	1.70
1000 hr	27.80	32.51	8.93	15.51	9.65	15.00	29.86	29.49	46.71
Shrubs									
Foliage	1.13	0.74	0.40	0.41	0.42	0.44	0.31	0.44	0.20
Branchwood	3.76	5.39	1.35	1.75	1.43	1.60	1.08	1.24	0.68

Source: From Kessell (1979:117)

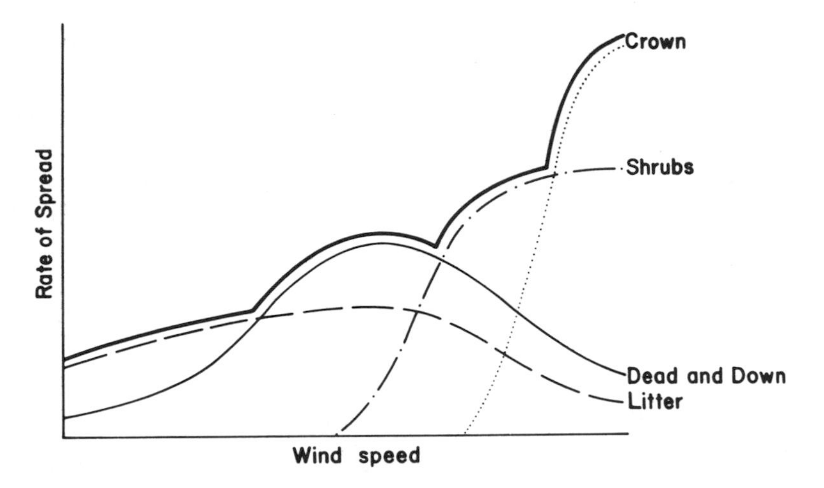

Figure 3.13. Idealized relationship between the rate of fire spread and wind speed for various fuel strata; the fire speed (heavy solid line) equals the fastest speed in any stratum. (From Kessell 1979:197)

litter fuels (as in Table 3.1); and, finally, from this and weather data, the fire model is used to estimate fire spread rate (as in Figure 3.13). (5) Succession models predict the vegetation likely to cover a site as a function of time, both for burned and unburned sites (also see Noble & Slatyer 1980). (6) Vertebrate habitat utilization models indicate the value of various vegetation covers to vertebrate species, for feeding and for reproduction. Some animals require a diversity of habitats within an area of acceptable extent. Together with the site inventory, this information helps to indicate, for example, whether the grizzly bears in a given region of the park need additional, recently burned habitat with abundant huckleberries. (7) Water sources, access trails, buildings, and so on are also indicated in the site inventory. (8) Computer programs integrate the preceding data and models, providing a fire officer with useful information and predictions. The coordinates of a new fire may be supplied to the computer model, for example, and a report produced in a minute, giving the estimated rate of spread of the fire, the fire control options by various means suited to various access possibilities, the likely course of the fire if not controlled given the present and forecasted weather conditions, the projected vegetation cover with and without fire, and the implications for wildlife habitat. Such extensive, rapid, computer output helps fire officers in making the complex fast decisions required in fire management.

The expense of this fire management system is a small fraction of the loss incurred in a single large fire and, hence, appears quite reasonable. This complex system has many components, but direct gradient analysis is crucial because it provides detailed vegetation and fuel quantity estimates from the minimal site inventory descriptions necessitated by cost considerations. This application of direct gradient analysis is possible in the first place because of the typical structure of ecological communities; a few major environmental gradients largely determine the population levels of numerous plant and animal species and the values of numerous subordinate environmental factors.

Data set properties

A fundamental endeavor of ecologists is "abstracting certain generalities from the vast array of particulars" (Colwell & Fuentes 1975:281; also see Williams 1962 and McIntosh 1980). In this section, generalities will be sought about species and community distributions along environmental gradients, based on examples in the preceding section and additional references. The resulting generalities serve to provide an empirical basis for models of community structure. In turn, this provides a theoretical basis for designing appropriate, effective algorithms for ordination and classification techniques.

The most striking feature of the direct gradient analysis examples is the general similarity of species distributions along environmental gradients, whether the species are fungi, algae, higher plants, mollusks, insects, or mammals and whether the environmental gradient is depth in soil, elevation on a mountain, distance from a pollution source in a stream, or depth in a core of fossil samples. In these varied cases, most species reach a maximum at some point on the environmental gradient and decline to either side, producing a species response curve of an approximately bell-shaped or Gaussian form. The species are individualistic (each different from others) in the locations and heights of their modes, as well as the widths of their dispersions along the gradient. Because the species are individualistic instead of grouped into associations, community composition changes continuously along the gradient. These are the salient features of the *Gaussian model* of community structure, as will be discussed in greater detail in this section.

The Gaussian model of community structure is a statement in mathematical terms of the general features of community data. It may be presented under 12 points (Gauch & Whittaker 1972*a*, 1976).

(1) The abundances of each species along an environmental gradient generally form a bell-shaped, unimodal curve approximating the Gaussian (normal) curve. The Gaussian curve has three parameters: the mode μ, the maximum value Y_0, and the dispersion σ in units of standard deviation. If X represents the position along the environmental gradient, Y the species' abundance, and exponentiation uses the base of natural logarithms (≈ 2.718), the Gaussian equation is

$$Y = Y_0 \exp\left[-(X - \mu)^2/2\sigma^2\right]$$

Given two environmental gradients, a species' distribution generally approximates a Gaussian curve along each gradient, together constituting a Gaussian response surface over the plane defined by the two gradients. The bivariate Gaussian distribution requires modes and standard deviations on two axes, a single maximum, and an angle of rotation specifying the orientation of the surface with respect to the environmental gradient axes, for a total of six parameters (Gauch & Whittaker 1976). The Gaussian distribution is readily extended to three or more dimensions (Gauch, Whittaker, & Singer 1981).

Commonly, the communities in a study are affected by only one or a few major environmental gradients. Additional secondary gradients may have a small effect on all the samples or a notable effect on only a small fraction of the sample set. For example, the vegetation of a mountain range may be affected predominantly by major gradients of elevation and moisture status. Secondary gradients might include a pervasive, but small, effect of soil calcium level and, for only alpine samples, a strong effect of wind and snow depth.

The Gaussian species response model is only an approximation. For example, Whittaker (1956) plotted 44 tree distributions in the Great Smoky Mountains in a coenoplane involving gradients of elevation and moisture status. Of these 44 species, 38 were unimodal (essentially Gaussian), 5 were bimodal, and 1 (*Fagus grandifolia*) was trimodal. Austin (1976*a*, 1980*b*) and Austin & Austin (1980) note that bimodal and skewed species responses are not rare (also see Groenewoud 1976 and Prentice 1980*b*). The Gaussian model is presented as and should be received as an approximation – a generality from a vast array of particulars. At its level of detail, however, no alternative model appears better. The situation is analogous to curve fitting of, say, 50 points; a polynomial with 50 adjustable constants can fit the data perfectly, but a polynomial or other form with, say, three constants is likely to be imperfect. Here the Gaussian curve has the advantages that it fits the empirical data as well as any curve with no more than three constants and it has good mathematical tractability for theoretical work. The Gaussian model is an approximation, but it is an excellent approximation. Some approximation must be made in order to speak of the general properties of communities because the alternative, without approximation, is to speak on the level of raw data in their full particularity.

(2) The modes of species distributions are scattered individualistically along an environmental gradient. The species do not tend to be concentrated into groups of highly associated species with little overlap. Minor species are scattered at random, but major species may show relatively regular spacing, which reduces the implied competition between them (Whittaker, Levin, & Root 1973; Whittaker 1978*b*).

(3) Maximum abundances in any single sample for each species may form lognormal or lograndom distributions. Species maxima may be grouped into octaves, where each octave is a doubling of the maximum value (for example, seven octaves may be used for maxima with values 1 to 2, 2 to 4, 4 to 8, 8 to 16, 16 to 32, 32 to 64, and 64 to 128, with a species on a dividing point contributing a half count to both octaves; Preston 1948). For temperate forest trees, the numbers of species in the octaves are often approximately equal. This distribution is termed *lograndom* because the octaves are a logarithmic scale and random allocation leads to approximately equal octave counts. When maxima were examined for a large number of forest herbs or desert shrubs and herbs, the numbers of species in the octaves were uneven, being approximately normal (Gaussian). This distribution is termed *lognormal* because it involves a normal curve with a logarithmic scaling of maximum abundances. Especially for communities having a large number of species (greater than 25 or 50), the lognormal is common. With smaller numbers of species, the lograndom and lognormal are rarely, if ever, empirically distinguishable. The lograndom and lognormal distributions of species maxima may be considered as alternatives for different circumstances, with the lognormal most common.

(4) Species dispersions vary, with some species restricted to narrow ranges along an environmental gradient and others occurring more broadly. For example, in Figure 3.3, the fungus *Pullularia* has a relatively narrow distribution, whereas *Penicillium* has a broad distribution. Coenoclines involving numerous species have been found to have a set of species dispersions tending to form a normal distribution, with a standard deviation of about 0.3 times the average value.

(5) Environmental gradients used as separate axes in direct gradient analysis may, in fact, not be entirely independent in their effects on the communities. Instead, there may be a partial correlation between gradients. For example, in Figure 3.6*A*, the bivariate Gaussian distribution for *Sassafras albidum* in the Great Smoky Mountains is elongated at an angle of about 50° from the topographic moisture conditions axis on the abscissa (in contrast to the alternatives of being elongated in no direction or in the direction of the abscissa or the ordinate). Consequently, moisture and elevation are partially correlated in their effects on *S. albidum* abundance. In this case, numerous additional tree species have a similar orientation in their distributions, so an overall partial correlation exists between moisture and elevation. Such partial correlation occurs frequently because of the interacting effects of the environmental gradients on the physiology of organisms, trends in the sample set of one environmental gradient wih respect to another, or interacting effects in terms of some more general environmental gradient affected by both of the original gradients. In this example, a trend of decrease in climatic humidity toward lower elevations is probably the major cause of partial correlation.

Partial correlation in community effects of environmental gradients may be accommodated mathematically by the use of nonzero values for the rotation parameter in the bivariate Gaussian distribution (Gauch & Whittaker 1976). In the example of trees in the Great Smoky Mountains in relation to moisture and elevation, the rotation angles for these 44 species show an average value of about 20° (from the moisture axis toward the elevation axis), with normally distributed deviations from this average.

In other cases, the rotation angles may be uniformly small; in other words, the environmental axes may be nearly independent. In still other cases, individual species may show marked elongations in directions other than those of the axes, but there may be no tendency for the rotation angles to point in a consistent direction. This case implies that the interactions of the environmental factors do affect species, but the effects differ from one species to another.

There does not appear to be any basis for consistent expectation of any of the three possibilities: uniform orientation, random distribution, or normal distribution of rotation angles. These possibilities imply, respectively, no interaction of environmental gradients on species distributions, present but inconsistent interactions, and consistent interactions.

(6) Given the parameters describing each species (such as the three parameters for the Gaussian curve or the six for the bivariate Gaussian surface), one may ask whether correlations relate some of these. In particular, one might suspect a correlation to exist between the maximum abundances and the dispersion widths of species or between the dispersion widths along one gradient and the dispersion widths along another gradient. Correlations between these or any other pairs of parameters do not appear to characterize field data.

(7) A gradient of community composition (coenocline) can be characterized by the amount of *species turnover* from one end to the other. This property is termed *beta diversity* by Whittaker (1967).

Species turnover can be conveniently expressed by either of two units. One unit is the *half-change* (HC), defined as the ecological separation at which the similarity between two samples is half of the value for replicate samples (where similarity is computed by any measure of between-sample similarity, such as percentage similarity). The other convenient unit is the number of *average standard deviations of species turnover* along a gradient (SD; Hill 1979*a*; Hill & Gauch 1980). This unit is identical to the Z unit of Gauch & Whittaker (1972*a*), but this latter symbol carries no connotation as to the unit's meaning, so we prefer SD. As an example, if the average dispersion width of species along an elevation gradient is 300 m and this ecocline spans 1500 m, then the beta diversity is 5 SD units. If data have no noise, Gauch (1973*a*) shows that 1 HC is equivalent to 1.349 SD. Noise lowers this factor of 1.349, and for typical field data, this factor is approximately 1 (Hill 1979*a*:8).

Visual inspection suffices for gross estimation or comparison of species turnover. Its more quantitative measurement has been undertaken in the GRADBETA computer program by Wilson & Mohler (1981) and in DECORANA by Hill (1979*a*); also see a graphical technique by Whittaker (1960). There are, nevertheless, difficult problems. One problem is that noise decreases the similarity between adjacent samples along the ecocline and, hence, inflates species turnover (so that species turnover estimates decrease as one averages or smooths the data, but averaging and smoothing are incidental matters, which should not affect species turnover measurement). Another problem occurs when more than one community gradient is present and it is necessary (but often difficult) to pick out sample subsets that traverse only one coenocline at a time (while holding constant the locations on other gradients). Subsequently, the difficult task remains of piecing together the overall pattern from these individual slices. Fortunately, many purposes of characterization and comparison involve gross variations in species turnover, and consequently, modest accuracy is quite sufficient.

(8) Community studies vary in the number of species encountered in a sample (alpha diversity) and in the total number of species encountered in the study (gamma diversity, Whittaker 1960, 1967). Alpha diversity may change from one part of the community pattern to another. For example, in many mountain ranges, the alpha diversity of trees decreases as elevation increases (but compare Peet 1978).

(9) Gaussian species response surfaces may be modified by competition. Species are not distributed merely in accordance with their physiological tolerances; rather, distributions are further limited and modified by competition with other species (Cormack 1979). Competition is a frequent cause of bimodal species responses. For example, in ravines in the Finger Lakes region of New York, eastern hemlock (*Tsuga canadensis*) is a strong competitor at middle heights (Lewin 1974). Because of this dominance of *Tsuga,* other tree species show bimodal distributions.

A simple way to model competition is to assume that the total for all species abundances at each position on the coenocline must equal a constant (or some simple function of coenocline position), representing the environmental carrying capacity, and reduce (or increase) all species abundances for that position by the same proportion to result in the desired total.

(10) The flanks of Gaussian surfaces extend indefinitely. In real field data, there are coenocline distances beyond which a species is never found, and short of this, there is also a cutoff point beyond which the species is too rare to be found with any regularity within sample plots of a given size. The point at which a species is absent (or very rare) in real field data is analogous to mathematically truncating the flank of a Gaussian surface at some small value (such as 0.1% or 1% of the maximum species abundance).

(11) Field data are noisy. By noise is meant variations in one species's abundances not correlated with variations in other species' abundances, as discussed in the introduction (Poore 1956; Gauch 1981). The noise in community data can be simulated by various equations employing random numbers (Gauch & Whittaker 1972*a*; Gauch 1981).

Field data vary in noise level. The noise level is best estimated by calculating a similarity measure for pairs of replicate samples collected at the same position along recognized environmental gradients and taking an average. Ordinarily replicate samples are taken at physically close sample sites to better guarantee uniformity of physical conditions. For field data, the percentage similarity of replicate samples is not 100% (no sample noise), but rather 80 to 90% for low noise, 70 to 80% for medium noise, and 50 to 70% for high noise.

(12) Patterns of sample placement within the community gradients present may vary, as discussed in the previous chapter on sampling methods. Various sample placement procedures can be modeled as (a) random placement, (b) placement in a regular transect, grid, or cube, and (c) deliberate placement in order to produce sample clusters or gaps, even or uneven representation, or any desired pattern.

The preceding 12 points describe the general structure observed for the distribution of species populations along environmental gradients. The Gaussian model of communities is incorporated in the community computer simulations of Gauch & Whittaker (1972*a*, 1976).

Frequently, community-level variables (such as community type and life-form) also have Gaussian distributions along environmental gradients. This may occur because species with similar distributions (in relation to the overall community gradients) combine to contribute to a given value or state of a community-level variable; in effect, several relatively similar Gaussian curves are summed to yield one new Gaussian curve. This explanation fits the community-type variables in Figure 3.9 and the geographic origin and life-form variables in Figure 3.10. In other cases, the community-level variable integrates all species nondifferentially, but the variable itself has systematic trends along environmental gradients that are approximately Gaussian in form. This is often the case for gradients in species diversity and primary productivity. Although the Gaussian model principally concerns species distributions, it has numerous implications for distributions of community-level variables as well.

Simulated data

Direct gradient analysis is used to characterize the distributions of species populations along environmental gradients, as summarized in the Gaussian model of community structure. In turn, the Gaussian model may be embodied in computer programs simulating community data. Simulated data are valuable for testing ordination and classification techniques, as will be evident in the following two chapters.

Swan (1970) first constructed simulated coenoclines using a simple Gaussian model. For species abundance curves in response to a hypothetical environmental gradient, Swan used bell-shaped Gaussian curves, along a single gradient, without noise, with no variation in spacing, height, or peakedness (within a single data set). He did vary the amount of species turnover, however, by producing five data sets using curves of different peakedness.

Although his simulated coenoclines were simple, analysis of these simulated data brought critical problems into focus to a degree hitherto unrealized by study of field data. Confirming earlier work (for example, Whittaker 1960, 1967), Swan showed that measures of the similarity between samples are nonlinear functions of gradient separation. Swan's work, however, had an exactness and simplicity that brought this nonlinearity into sharper focus. This problem is now called the *nonlinearity* problem. He also characterized the distortion in ordination (using a technique similar to principal components analysis) that results from nonlinearity: as species turnover increases, the original, one-dimensional coenocline is distorted into an increasingly arched configuration. This problem is commonly called the *arch,* or *horseshoe,* problem. Subsequently, Noy-Meir & Austin (1970) subjected Swan's simulated data to principal components analysis and confirmed the existence, for this ordination technique, of the same problems. In retrospect, numerous principal components ordinations of field data show this arch. Field data are noisy and complex, however, and the problem is evident only under certain circumstances (namely with a single, long, community gradient). Consequently, the arch problem was not clearly recognized prior to Swan's work.

Comparisons of Swan's simulated coenoclines with actual coenoclines reveal the extreme simplicity of these simulated data; they simulate only the coarsest, most general features of actual coenoclines. Nevertheless, Swan's simulation study defined the nonlinearity and arch problems that have turned out to be major problems for ordination methodology. The fruitfulness of simulation studies is due to two advantages: (1) the expected result (null hypothesis) is known exactly and (2) parameters (such as species turnover and noise level) can be varied readily with little expense. By contrast, expected results with field data are known only approximately at best, and each field data set requires so much time and expense that experimental designs involving variations in several data set properties are not feasible. On the other hand, the advantage of field data is realism. Only field data are fully realistic and no simulation can fully capture their complexity. The advantages of field and simulated data are different and complementary. Methodological research in community ecology benefits from both kinds of data and would be greatly hampered by exclusive use of only one kind.

There was interest in developing more realistic simulations than the original coenocline simulation by Swan. Gauch & Whittaker (1972*a*) presented a coenocline simulation in which heights and widths of Gaussian curves varied, modes were scattered more complexly, and noise could be incorporated (Figure 3.14). Study of these simulated data sets at a number of laboratories helped to evaluate the performance of many multivariate techniques under a variety of circumstances of data set properties (as will be discussed in the following chapters). Coenocline simulation was extended to two community gradients, using the coenoplane (Austin & Noy-Meir 1971; Gauch & Whittaker 1976). The assumptions in the coenoplane model of Gauch & Whittaker are given in the 12 points of the preceding section on data set properties. More recently, three- and four-dimensional simulated data sets have been used (Hill & Gauch 1980; Gauch & Whittaker 1981; Gauch, Whittaker, & Singer 1981). Additional applications of Gaussian simulated data to testing of multivariate techniques include those by Austin (1972), LaFrance (1972), Aart (1973), Ihm & Groenewoud (1975), Kessell & Whittaker (1976), Phillips (1978), Robertson (1979), Feoli & Feoli Chiapella (1980), Fasham (1977), Johnson & Goodall (1979), and Prentice (1980*b*). A pioneering test by Austin & Greig-Smith (1968) used non-Gaussian simulated data. Figure 3.15 illustrates the effectiveness of simulated data for revealing differences in the accuracy of the ordination methods presented in the next chapter.

Discussion

The concerns of this chapter, the methods and results of direct gradient analysis, have been treated on a descriptive level. This level is also appropriate for the ordination and classification concerns to follow. These results also invite questions on other levels, however, such as origins. What are the origins and causes of observed community structure? Such questions do not bear directly on the concerns of this book and cannot be pursued at length, but a few comments and references seem in order.

Community structure as observed through direct gradient analysis

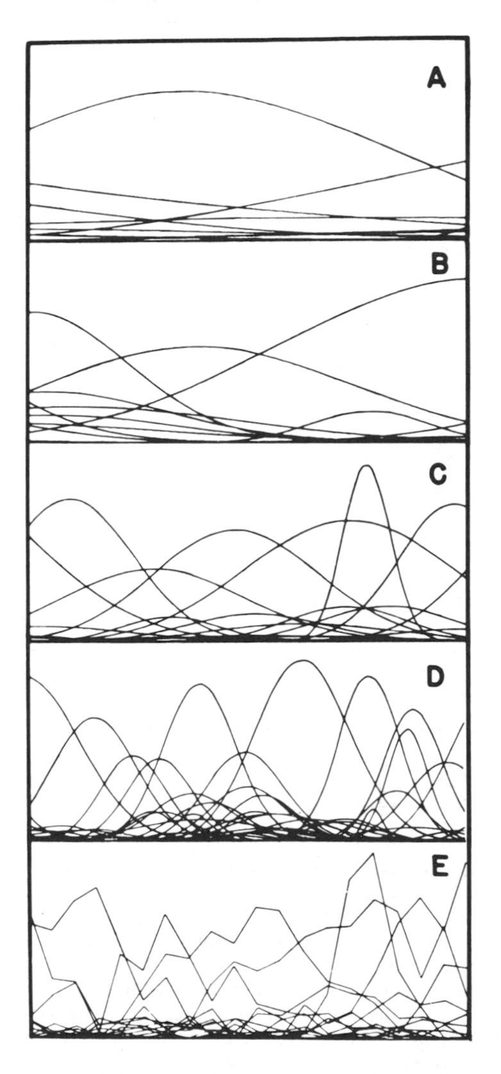

Figure 3.14. Five simulated coenoclines. Coenoclines (*A*)–(*D*) have species turnovers of 1.1, 2.2, 4.4, and 8.9 half-changes. Coenocline (*E*) is identical to (*D*) except that sampling errors have been introduced (producing a percentage similarity between replicate noisy samples of 77%, a medium or typical noise level for field data). (From Gauch & Whittaker 1972*a*:448. Reprinted by permission of the publisher. Copyright Ecological Society of America)

has two major origins: most importantly, biological processes and secondarily, sampling imperfections. The many species present in a community are growing and reproducing according to their individual genetics, physiological requirements, life cycles, and strategies. They are organized in communities in a complex way, which allows or fosters

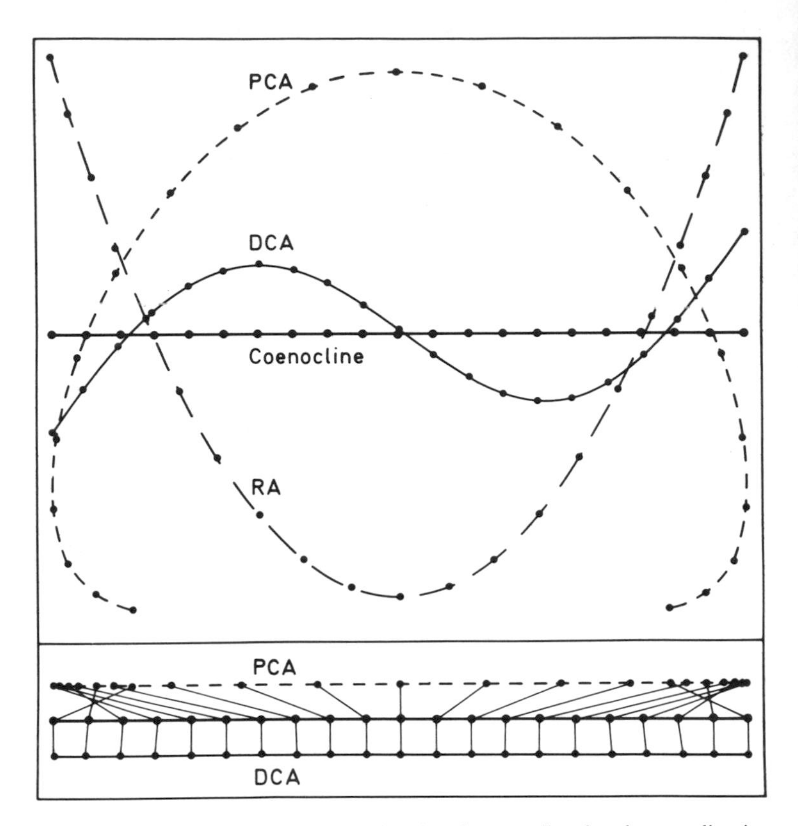

Figure 3.15. Representation of a simulated coenocline by three ordination techniques: centered principal components analysis (- - -), detrended correspondence analysis (——), and reciprocal averaging (— — —). The simulated coenocline is essentially like Swan's (1970), having 21 evenly spaced Gaussian curves representing species, 21 evenly spaced samples depicted by the dots on the coenocline lines, and a gradient length of 5 half-changes (Gauch 1977: 172, data set 1). The upper panel shows two-dimensional solutions. The first ordination axes are shown on the abscissa and are scaled to the same width in order to facilitate comparison; the second axes are shown on the ordinate and are scaled in proper proportion to their corresponding first axes. The polarity of the second axes is arbitrary but RA and PCA are presented with opposite polarities in order to avoid cluttering. Note that the DCA ordination shows the least distortion of the original coenocline, RA distorts it into an arch, and PCA both arches and involutes the original one-dimensional configuration. The lower panel shows that the first axis of PCA recovers the coenocline poorly, whereas the first axis of DCA is almost ideal. The first axis of RA (not shown) is compressed at the axis end but at least has the correct sample sequence (similar to Figure 4.11).

their mutual coexistence. The crucial questions are: (1) What present processes are involved in the ongoing stabilization or modification of species populations? (2) What past, long-term processes were involved over hundreds or thousands of generations that affected the presence of species as well as their genetic makeup and places and roles in the community? These are difficult questions. Interesting results and concepts may be found in the work of Whittaker (1970, 1975), Terborgh (1971), Colwell & Fuentes (1975), Whittaker & Levin (1975), Connell (1978, 1980), Simberloff (1978, 1980), Grime (1979), and Hubbell (1979).

Community data are structured, secondarily, by the sampling process itself. This structuring is minor when compared with the biological structuring because community observations are effective, but imperfection must be recognized in order to avoid viewing artifacts as facts. The commonness of the Gaussian species response curve may partially result from sampling errors (M. O. Hill, personal communication). Given a mildly skewed curve or slight bimodality, little error in the locations of data points is needed to make the configuration indistinguishable from a Gaussian curve. This critique does not completely invalidate the Gaussian model. It merely implies that it fits the data as frequently as it does mainly because it is approximately correct but additionally, because sometimes the data are not precise enough to reveal departures from Gaussian form.

The next two chapters will be concerned with ordination and classification techniques. Direct gradient analysis contributes substantially to the foundation upon which ordination and classification stand in a manner that should be recognized clearly.

(1) Direct gradient analyses of a great diversity of communities reveal several general features in common, despite the particularities of each community. This result implies that general-purpose multivariate analyses are possible and that they can account for a respectable portion of the structure in community data. By contrast, were general features lacking, each community would have to be approached as a unique case with its own custom-designed analytical methods.

(2) Direct gradient analyses show that despite the superficial high-dimensionality of samples-by-species data matrices, the intrinsic dimensionality of the data is ordinarily low because species distributions are structured by a modest number of important environmental gradients. Intrinsic low dimensionality is precisely what makes multivariate analysis worthwhile.

(3) Models of community structure are based on direct gradient analysis. A fairly detailed Gaussian community model with 12 data set properties was presented in this chapter. A coarser description of community data with only four aspects was presented earlier in the introduction. Community models are the foundation for designing multivariate analyses having matching assumptions. When data properties and analysis model match, effective studies result (Austin 1980*b*). When they do not match, (a) significant data structure is not recognized, (b) spurious structure may emerge, (c) insignificant noise may persist, obscuring or diluting desired information, and (d) desired information may be presented in an unnecessarily complex, indirect, ineffective form.

(4) Direct gradient analysis provides well-known data sets by two means. (a) Field data sets may be analyzed by direct gradient analysis for cases in which major environmental factors are evident. (b) Simulated data sets with precisely known properties can be constructed in accord with general community data set properties, as appreciated from numerous direct gradient analyses. Ideally, the match between data structure and analysis model is evaluated by direct mathematical reasoning, but in practice, the complexity and mathematical intractability of aspects of this evaluation make it necessary to elucidate some matters by trials with test data sets.

For these reasons, ordination and classification methods are based on direct gradient analysis. On balance, it should be remembered, however, that direct gradient analysis has limited applicability; only with evident environmental gradients is it applicable. Also, its level of objectivity is lower than for ordination and classification. Furthermore, some research purposes dictate a distinctively community-centered approach, in contrast to the fundamentally environment-centered approach of direct gradient analysis. Consequently, as was emphasized in the introduction, the full spectrum of community research employs a methodological triad of direct gradient analysis, ordination, and classification. Particular research questions may require a certain one of these methods or a certain combination of them.

4

Ordination

Community ecologists often analyze data by a methodological triad consisting of direct gradient analysis, ordination, and classification. These three methods have the common goal of organizing data for purposes of description, discussion, understanding, and management of communities. They vary in strategy. Direct gradient analysis portrays species and community variables along recognized environmental gradients. By contrast, ordination and classification techniques organize community data on species abundances exclusively, apart from environmental data, leaving environmental interpretation to a subsequent, independent step (with a few exceptions, as will be noted later). The result of ordination is the arrangement of species and samples in a low-dimensional space such that similar entities are close by and dissimilar entities far apart. The result of classification is the assignment of species and samples to classes; the classes may or may not be arranged in a hierarchy. These three approaches are complementary, as was shown by examples in the first chapter.

Work in organizing data matrices pertinent to ordination began early in this century, and substantial ordination work was done around 1950, using simple algorithms and hand calculations (for reviews, see Becking 1957 and Whittaker 1967). By 1970, computers were available to most ecologists, and a great number of ordination techniques had been developed. Ordination techniques were introduced more rapidly than they were tested, however, until the 1970s, when ordination comparisons were made leading to reliable recommendations and preferences. Also in the 1970s, communities were modeled explicitly, and this aided the development of more realistic and effective ordination techniques. Progress has been rapid. The newest ordination techniques appear to analyze community data nearly as well as is possible, given inherent limitations, and, at the same time, make entirely reasonable demands upon computer resources.

The major resources for this rapid progress in ordination methodology were realistic community models from direct gradient analyses and

Table 4.1. *Samples-by-species data matrix*

	Samples	
Species	a	b
1	6	20
2	5	10
3	7.5	15

access to computers. The major motivations for this progress were the speed, low cost, and effectiveness of computerized ordination techniques, together with an increased level of objectivity.

Five conceptual spaces

Before describing the calculations involved in several common ordination techniques, five conceptual spaces require presentation. These five spaces serve as the best point of departure for explaining most ordination algorithms. They also relate effectively to fundamental considerations about the role and purpose of ordination. The geometric concepts afforded by these spaces are more lucid to most readers than are their equivalents in matrix algebra (but see Williams 1976:3–28 and Jöreskog, Klovan, & Reyment 1976:8–52 for remarkably lucid introductions to matrix algebra).

The primary data in community ecology are contained in samples-by-species abundances matrices (as in Table 1.1). The data of a samples-by-species matrix can be conceptualized as a *species space* in which the species are axes of multidimensional space and the samples are points located by their abundances for each species. Consider the simple samples-by-species data matrix, with three species and two samples, shown in Table 4.1. For example, sample b for species 3 has an abundance of 15. These data can be presented by a matrix, as here, or equivalently by a geometric, spatial representation using species space. For this matrix, the corresponding species space is three-dimensional and contains two sample points, as in Figure 4.1*A* (adapted from Groenewoud 1965; also see Gittins 1969; Crovello 1970; Allen & Skagen 1973; Beals 1973; Jöreskog, Klovan, & Reyment 1976:11–12; Pielou 1977: 332; Orlóci 1978*a*:43; Noy-Meir 1979). Samples with similar species composition occupy nearby positions in species space.

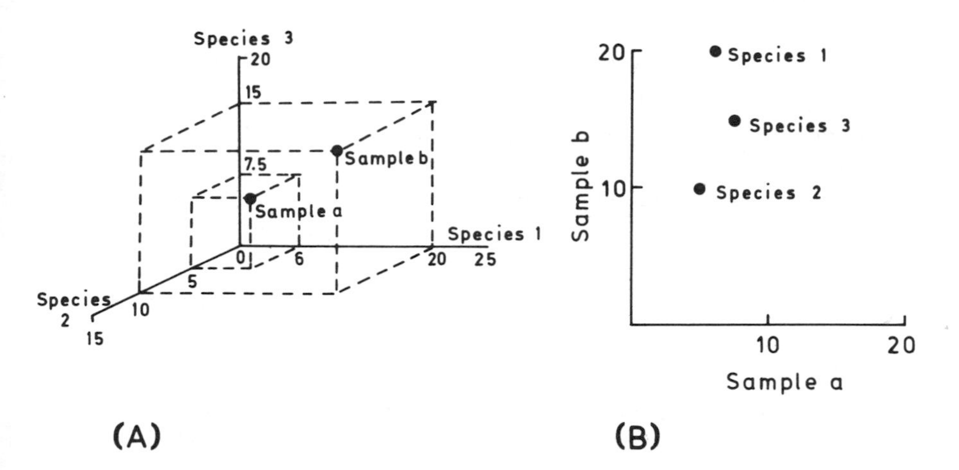

Figure 4.1. Species space (*A*) and samples space (*B*). Sample a has abundances of 6, 5, and 7.5 for species 1–3; sample b abundances of 20, 10, and 15. (*A*, after Groenewoud 1965)

There is a converse *samples space* in which samples are axes and species are points placed along the axes according to the species' importances in each sample. Figure 4.1*B* shows the same data as Figure 4.1*A* but uses samples space. Species with similar distributions in the sample set occupy nearby positions in samples space. For example, species 1 and 3 are more similar in their distributions than are species 1 and 2.

Both species space and samples space contain identical information, namely, the abundance values in the samples-by-species data matrix. They contain this information fully. Species and samples spaces differ in their use of axes that represent species and samples, respectively, and their use of points representing samples and species, respectively. Here these spaces are named for the entity constituting their axes; species space has species for axes, and samples space has samples for axes.

A secondary matrix can be computed from the primary samples-by-species data matrix having a dissimilarity (distance) value for every pair of species (or for every pair of samples). If species are compared, the result is a species-by-species dissimilarities matrix; if samples are compared, a samples-by-samples dissimilarities matrix results. Given N species (or samples), there are $[N(N - 1)]/2$ comparisons; 10 species have 45 pairwise comparisons. These dissimilarity values can be arranged in a square, symmetric matrix, with diagonal elements for self-comparisons of zero dissimilarity.

There are numerous dissimilarity (distance) measures for comparing sample pairs or species pairs (Jardine & Sibson 1971:3–36; Orlóci 1972, 1978*a*:42–101, 1978*b*; Gauch 1973*b*; Clifford & Williams 1976; Goodall 1978*b*; Maarel 1979*a*; Pielou 1979; Green 1980). Three common distance measures are *percentage dissimilarity* (PD, also sometimes termed *percentage distance* or *percentage difference*), *complemented coefficient of community* (CD), and *Euclidean distance* (ED). The calculation of PD and CD, but not ED, begins with the computation of a similarity. This similarity has to be subtracted from the similarity among replicate samples, the *internal association,* in order to convert it to a distance (Bray & Curtis 1957; Gauch 1973*a*). Because overestimation of the internal association is less problematic than underestimation (Gauch 1973*a*), the value 100 is often suitable if one cannot supply an empirical estimate of the replicate similarity. Negative PD or CD distances will result if the internal association is set lower than the similarity between the most similar samples; the value 100 for internal association cannot lead to this problem. Ricklefs & Lau (1980) discuss the bias and dispersion properties of PD and ED. The equations for similarity measures (*percentage similarity* PS and *coefficient of community* CC) and distance measures (Euclidean distance ED, percentage dissimilarity PD, and complemented coefficient of community CD) between samples j and k are

$$\mathrm{ED}_{jk} = \left[\sum_{i=1}^{I} (A_{ij} - A_{ik})^2 \right]^{1/2}$$

$$\mathrm{PD}_{jk} = \mathrm{IA} - \mathrm{PS}_{jk}, \quad \text{where} \quad \mathrm{PS}_{jk} = \frac{200 \sum_{i=1}^{I} \min(A_{ij}, A_{ik})}{\sum_{i=1}^{I} (A_{ij} + A_{ik})}$$

and

$$\mathrm{CD}_{jk} = \mathrm{IA} - \mathrm{CC}_{jk}, \quad \text{where} \quad \mathrm{CC}_{jk} = \frac{200 S_c}{S_j + S_k}$$

The summations are over all species I, A_{ij} and A_{ik} are the abundances of species i in samples j and k, IA the internal association, S_j and S_k the numbers of species in samples j and k, and S_c the number of species in common.

If samples are relativized to total 100, the denominator of the equation for PS becomes 200, and the equation may be simplified accordingly. To obtain similar equations for species (rather than samples), merely reverse the terms species and samples in the preceding paragraph. (The concept of replicate species is not quite comparable to that of replicate samples because species are individualistic. The greatest observed species similarity value or else 100 should suffice, however, for an IA estimate for species distance calculations.)

These three dissimilarity measures (ED, PD, and CD) differ in the features of community data emphasized. By squaring abundance values, ED emphasizes the larger abundance values in the samples-by-species data matrix; that is, ED values are determined mostly by the dominant species. By considering only species presence or absence, CD has the opposite emphasis, giving minor and major species the same emphasis. The measure PD is intermediate, with a linear weighting of species abundances. In most ordination applications, the intermediate weighting of PD seems best, emphasizing dominants somewhat but still considering minor species (Gauch 1973*a*). The dissimilarity values of PD conform well to the mental scaling of dissimilarities originating from ecologists' field observations.

A secondary species-by-species percentage dissimilarity matrix may be computed from the data shown in Figure 4.1. Species 1 and 2 have a percentage similarity $PS = [200 \times (5 + 10)]/(6 + 5 + 20 + 10) \approx 73$. Assuming an internal association IA of 100, their percentage dissimilarity is $PD = 100 - 73 = 27$. Likewise, species 1 and 3 have $PD = 13$, and species 2 and 3 have $PD = 20$. These PD values may be entered into a species-by-species PD matrix as shown in Table 4.2. This matrix is square and symmetric. (A corresponding samples-by-samples PD matrix may be computed, but since in this example there are only two samples, it is quite small, merely having 0, 42 in its first row and 42, 0 in its second row.)

A dissimilarity matrix likewise may be conceived geometrically. *Samples dissimilarity space* uses sample dissimilarities as axes and samples as points. That is, axis 1 is the dissimilarity to sample 1, axis 2 the dissimilarity to sample 2, and so on; sample points are located in this space by their dissimilarities to each sample.

The corresponding *species dissimilarity space* uses species dissimilarities as axes and species as points. The species-by-species PD matrix shown in Table 4.2 (based on the same data as Figure 4.1) defines the species dissimilarity space shown in Figure 4.2. (The corresponding samples dissimilarity space may be conceived for this small data set, having merely two axes for its two samples, with a point for sample 1

Table 4.2 *Species-by-species percentage dissimilarity matrix*

	Species		
Species	1	2	3
1	0	27	13
2	27	0	20
3	13	20	0

at 42 on the second axis and a point for sample 2 at 42 on the first axis.) Ordinarily, the configuration of species points in samples space is preserved in species dissimilarity space, so both spaces imply similar configurations of species points. This is the case for Figures 4.1*B* and 4.2, both having three species points in a triangle with the points for species 1 and 3 closest. When the sample set is relatively homogeneous, these configurations usually correspond closely, but for sets with a great diversity of samples, the discrepancy between configurations may be substantial.

Dissimilarity spaces have the same number of axes as points. For example, a data set with 85 species and 315 samples has an 85-dimensional species dissimilarity space with 85 points and a 315-dimensional samples dissimilarity space with 315 points. By contrast, the original data can be represented by 315 sample points in 85-dimensional species space or by 85 species points in 315-dimensional samples space.

Species space and samples space each represent the full information content of the abundance values in a samples-by-species data matrix. These spaces are simply geometric equivalents of the data matrix. In contrast, a species dissimilarity space contains information on species relationships but no information on samples (not even the number of samples). Likewise, samples dissimilarity space contains only information on samples. Hence the information content of a dissimilarity space is impoverished as compared with the original species space or samples space. Furthermore, dissimilarity values are less economical for expressing community information than are the original data. Given I species and J samples, the original data matrix contains IJ elements. The species and samples dissimilarity matrices together contain $\{[I(I - 1)]/2\} + \{[J(J - 1)]/2\}$ elements. If I nearly equals J, these two numbers of elements are about equal, but otherwise, the original data matrix

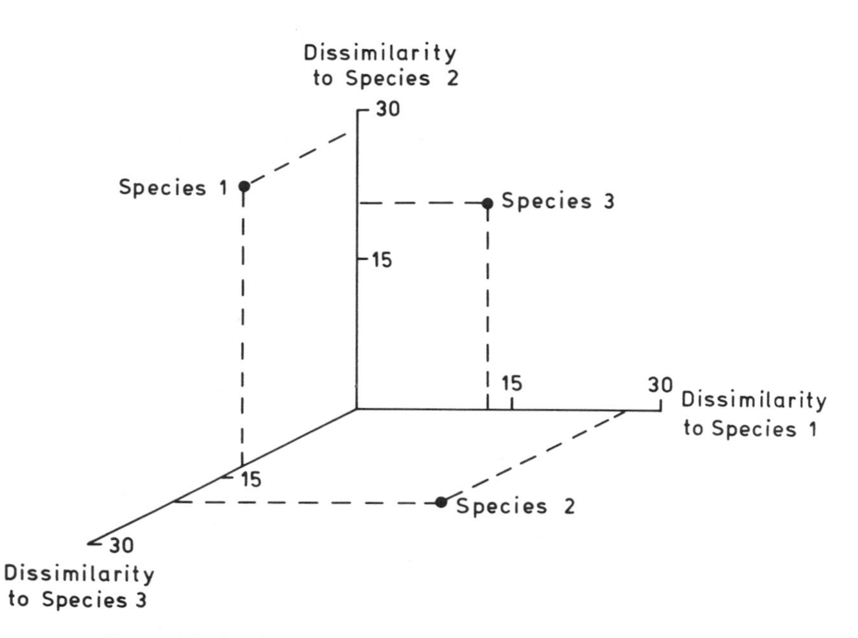

Figure 4.2. Species dissimilarity space using the same data as in Figure 4.1, with dissimilarity measured in units of percentage dissimilarity.

has fewer elements. For the example with 85 species and 315 samples, the original data matrix has 26 775 elements, but the dissimilarity matrices have 53 025 elements. Finally, even given both dissimilarity spaces (for species and for samples), the available information content is still less than that of the original data matrix because dissimilarity matrices are computable readily from the original data but the reverse is not possible.

To summarize thus far, the samples-by-species data matrix may be conceived geometrically by four spaces: species space, samples space, species dissimilarity space, and samples dissimilarity space. The first two both represent the entire information of the original data. The latter two both contain information on only species or only samples, respectively, but even together do not fully represent the information content of the original data. A dissimilarity matrix may be computed from the original data by any of numerous dissimilarity measures (such as percentage dissimilarity PD) and its values conceived geometrically by a dissimilarity space.

These four spaces are the starting point for most ordination and classification techniques. The utility of these four spaces for multivariate analysis is due to three properties. (1) Species (or sample) relationships are manifested in these spaces in that points for similar species are near each other, whereas dissimilar species are far apart. For example, in Figure 4.1, the most similar pair of species is species 1 and 3, and they are the closest pair of species points in samples space (Figure 4.1*B*). In this case, species relationships are maintained properly (although approximately) in the transformation to species dissimilarity space (Figure 4.2). (2) Some combinations of species abundances are more frequent or possible than others, so the overall distribution of sample points is not random or round but, rather, concentrated in certain places and directions (Becking 1957; Orlóci 1978*a*:103). Likewise, species points show definite structure because of patterns in species distributions. Consequently, the points may occupy a lower-dimensional subspace within a larger-dimensional space. This property, as will be discussed further, is the basic property that makes possible the dimensionality reduction (data summarization) efforts of ordination techniques. (3) Graphical explanations of ordination algorithms using these spaces have heuristic value, being more easily communicated to most readers than corresponding complex matrix algebra (Noy-Meir 1979).

Species space represents the abundance values of the original species-by-samples data matrix fully (as does samples space). The only problem is that species space is of many dimensions, typically dozens to hundreds of dimensions, and hence impossible to visualize or inspect. The simplest solution to this problem is to project the original high-dimensional species space onto a space of fewer dimensions, say, two or three dimensions, so that the distribution of sample points can be inspected. When this is done, structure along the chosen two or three dimensions will be visible and structure in other dimensions will be lost. If the sample points lie at random, thousands of different projections are possible and none is especially better or worse than others; consequently, projection into fewer dimensions will be of little value. On the other hand, if the structure of sample points is concentrated within two or three dimensions (because other dimensions are correlated with these), then projections along these dimensions will show much more of the original structure than less-favored projections. In this case, certain projections will be of value and will offer, in two or three dimensions, a fairly accurate look at the structure of the points in their original space. In practice, this is frequently the case, and consequently, ordination is fruitful.

A geometric interpretation of the derivation of samples (or species) dissimilarity space from samples (or species) space is possible (Gittins

1969; Aart 1973). The geometric interpretation is mathematically equivalent to the preceding algebraic derivation in terms of dissimilarity measures and secondary matrices, but the geometric viewpoint may be more lucid. In Figure 4.1*A*, the three-dimensional species space contains two samples represented as points. An alternative, equivalent conception to this point representation of the samples uses a vector representation, with vectors going from the origin to the sample points. It is then possible to describe much (but not all) of the information about sample relationships in terms of the angles between pairs of vectors. Furthermore, the angles may be expressed by their cosines. For example, the angle between the vectors from the origin to samples a and b in Figure 4.1*A* is 14° and 0.97 is its corresponding cosine. This cosine has a statistical meaning because it is identically the correlation coefficient between the two samples. Consequently, cosines of angles between vectors for sample pairs are analogous to a samples-by-samples dissimilarity matrix (and thus a sample dissimilarity space) for which the dissimilarity measure used is the correlation coefficent after subtraction from 1 (where subtraction changes the correlation coefficient from a measure of similarity into a measure of dissimilarity). For distance measures other than the one-complement of the correlation coefficient, geometric interpretation is not quite as simple, but the general feature remains that transformation from samples (or species) space to samples (or species) dissimilarity space involves some pairwise comparison of samples (or species).

The fifth and final conceptual space is *ecological space,* having environmental gradients as axes (Whittaker 1967; Austin 1976*a,* 1979; Whittaker & Gauch 1978; Pignatti 1980). The points plotted within ecological space may be of many kinds, including samples, species, and community-level variables, as exemplified by figures in the preceding chapter of direct gradient analyses using environmental factors as axes.

A fundamental point emerging from direct gradient analyses is that a single environmental gradient simultaneously affects many species in their distributions, and consequently, species abundance data are partially redundant (in that a high or low value of a single species provides considerable information on values to expect for other species). Consequently, the sample points in species space occupy a relatively low-dimensional subspace within species space, and the configuration of sample points is simpler than might otherwise be anticipated by the high dimensionality of species space. By contrast, ecological space is low dimensional from the outset. The remarkable feature of community data is that the intrinsically low-dimensional sample configuration complexly embedded within high-dimensional species space often reflects with much fidelity the sample configuration in low-dimensional ecological space, despite a vast difference in superficial dimensionality. Ecological space is the desired output from ordination.

Table 4.3 summarizes these five conceptual spaces, giving their axes, points, information content, ordination role, and dimensionality. The names of the spaces are based on the nature of their axes. The first four spaces are alternative spaces for representing data input to ordination and the fifth is the desired ordination output.

The purposes of ordination

From a wider perspective encompassing all multivariate analyses, the introduction stated three purposes: summarizing community data, relating community variation to environmental gradients, and understanding community structure so that increasingly effective multivariate techniques may be developed. This wider perspective includes ordination, and these are the purposes of ordination in general terms. The more specific purposes of ordination will be detailed here.

Ordination serves to summarize community data by producing a low-dimensional ordination space (of typically one to three dimensions) in which similar species and samples are close together and dissimilar entities far apart. Field data must be high dimensional because of the large numbers of species and samples, whereas the final results must be low dimensional because of the human limitations. Ordination is a means for bridging this gap, for producing effective, low-dimensional summaries from field data by relatively convenient and objective means.

A basic goal of many ordination studies is to derive an ecological space from the input data. A two-step procedure is standard. First, ordination is used to summarize community patterns. Second, the community patterns are compared with environmental information in order to produce an environmental interpretation of the ordination (community) results. Consequently, ordination may be viewed as the first half of the overall process for deriving an ecological space from the input data. Although ordination of community data leaves environmental interpretation to a subsequent step, the needed comparison of community and environmental patterns could hardly succeed apart from the preliminary elucidation of community patterns by ordination.

The understanding of communities that ordination work has fostered

Table 4.3. *Properties of five conceptual spaces*

Space (axes)	Points	Information content	Ordination role	Dimensionality
Species	Samples	Raw data completely	Input	High
Samples	Species	Raw data completely	Input	High
Species dissimilarity	Species	Species relationships	Input	High
Samples dissimilarity	Samples	Sample relationships	Input	High
Ecological	Species, samples, or community-level variables	Environmental interpretation of community data	Output	Low

aids the development of appropriate, effective, multivariate analysis techniques. Direct gradient analysis is more important than ordination for this role, however, because environmental data are integrated in the results by a simpler and more direct process.

The purpose of ordination is further clarified by contrasting ordination with alternative methodologies, particularly direct gradient analysis and classification. Direct gradient analysis is useful when important environmental factors are readily appreciated and measured and when research purposes call for a direct, integrated use of environmental data. Classification is useful for providing research results expressed in terms of community types and for providing a workable number of community types for remembering and communicating results. Often ordination tackles certain aspects of research problems in an overall program using ordination, classification, and direct gradient analysis.

Ordination techniques

In this section the algorithms for several ordination techniques will be considered. Techniques will be presented in order of increasing mathematical complexity, which generally corresponds with chronological order. A deliberate attempt will be made to present relatively few preferred techniques. This choice reflects two considerations. First, compendia of available ordination techniques already exist (Orlóci 1978*a*; Whittaker 1978*c*). Second, tests have shown numerous techniques to be either inferior to others or else essentially similar to others, relegating such to historical interest for ordination specialists. Most needs are served best by presenting several preferred ordination techniques with clear recommendations for choices among them. This strategy seems adequate in most cases, given the effectiveness of the better techniques, and seems desirable to foster comparability of community studies by emphasizing relatively standard methods.

Weighted averages

The use of weighted averages is the simplest ordination technique, dating back to early work by Ellenberg (1948), Whittaker (1948), Curtis & McIntosh (1951), and Rowe (1956). Weighted averages ordination is used to produce sample ordination scores from an ecologist's previous knowledge of the species or species ordination scores from previous knowledge of the samples.

The simple algorithm of weighted averages is best explained by reference to the samples-by-species data matrix (unlike most ordinations, for which species space or another such space offers the clearest point of departure). Let A_{ij} be the abundance of species i in sample j and let W_i be the weight for species i. The weights indicate the position of each species on an ecological scale perceived by the investigator (for example, a moisture preference scale from 1 for dry to 6 for wet, as in the Nelson County vegetation example considered earlier in Figures 3.1 and 3.2). The weighted averages algorithm then computes an ordination score S_j for each sample j as

$$S_j = \frac{\sum A_{ij} W_i}{\sum A_{ij}}$$

where the summations are over all species I. The reverse calculation, although less common, is equally simple: given weights for samples, weighted averages may be computed for species.

Calculation of a weighted average merely involves simple arithmetic. The real challenge is to gain adequate ecological insight to produce useful species weights (or sample weights) in the first place.

Curtis & McIntosh (1951) developed species weights for upland forest trees in southern Wisconsin by a logical sequence of steps, which, in its general features, has broad applicability. They began by analyzing just the 4 most common of the 22 tree species and interpreting the results ecologically (as shown earlier in Figure 3.8). This initial framework was found suitable for revealing patterns in the distributions of other tree species not included in the initial analysis, as well as distributions of herbs, shrubs, and environmental parameters. The major community gradient that they found was a pioneer-to-climax forest gradient, with a corresponding shift from *Quercus velutina* (black oak) dominance to *Acer saccharum* (sugar maple) dominance, and an environmental shift from high light, variable moisture, and immature soil conditions (with low pH and low nutrients) to low light, medium moisture, and mature soil conditions. The maximum abundance of each of the 22 tree species along this succession gradient was determined, and each species was assigned a weight, called a *climax adaptation number,* from 1 for pioneer species to 10 for climax species.

Using weighted averages of the climax adaptation number, a sample ordination score may be computed for any sample, utilizing all of the tree species information to express the position of the sample on a

Table 4.4. *Ten composite forest samples with 14 tree species in southern Wisconsin upland forests, and species climax adaptation weights*

			Sample[b]									
Scientific name	Common name	Weight[a]	1	2	3	4	5	6	7	8	9	10
Quercus macrocarpa	Bur oak	1.0	9	8	3	5	6		5			
Quercus velutina	Black oak	2.0	8	9	8	7						
Carya ovata	Shagbark hickory	3.5	6	6	2	7		2				
Prunus serotina	Black cherry	3.5	3	5	6	6	6	4	5		4	1
Quercus alba	White oak	4.0	5	4	9	9	7	7	4	6		2
Juglans nigra	Black walnut	5.0	2				3	5	6	4	3	
Quercus rubra	Red oak	6.0	3	4		6	9	8	7	6	4	3
Juglans cinerea	Butternut	7.0			5		2			2		2
Ulmus americana	American elm	7.5	2	2	4	5	6		5		2	5
Tilia americana	Basswood	8.0					2	7	6	6	7	6
Ulmus rubra	Slippery elm	8.0	4		2	2	5	7	8	8	8	7
Carya cordiformis	Yellowbud hickory	8.5						5	6	4		3
Ostrya virginiana	Ironwood	9.0							7	4	6	5
Acer saccharum	Sugar maple	10.0						5	4	8	8	9

[a]The species weight is the climax adaptation number indicating preference for pioneer (1.0) to climax (10.0) forest conditions (from Curtis & McIntosh 1951, except for the *Ulmus americana* weight from Curtis 1959:519).
[b]The composite samples are based on data of Peet & Loucks (1977), obtained by summing the abundances for each set of three replicate samples, deleting species occurring in fewer than four composites, converting matrix values to the percentages of the largest matrix entry, and, finally, applying the octave transformation to the result in order to obtain one-digit matrix entries. The composite samples are in a sequence from pioneer (1) to climax (10) forests.

succession gradient. The weighted averages ordination procedure may be illustrated using ten composite samples of southern Wisconsin upland forests and species weights given in Table 4.4. There are 14 tree species listed with scientific names, common names, and a weight, which is the climax adaptation number, indicating a preference for conditions in pioneer (1.0) to climax (10.0) forests; for example, *Quercus macrocarpa* and *Q. velutina* characterize pioneer forests, whereas *Ostrya virginiana* and *Acer saccharum* dominate climax forests. The ten composite samples likewise are graded from pioneer (1) to climax (10) forests. The weighted averages ordination score for sample 1 is

$$S_1 = \frac{\begin{array}{c}(9 \cdot 1.0) + (8 \cdot 2.0) + (6 \cdot 3.5) + (3 \cdot 3.5) + (5 \cdot 4.0) \\ + (2 \cdot 5.0) + (3 \cdot 6.0) + (2 \cdot 7.5) + (4 \cdot 8.0)\end{array}}{9 + 8 + 6 + 3 + 5 + 2 + 3 + 2 + 4}$$

or 3.6, indicating a relatively pioneer forest. Proceeding in the same manner, the ordination scores for composite samples 2 to 10 are: 3.1, 4.2, 4.0, 5.2, 6.5, 6.6, 7.4, 7.7, and 8.0. The general trend from pioneer-to-climax forests is evident.

This example of weighted averages illustrates the derivation of sample ordination scores from species weights concerning successional status. Species weight for these trees along other gradients may be derived from field observations (Peet & Loucks 1977), and these newly derived weights used to ordinate samples along additional gradients. Likewise, the procedure may be reversed, supplying sample weights for some ecological property of the samples (such as percent sand in the A_1 soil horizon), and the species ordinated by weighted averages along this gradient. A given data matrix may be subjected to numerous different weighted averages ordinations, one ordination for each set of species or sample weights that the ecologist can supply, in order to obtain an ordination with two or more dimensions. An alternative approach by Grime (1979:56–78), triangular ordination, uses three poles in an equilateral triangle to ordinate species or samples with respect to competition, stress, and disturbance.

A remarkable feature of the early study by Curtis & McIntosh (1951) is that an understanding of only the four most common tree species revealed the basic succession pattern in their forests; other tree, herb, and shrub species and environmental parameters could be incorporated into this pattern subsequently because of correlations. Later, their succession gradient also proved effective for organizing distributional data on herbs (Gilbert & Curtis 1953), shrubs (Loucks & Schnur 1976), lianas (Swan 1961), lichens (Hale 1955), soil fungi (Tresner, Backus, & Curtis 1954), and birds (Bond 1957). A logical, simple process of successive refinement produced useful species weights, and these weights served to locate forest samples on a succession gradient by calculating weighted averages. This example shows that weighted averages may be successful even in cases where the ecologist's initial understanding of the material is meager.

The reasoning used by Curtis & McIntosh (1951) to arrive at species weights was effective, but a considerable subjective element must be admitted. A further limitation was the definition of a single gradient (of succession); little imagination is required to suspect the existence of additional gradients, even if they are less important. Subsequent studies of upland forests in southern Wisconsin have shown the additional gradients of water drainage, air temperature, and soil nutrient status (Bray & Curtis 1957; Forsythe & Loucks 1972; Peet & Loucks 1977). Some refinement of a weighted averages study is possible by producing several sets of weights for the same species, each set reflecting a different ecological factor, such as successional status, moisture preference, and temperature requirements (Ellenberg 1948; Waring & Major 1964; Sobolev 1975; Peet & Loucks 1977; Persson 1981). Greater objectivity requires use of other ordination techniques, as weighted averages is inherently a relatively subjective technique, using an ecologist's knowledge of species to ordinate samples or knowledge of samples to ordinate species.

An intriguing modification of weighted averages was discovered by Goff & Cottam (1967). Termed *index iteration,* in this technique the investigator supplies no weights, and ordination scores are produced by an entirely objective computation producing a unique solution. A samples-by-samples (or species-by-species) similarity matrix is produced; the species are given arbitrary ordination scores (such as sequential integers); weighted averages are computed using these scores; and this last step is iterated until the scores stabilize. After solving for one ordination axis, further ordination axes can be extracted for any number of dimensions, again continuing in a thoroughly objective manner requiring no weights or decisions from the investigator. More recently, it has been shown that an analogous procedure can be applied to the original samples-by-species data matrix, producing both sample and species ordination scores simultaneously for any number of dimensions (Hill 1973, 1974). This ordination technique is called *reciprocal averaging* and will be discussed later in this chapter. Reciprocal averaging can also be applied to a secondary samples-by-samples (or species-by-species) matrix, and it is then termed *secondary reciprocal averaging* (Gauch, Whittaker, & Wentworth 1977; this is identical to index iteration). Reciprocal averaging has a relative objectivity much greater than weighted averages. The amount of calculation required, however, is much greater than weighted averages (in that about 100 weighted averages iterations are required, more or less, depending on the data set, the desired precision and number of axes, and the details of the computational procedure used), so access to a computer is required.

Weighted averages ordinations customarily incorporate environmental or ecological gradients relatively directly. Usually the species weights (or sample weights) reflect successional status, moisture conditions, or a similar, relatively concrete, ecological gradient. Consequently, the environmental interpretation of results is rather straightforward, almost as straightforward as in direct gradient analysis (Peet & Loucks 1977). In contrast, precisely because more objective methods (like reciprocal averaging) do not utilize the ecologist's perceptions of environmental gradients, objective methods leave environmental interpretation to a subsequent and potentially difficult step.

Goff & Cottam (1967) note that weighted averages may be applied to abundance data (as already discussed) and also to presence and absence data (which is equivalent to transforming any positive abundance to the same value, say, unity, prior to calculating weighted averages). As immature or weak individuals are frequently found far beyond their preferred habitats, mere presence is much less indicative of a species's ecological preferences than is quantitative abundance. Consequently, Goff & Cottam recommended general use of quantitative data when computing weighted averages.

Recognizing that some species are bimodal or insensitive in their distributions along a given environmental gradient, they may be excluded from weighted averages in order to improve the accuracy of ordination (Ellenberg 1952; Rowe 1956; Waring & Major 1964; also see Mirkin & Rozenberg 1977). Alternatively, recognizing that the indicator value of species varies in degree on a continuous scale, species may be given weights expressing their indicator value for a particular gradient and then a double weighting used to give more accurate results than the customary single weighting (Goff & Cottam 1967).

Weighted averages may be applied to community data other than species abundances (Goff & Cottam 1967), just as direct gradient analysis was shown to be applicable to many kinds of data. Knight & Loucks (1969) studied 29 structural and functional features of Wisconsin forest vegetation (including life-form, leaf size, flowering season, pollination mechanism, bark thickness, branching pattern, shade tolerance, moisture optimum, leaf persistence, seed dispersal mechanism, surface fire susceptibility, twig diameter, and vegetative reproduction). Structural–functional species weights were derived, permitting weighted averages calculations of sample positions along structural–functional gradients. The resulting sample ordinations were informative about functional, successional, and environmental relationships, offering an interesting alternative to conventional study of species abundances. Another article concerning structural gradients is by Goff & Zedler (1968).

The robustness of Curtis & McIntosh's (1951) results, from three different techniques including weighted averages, is encouraging. Subsequently, Goff & Cottam (1967), with seven further analyses, and

Peet & Loucks (1977), with two further analyses, derived similar weights for these tree species. These analyses varied in field methods, transformations of the data, and ordination techniques and yet gave rather consistent results. The strength of the succession gradient may make this forest region in Wisconsin ideal for ordination studies, so universal robustness can hardly be claimed on the basis of this single example. Nevertheless, it is fair to say that, in some cases, weighted averages is, despite its simplicity, an adequate ordination technique to reveal the main community and environmental gradients and that further ordination analyses may only confirm patterns already demonstrated. In other cases, however, research purposes may require greater objectivity or ecological knowledge of the study area may be inadequate to provide species or sample weights. Even if weighted averages ordination is applicable and effective, additional ordination techniques may provide confirmation or may be complementary.

Polar ordination

Bray & Curtis (1957) devised an ordination technique that has been used widely in plant ecology (Cottam, Goff, & Whittaker 1978). Two samples serve in a special role as poles of an ordination axis, so the technique is commonly called *polar ordination.*

The vegetation studied by Bray & Curtis (1957) was, like Curtis & McIntosh's (1951), the upland forests of southern Wisconsin. Bray & Curtis augmented earlier work by devising an ordination technique based on vegetational data alone (apart from environmental data), which allows the extraction of more than one gradient in community composition. These goals were motivated by their view of community structure and, hence, of appropriate data analysis. The model of community structure of Bray & Curtis is not one of simple cause and effect, with environmental factors causing structure in species distributions. Rather, they emphasize a complex interaction of environmental and vegetational factors, with a species's success affected by both environmental factors and competition from other plants, and environmental factors affected both by other environmental factors and influences of the vegetation upon the site. The goal of ordination, given such a data structure, is to map the features (species, samples, environmental factors) in an ordination space in which "the relative proximity of different features and their varying spatial patterns [indicate] the degree to which the features may participate in a mutually determined complex of factors" (Bray & Curtis 1957:327). More than one ordination axis is extracted to depict the data's complex structure more faithfully than is possible with only one axis.

The relatively simple algorithm of polar ordination begins with the computation of a samples-by-samples distance (dissimilarity) matrix. Bray & Curtis used percentage dissimilarity (as defined earlier in this chapter). For example, in Table 4.4, the percentage similarity of samples 1 and 5 is

$$\frac{200 \cdot (6 + 3 + 5 + 2 + 3 + 2 + 4)}{9 + 6 + 8 + 0 + 6 + 0 + 3 + 6 + 5 + 7 + 2 + 3 + 3 + 9 + 0 + 2 + 2 + 6 + 0 + 2 + 4 + 5}$$

$$= \frac{200 \cdot 25}{88} \approx 57$$

Assuming, for simplicity, an internal association (replicate similarity) of 100, the percentage dissimilarity of samples 1 and 5 is $100 - 57 = 43$. Likewise, samples 5 and 10 have a distance of 55, and samples 1 and 10 a distance of 72.

The second step is selection of two samples to serve as poles. The samples-by-samples distance matrix is searched for its greatest value, to find that sample pair separated by the greatest distance. These samples then serve as poles, or endpoints, of the ordination axis, defining, it is hoped, a major direction of community variation. As Bray & Curtis note, however, this endpoint selection can be problematic because (1) the largest distance encountered may be achieved by more than one sample pair, making the selection ambiguous and (2) this procedure may select outliers, which will fail in the next step to spread samples in the ordination space. An alternative procedure for selecting endpoints is deliberate endpoint selection based on perceived ecological relationships, for example, choosing a sample from a wet and from a dry site. Deliberate endpoint selection avoids the preceding two problems and also requires computation of far fewer sample distances. Given J samples, the entire distances matrix required for objective endpoint selection has $(J^2 - J)/2$ elements, but the two sets of distances from the two endpoints required for ordination with deliberate endpoint selection involve only $2J$ elements; for 100 samples, these options require 4950 versus 200 distance calculations. The original axis selection procedure is objective and useful (in cases where no problem is encountered); on the other hand, deliberate endpoint selection makes subsequent interpretation of results relatively direct. Both selection procedures are useful, with these complementary advantages.

The third and final step is computation of ordination values X_j for each sample j and, optionally, distances off the ordination axis E_j.

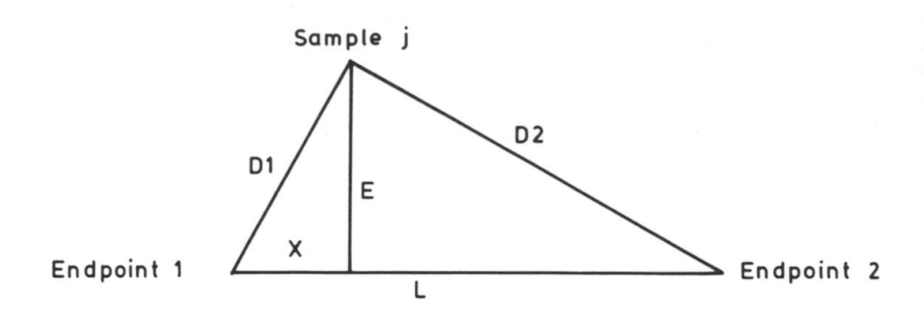

Figure 4.3. Diagram used in computing polar ordination values of X_j. See text for explanation.

Each computation involves the two endpoints and a sample *j*, with the associated three interpoint distances used to define a triangle; as shown in Figure 4.3, where L is the baseline distance between endpoints, $D1$ the distance of sample *j* to the first endpoint, and $D2$ the distance of sample *j* to the second endpoint. The location of sample *j* is projected perpendicularly onto the baseline to define the ordination value X and distance off the axis E, as shown in the figure. Although Bray & Curtis (1957) solved the problem graphically, it is easily translated into algebraic terms (Beals 1960):

$$X = \frac{L^2 + (D1)^2 - (D2)^2}{2L}$$

and

$$E = [(D1)^2 - X^2]^{1/2}$$

Because each computation involves only three points, it can be represented by the simple triangle in Figure 4.3 or the preceding equations. The entire ordination of all the samples may be conceptualized in samples dissimilarity space: given a cloud of sample points in samples dissimilarity space, two endpoints are chosen that are most distant (and, it is hoped, thereby represent poles along the major community gradient), and all sample points are projected perpendicularly onto the ordination axis joining the two endpoints.

A sample similar to the first endpoint and dissimilar to the second endpoint will fall near the first endpoint, and likewise, samples similar only to the second endpoint will ordinate near it. Samples dissimilar to both endpoints will ordinate around the center of the ordination axis. Samples of similar species composition will ordinate in any case near each other.

Only X is actually an ordination score. The distance off the axis E measures the extent to which the ordination score fails to define the location of a sample's point in the multidimensional samples dissimilarity space: A small value of E (in relation to the length of the ordination axis, that is, the spread in X values) indicates that the point is near the ordination axis, but a large value of E indicates that the point is far off the axis (because of community gradients additional to the community gradient captured in the ordination scores X). Often, only X is calculated. Unless an ordination is clearly adequate, however, it is desirable to compute E values. If E values are large (averaging 25 to 40% or more of the range in ordination values X), much of the data structure is not captured by the ordination axis and additional ordination axes may be useful (as explained shortly).

The samples of Table 4.4 may be arranged by polar ordination. Because these samples are given in order from most pioneer (1) to climax (10), as perceived by Peet & Loucks (1977), samples 1 and 10 may be selected as endpoints for an ordination axis intended to reflect successional status. Given distances previously computed between sample points 1 and 5 of 43, 5 and 10 of 55, and 1 and 10 of 72, the ordination score for sample 5 is

$$X = \frac{(72)^2 + (43)^2 - (55)^2}{2 \cdot 72} \approx 28$$

This calculation may be performed for all ten samples and the ordination scores rescaled into the range 0 to 100 for convenience in graphing. The resultant polar ordination of the samples, given in Figure 4.4, shows the pioneer-to-climax community gradient effectively. The sample sequence is nearly, but not precisely, 1 to 10, so the original sample assessments by Peet & Loucks (1977) and this polar ordination differ slightly in detail. Because the data of Table 4.4 are an abbreviated version of those presented by Peet & Loucks, the two analyses are not strictly comparable, and it would not be meaningful to ask which analysis is preferable given such slight difference. Sample 2 is placed at one end of the ordination despite its not being an endpoint; this phenomenon is not uncommon in polar ordination analyses and is not ordinarily problematic. If instead of this deliberate choice of samples 1 and 10 as endpoints, one uses the objective procedure of Bray & Curtis, samples 2 and 8 are selected as most distant from each other. The resultant ordination (not shown) is very similar, in general, to the sample ordination shown in Figure 4.4.

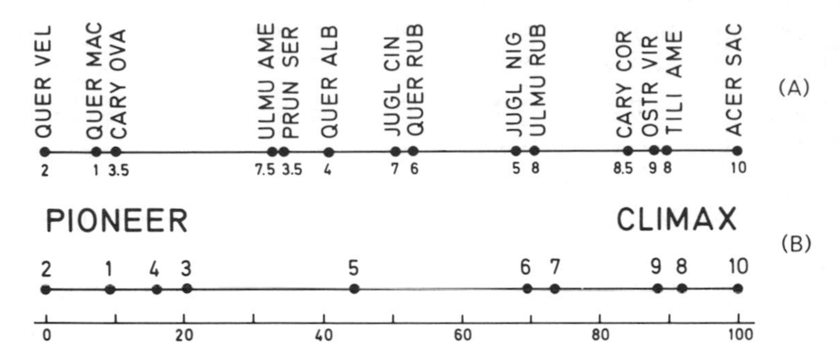

Figure 4.4. Polar ordinations of southern Wisconsin upland forest tree species (A) and samples (B). The data and species names are given in Table 4.4. The species polar ordination uses *Quercus macrocarpa* and *Acer saccharum* as endpoint species; the species show a clear, general trend from pioneer to climax as reflected in the climax adaptation numbers running from low on the left to high on the right. The sample polar ordination uses samples 1 and 10 as endpoint samples; again a general trend from pioneer to climax status is evident, the samples having been numbered originally from 1 to 10 in order of increasingly climax composition as perceived by Peet & Loucks (1977). These two ordinations involve independent calculations; they are not linked mathematically. Both are scaled from 0 to 100 for convenience in graphing. Both ordinations happen to express pioneer-to-climax status because in both cases endpoints that emphasize this community gradient were selected.

Polar ordination may be applied to species as easily as to samples. The computational steps are essentially the same: First a species-by-species dissimilarities matrix is computed; second, two species are chosen as endpoints to define the ordination axis (by being most dissimilar or by subjective choice); and third, all species points are projected perpendicularly onto the ordination axis. The ordination scores may be rescaled into the range 0 to 100 for convenience. The species ordination resulting from using a pioneer species, *Quercus macrocarpa,* and a climax species, *Acer saccharum,* as endpoints is also shown in Figure 4.4. The climax adaptation numbers of these species are written below the species codes to show the general trend in successional status. In Figure 4.4 only one ordination axis is shown because this small data set does not clearly reveal additional community gradients.

It is important to realize that, although in Figure 4.4 both species and sample ordinations are shown and although both ordinations show the same succession gradient because of the pioneer-to-climax endpoints chosen in both cases, the two ordinations are separate calculations and have no direct mathematical relationship to each other. The unfortunate consequences of this independence are that (1) having computed species and sample polar ordinations for a data set, the two ordinations must be interpreted separately and (2) the two ordinations may not bear any straightforward relationship to each other. This unfortunate state of affairs is in marked contrast to eigenvector ordination techniques, which will be discussed subsequently, in which species and sample scores are unified in a single ordination space and a single interpretation task applies to the whole. Likewise, in weighted averages, the weights and ordination scores are related directly.

An alternative to separate species and sample polar ordinations is to produce only a sample polar ordination and then show species optima within this ordination framework. This approach has the advantage of integrating sample results and species results and, in fact, was used by Bray & Curtis (1957, Figure 8).

A second ordination axis may be derived by ordinating samples (or species) along another axis defined by a new pair of endpoint samples. Bray & Curtis give the criteria that the second axis endpoints (1) be separated by the greatest interpoint distance and (2) be separated by the least projected distance on the first axis. This selection procedure is likely to result in a second ordination axis of relatively great extent, which is different from the first axis. Obviously, because two independent criteria are involved, one sample pair might be optimal for the first criterion but a different sample pair optimal for the second criterion. The Bray & Curtis second axis selection procedure does not specify exactly how to weight or blend these two criteria; it is formulated somewhat vaguely. A rigorous formulation suitable for automatic computer processing is given by Gauch (1977:55). Third and higher axes may be derived by continuing this procedure, using endpoints with large interpoint distances but small projected distances on all previous axes. Bray & Curtis derive three ordination axes for their data set. Geometrically, the axes of a multidimensional polar ordination are usually not exactly perpendicular to each other (in the samples dissimilarity space), but they are approximately perpendicular and, for simplicity, are customarily so drawn in ordination graphs. Orlóci (1978*a*:133–7) gives a correction for axis nonperpendicularity, but few investigators have considered the correction great enough to justify the required increase in computation.

A two-dimensional polar ordination of the 59 upland forest samples of Bray & Curtis (1957) is shown in Figure 4.5. Besides merely showing sample locations in the ordination space, this figure shows the abundance of *Tilia americana* in each sample. A definite concentration

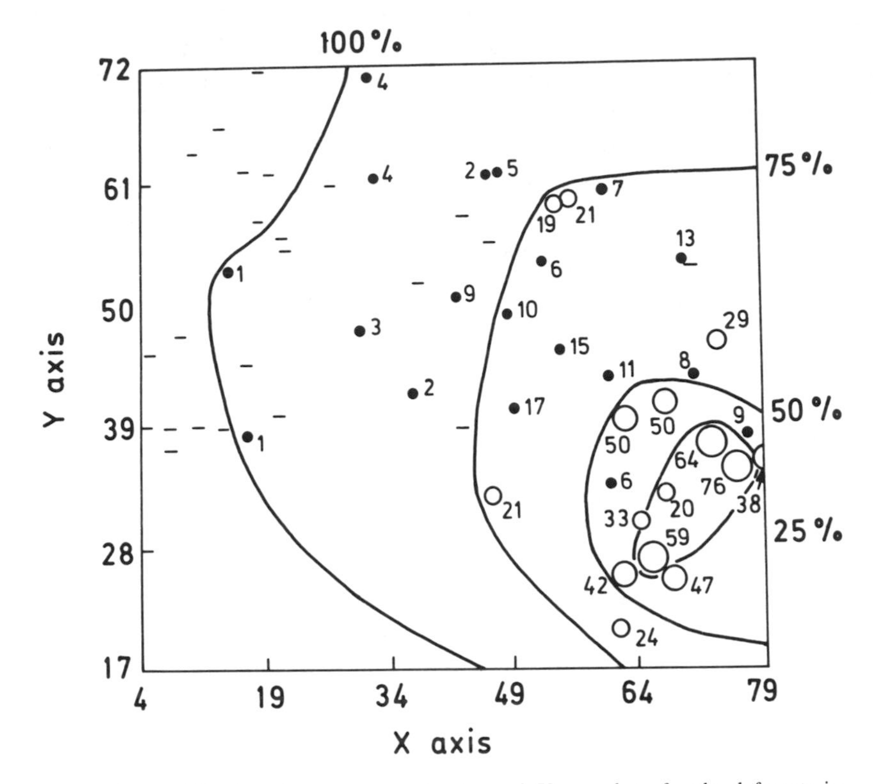

Figure 4.5. Two-dimensional ordination of 59 samples of upland forests in southern Wisconsin. At each sample point, the abundance of *Tilia americana* (basswood) is given (with a dash for absence, or a number in units of hundreds of square inches of basal area per acre at breast height); contour lines by quartile are drawn to show the population concentration in the lower right portion of the figure. (After Bray & Curtis 1957:333. Reproduced by permission of the publisher. Copyright Ecological Society of America.)

of abundance in the lower right portion of the figure is evident. Analogous graphs for numerous additional species in Bray & Curtis (1957) show each species concentrated in some portion of the ordination space, tapering off in all directions with an approximately bell-shaped distribution. The species are individualistic, each having a unique distribution and optimum, and community variation is continuous. In addition to the 26 main tree, shrub, and herb species used to construct the ordination, rare species not used in producing the ordination likewise show orderly patterns in the sample ordination space. Consequently, the ordination is judged successful in summarizing community composition and gradients.

Ecological interpretation of their three ordination axes is possible for Bray & Curtis (1957) because of knowledge of the species' habitats and assessment at each site of canopy cover and nine soil parameters: A_1 soil horizon depth, pH, water-retaining capacity, organic matter content, concentrations of Ca, K, P, and ammonia, and the ratio Ca/K. The first axis is correlated with canopy cover, A_1 depth, organic matter, pH, Ca, P, and Ca/K ratio – all features of an increasingly shaded and mesic environment. It is interpreted as a succession gradient, essentially as found earlier by Curtis & McIntosh (1951). The second axis shows correlation with soil water-retaining capacity and ammonia and, hence, indicates poor drainage and poor soil aeration. The third axis shows correlation with K, Ca/K ratio, and organic matter and features three species that benefit from recent, small openings in the forest (gap phase species, such as *Prunus serotina, Populus gradidentata,* and *Carya ovata*). The three ordination axes show little to negligible correlations with each other and are judged to convey essentially independent information.

A refreshing element in Bray & Curtis' classic paper is the precise criteria for assessing ordination results and the effective, yet realistic role given to ordination. Two assessment criteria are used. First, the ordination should effectively summarize community composition and gradients (although not entirely or perfectly because of noisy data). Second and more importantly, the community gradients found in the ordination should relate to gradients in environmental features. Their results pass these two tests and significantly increase ecological knowledge of Wisconsin upland forests. The role of ordination is to summarize community patterns; this background is further useful for generating hypotheses about community structure and about plant and environment interactions, which can be tested subsequently.

The success of Bray & Curtis' research may be attributed to two factors. First, a consistent conceptual framework prevails from the original statement of community structure, to design of the ordination, to assessment of results. Second, a substantial, quantitative, field data set contains enough information to support detailed investigation. Bray & Curtis were also successful in perceiving the limitations of their study. Because so many environmental parameters change along the complex-gradient, which may be termed a succession gradient in general, the interpretation of this gradient lacks precision. Although such imprecision may be natural to such investigations to some degree, Bray & Curtis suggest a somewhat more precise interpretation of the first axis succession gradient as recovery from past major disturbance by fire and say that further research may confirm or contradict this hypothesis.

Few techniques in ecology other than polar ordination could claim as extensive a history of subsequently suggested refinements and variations (Bannister 1968; Orlóci 1966, 1974*b*, 1978*a*:133; Gauch 1973*a*; Gauch & Scruggs 1979; see additional citations in these references). For example, Bray & Curtis (1957) applied a preliminary data standardization to give all species relatively equal emphasis (and likewise all samples). Not infrequently, this standardization improves ordination results, but it is not always used. Likewise, other distance measures may be used, but percentage dissimilarity (used by Bray & Curtis) appears generally best (Gauch 1973*a*). Concerning endpoint selection, both the original automatic endpoint selection procedure and deliberate endpoint selection based on ecological judgment are useful and complementary approaches. Many suggested refinements seem to be of little practical value, however, and, indeed, have not found wide adoption. Suggested refinements that change ordination values no more than 5 to 10% in typical cases may be judged generally insignificant in view of the inherent subjectivity and noisiness of community data. Likewise, certain refinements that give marginally superior results given data sets with specific data properties may be of little value when the assessment or estimation of these data properties is itself problematic.

Polar ordination has partial similarities, mathematically and functionally, to weighted averages and to eigenvector ordinations. These similarities must be appreciated in order to see relationships among ordination techniques properly. Unfortunately, the great differences in computational procedures and sophistication tend to hide underlying similarities. Polar ordination contrasts two polar samples, and approximately the same result can be achieved through weighted averages by designing a set of species weights having high values for species abundant in the first endpoint sample and low values for species abundant in the second endpoint sample. The selection of two endpoint samples in polar ordination comes close in function and in results to being merely a simple means for defining a set of species weights for weighted averages ordination of the samples. Indeed, for the data of Table 4.4, the polar ordination (Fig. 4.4) and the weighted averages ordination (given earlier) are very similar; they are even identical as to rank order of the samples. Likewise, principal components analysis of a variance–covariance matrix (as explained shortly) uses an identical samples dissimilarity space as does polar ordination using Euclidean distances. To the degree that the axis defined by the endpoint pair is parallel to that defined by the principal components axis, the ordination results will be the same. Given even a modest number of samples (say, 15 or 20), the probability that some sample pair has very nearly the same orientation as the principal components axis is high, especially as elongation of the sample point cloud in this major direction of community variation enhances the probability. If, instead, a distance measure other than Euclidean distance is used for polar ordination, the configuration of sample points is ordinarily still like that using Euclidean distance to a first approximation (especially if the samples are relatively homogeneous), so a moderate similarity between polar ordination and principal components analysis may remain. Furthermore, reciprocal averaging ordination and detrended correspondence analysis are, like principal components analysis, also eigenvector methods with related algorithms. Consequently, their results often approximate results from polar ordination. It would be a mistake to consider weighted averages, polar ordination, principal components analysis, reciprocal averaging, detrended correspondence analysis, and other ordination techniques to have entirely different algorithms or to give entirely unrelated results. It would be equally erroneous to overemphasize the similarities among these ordination techniques. The differences in function are significant, and differences in the quality of ordination results may be great for difficult data sets.

Polar ordination remains of interest because of the comparability of studies employing the same technique, the simplicity and understandability of the calculations, the options for relatively objective or subjective ordination axes, and its frequent effectiveness. In many studies, polar ordination alone has sufficed for effective community analysis, but a safer approach considers polar ordination a useful technique in concert with a few additional, complementary, ordination techniques.

Principal components analysis

Principal components analysis (PCA) was first applied to ecological data by Goodall (1954*a*), although the method was invented earlier by Pearson (1901) and Hotelling (1933). Goodall analyzed species-by-species matrices concerning Australian scrub vegetation. The heavy computational demands of PCA necessitated the use of a computer, even for his data set of only 14 species. The intriguing novelty of PCA was that it gave ecologists their first ordination technique in which ordination scores were derived from the data matrix alone; the investigator did not supply weights, endpoint selections, or anything else. Thus PCA was welcomed as an especially objective ordination technique. A further advantage was the simultaneous production of species and sample ordination scores in one integrated analysis. These advantages relate directly to Goodall's particular perspective on the study of communities, a perspective that begins with the community itself and, subsequently, incorporates environmental and other information. "There is much to be said for the view that the complexes of environmental factors determining plant distribution can be indicated and measured better indirectly, through the plants themselves, than by direct physical measurements; this is, of course, the idea behind the use of 'phytometers' in agricultural meteorology." Unlike earlier methods that accommodated only one to a few indicator species, PCA enabled the "whole vegetation to be used . . . for the indication and indirect measurement of environmental complexes" (Goodall 1954*a*:322). Goodall expected this usage of more vegetational information to offer a considerable improvement upon earlier methods.

The intent and the name of PCA are expressed lucidly by Hotelling (1933:417):

> Consider n variables attaching to each individual of a population. These statistical variables $x_1, x_2, \ldots, x_n$ might for example be scores made by school children in tests of speed and skill in solving arithmetical problems or in reading; or they might be various physical properties of telephone poles, or the rates of exchange among various currencies. The x's will ordinarily be correlated. It is natural to ask whether some more fundamental set of independent variables exists, perhaps fewer in number than the x's, which determine the values the x's will take.

These more fundamental variables Hotelling termed *components,* and as they may vary in importance, interest attaches especially to the principal, or most important, components, hence the name *principal components analysis.*

To be strictly applicable, a data set must meet several assumptions of the PCA model, primarily that the components have normal distributions and be uncorrelated (Hotelling 1933; Dale 1976*b*). Field data sets rarely, if ever, meet the requirements precisely. For applications involving statistical testing of hypotheses, the assumptions must be met rather well; for merely descriptive purposes, larger departures from ideal data structure are tolerable (Greig-Smith 1980). Even for descriptive purposes, however, it must be remembered that PCA has an underlying mathematical model and, consequently, may be applicable to one data set but not another.

The algorithm of PCA may be introduced in geometric terms (Pearson 1901; Hotelling 1933; Gittins 1969; Allen & Skagen 1973; Williams 1976:47–58; Pielou 1977:332–7; see also Gower 1967*b*). Efficient projection of points in a multidimensional space into fewer dimensions is the function of PCA, such that "the arrangement of the points suffers the least possible distortion" (Pielou 1977:332). Figure 4.6 shows the derivation of PCA axes for a simplified species space with only two species. This figure shows three different coordinate systems: (1) The original axes of species space are marked "Species 1" and "Species 2," and the 25 sample points are placed according to their abundances of these two species. The lowest, left-most sample point, for example, represents a sample with an abundance of 17 for species 1 and 5 for species 2. (2) A second pair of axes, "X 1" and "X 2," is in the same direction as the original axes but with the origin moved to the centroid (which is simply the point located at the average for each species, namely, about 27 for species 1 and 18 for species 2). Movement of the coordinate system to the centroid is termed *centering*. (3) The third pair of axes, "PCA 1" and "PCA 2," involves a rigid rotation around the centroid such that the first PCA axis goes through the major extension of the cloud of points. More precisely, the variance is maximized of the ordination scores of the sample points as projected perpendicularly onto the first PCA axis. Because the total variance is the same for any of these three coordinate systems (since only rigid translations and rotations are involved), this maximization of variance along the axis is equivalent to minimizing the variance of the projection distances from the axis. In other words, the first PCA axis is in the direction that captures as much variance as possible along the ordination axis. A second PCA axis is then found orthogonal (perpendicular) to the first and accounts for maximal remaining variance, and so on, for as many PCA axes as desired. In Figure 4.6 the position of the PCA 2 axis is defined trivially by the orthogonality constraint, because this figure involves only two dimensions, but in the usual case with many dimensions, the direction of PCA axes is not trivial. The result of PCA ordination is a sequence of axes of diminishing importance.

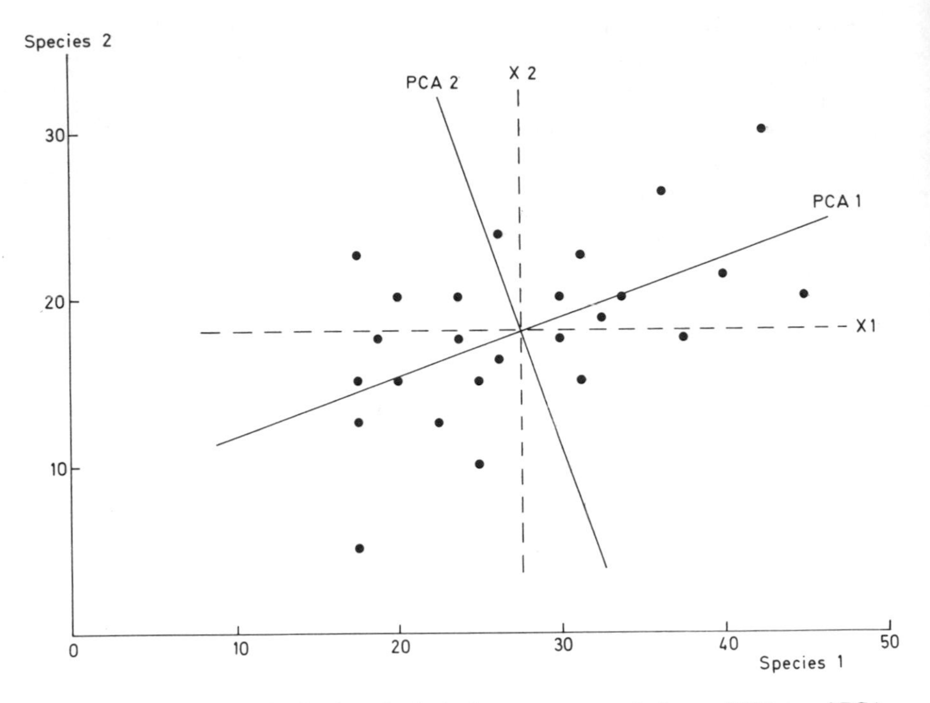

Figure 4.6. Derivation of principal components analysis axes (PCA 1 and PCA 2) from the original axes of species space (axes Species 1 and Species 2) by translation of the origin to the centroid (axes X 1 and X 2), followed by rotation around the centroid to maximize variances along the principal components analysis axes.

Four basic observations may be made in comparing the original species axes with the final PCA axes. (1) The arrangement of the points never changes; only the axes change. In particular, the distance between any pair of points is invariant. (2) The angular relations between points as viewed from the original are unaltered by the second transformation (rotation around the centroid), but they are changed by the first transformation (translation to the centroid). Hence correlation coefficients (which are cosines of the angles) are altered by moving the origin to the centroid, although not by rotation around the centroid. (3) The original axes have a simple meaning: abundances of individual species. The derived PCA axes have a complex meaning: linear combinations of species abundances. That is, location on a PCA axis is specified by a linear equation of the form a constant times species 1 abundance, plus another constant times species 2 abundance, plus . . . and so on, as is evident from Figure 4.6. Because in this example the PCA 1 axis is tilted from the species 1 axis by about 19° with a cosine of 0.946 and sine of 0.326, the PCA 1 ordination scores equal 0.946 times the centered species 1 abundance (the species 1 abundance minus 27), plus 0.326 times the centered species 2 abundance (the species 2 abundance minus 18; also see Jöreskog, Klovan, & Reyment 1976:33–4). The larger weight for species 1 than species 2 reflects the fact that the PCA 1 axis goes more nearly in the direction of the species 1 axis than the species 2 axis. (4) The PCA axes concentrate the variance or structure of the point configuration into relatively few axes, in contrast to the high dimensionality of the original species space. This dimensionality reduction facilitates comprehension of the relationships of the points to one another and constitutes the primary value of PCA. Broadly speaking, the use of PCA increases insight into relationships among points but at the same time introduces more complex axes.

The function of PCA ordination is appreciated best by considering the ordination process as a whole, starting with the original data and community model and ending with the final PCA result (Figure 4.7). The community model in Figure 4.7*A* exemplifies the usual Gaussian model: Three species are shown with Gaussian response curves along a hypothetical, one-dimensional, environmental gradient, and four samples are placed with even spacing along this gradient. The species abundance data are then used to place the four samples in three-dimensional species space, and the first PCA axis is indicated (Figure 4.7*B*). Finally, the sample points are projected into the first two dimensions of the PCA ordination (Figure 4.7*C*). Note that the original one-dimensional configuration of sample positions along an environment and community gradient has been recovered, in general, by the PCA result but has been distorted into an arch. This distortion is due to a mismatch between the community model and the underlying model (assumptions) of PCA where PCA assumes that variables (here, species abundances) change linearly along underlying gradients. These Gaussian responses are, however, not linear; they are not even monotonic. This nonlinearity problem was recognized by Goodall (1954*a*) in his original application of PCA and was brought into sharp focus later by Swan (1970) using simulated data with a Gaussian community model (also see Jeglum, Wehrhahn, & Swan 1971).

Figure 4.8 shows PCA results for the southern Wisconsin upland forests of Table 4.4. A pioneer-to-climax gradient for species and samples is evident, much like the polar ordination results shown earlier in Figure 4.4. This single succession gradient has been distorted some-

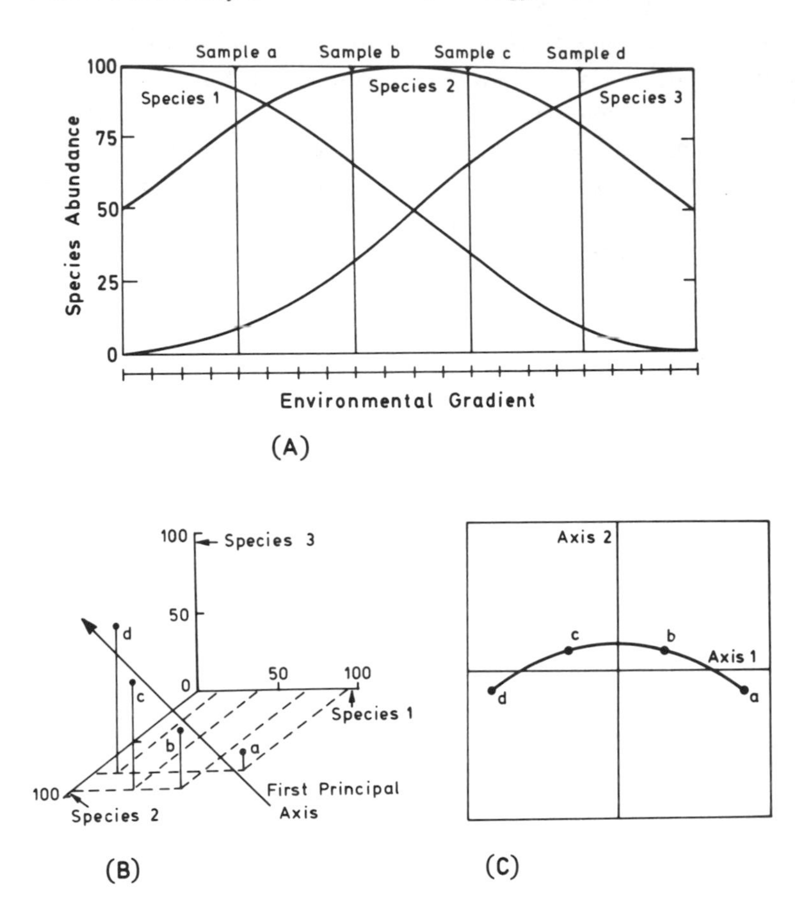

Figure 4.7. Gaussian species responses for a community model along a hypothetical environmental gradient (*A*); the representation in species space (*B*); and the resulting sample ordination by principal components analysis projection into two dimensions (*C*). (After Kelsey, Goff, & Fields 1976:14, 16, 18)

what by PCA, however, into an arched configuration. The second PCA axis in this case does not convey meaningful or independent information; it is merely a quadratic distortion of the valid first PCA axis, due to the inability of PCA to properly handle nonlinear species response curves (but see Peet & Loucks 1977 using their original data instead of the simplified summary of their data used here).

In the preceding discussion, PCA has been described geometrically. It can also be described succinctly in matrix algebra terms. To solve PCA, one begins by computing a secondary matrix from the samples-by-species data matrix. For simplicity, assume that a samples-by-

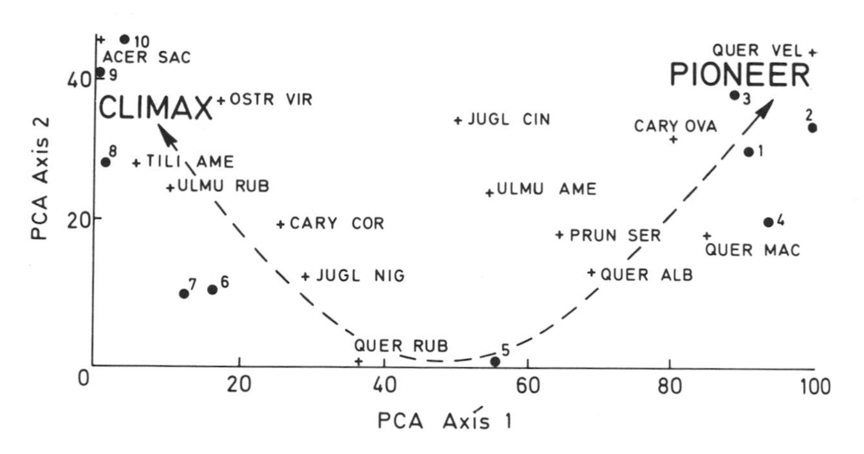

Figure 4.8. Principal components analysis (PCA) ordination of southern Wisconsin upland forests. The data and species names are given in Table 4.4. A pioneer-to-climax gradient for species and samples (indicated by the dashed line) is evident, although distorted somewhat into an arch.

samples secondary matrix is computed using variance–covariance (that is, computing the variance of a sample for comparison to itself and the covariance for comparisons between different samples). Eigenanalysis is then performed on this matrix. For each PCA axis, this calculation produces an *eigenvalue,* which is the variance accounted for by that axis. The axes are ranked by their eigenvalues, the first PCA axis having the greatest eigenvalue. If the cloud of points (in samples dissimilarity space) is concentrated in certain directions, as is usually the case, the bulk of the original total variance will be extracted in the early eigenvalues, and later eigenvalues will be small or zero. In a typical community study, the first three eigenvalues may account for 40 to 90% of the total variance. In some cases, however, particularly with large and noisy data sets, the first couple of PCA axes may account for as little as 5% of the total variance and yet be quite informative ecologically. On the other hand, in other cases, 90% of the variance may be accounted for, yet the result may be ecologically meaningless or severely distorted. In the end, the assessment of PCA results must be in terms of ecological utility; mere percentage of variance accounted for has not been found to be a reliable indicator of the quality of results (Austin & Greig-Smith 1968; Karr & Martin 1981; also see Robertson 1978). The other product of eigenanalysis is an *eigenvector* for each PCA axis. The number of values in an eigenvector equals the size of the secondary matrix analyzed; given a samples-by-samples secondary matrix with J samples, each eigenvector contains J values. These values are the ordination scores for the samples. They are cosines specifying the direction of the PCA axis (in species space). Finally, the data matrix may be multiplied by sample eigenvectors to obtain corresponding species eigenvectors. Principal components analysis has a useful samples–species duality: Both are ordinated at the same time. Either a samples-by-samples or species-by-species secondary matrix may be subjected to eigenanalysis and a simple matrix multiplication may be used to obtain the corresponding results. For computational efficiency, the smaller matrix of the two is ordinarily chosen for eigenanalysis.

Computational details of PCA may be found in the work of Hotelling (1933), Williams (1976:3–28), Pielou (1977:332–40), and Orlóci (1978*a*:109–29). There are many computational routes (Wilkinson & Reinsch 1971) possible to solve PCA, just as numerous possibilities exist for computing the square root of a number. The usual methods have computer requirements rising at approximately the second to third power of the number of samples or number of species, whichever is smaller. Consequently, such computer programs are limited in the amount of data that is practical. If, as is usually the case, only the first few eigenvalues and eigenvectors are of interest, PCA may be solved, however, by a direct iteration procedure, which even bypasses computation of a secondary matrix and implies computer requirements rising only linearly with the amount of data (Jennings 1967; Clint & Jennings 1970; Hill 1973; Hill & Smith 1976).

Several adjustments may be applied to the samples-by-species data matrix prior to computation of a secondary matrix for eigenanalysis. If none is applied, the PCA is noncentered and the secondary matrix is a dispersion matrix (Noy-Meir 1971, 1973*a*; Carleton 1979, 1980; Carleton & Maycock 1980). Subtraction of species means leads to centered PCA, using a variance–covariance matrix. Centered PCA was used for Figure 4.8. If species are centered and then standardized to unit variance, the analysis is termed *centered and standardized,* or simply *standardized,* PCA, using a correlation matrix. Geometrically, standardization to unit variance involves the expansion or contraction of each species axis of species space to unit variance. Otherwise, the geometry is the same as that for centered PCA as explained earlier (with Figures. 4.6 and 4.7). Noy-Meir, Walker, & Williams (1975) discuss the implications of these and other adjustments in terms of the investigator's purposes and the particular characteristics of individual communities. For ordination purposes, centered PCA and standardized PCA are commonly employed.

In the original application of PCA to ecological data, Goodall (1954*a*) tests hypotheses about vegetation structure, particularly whether or not his samples are homogeneous and whether the community variation is continuous or discontinuous. The relative objectivity of PCA gives special interest to its function in hypothesis testing. As Goodall observes, however, PCA has a nonlinearity problem, and furthermore, mathematical orthogonality of PCA axes does not necessarily imply biological independence. Because the underlying model of PCA is not fully realistic for community data, results are partially spurious; consequently, hypothesis testing is not entirely solid. Hotelling (1933) was also interested in PCA for testing hypotheses. Although of much interest, the function of ordination for hypothesis testing appears unfortunately limited at present (but see Strong 1980). Rather, the more common viewpoint stresses a function of ordination for hypothesis generation (Williams 1976). Unquestionably, the most common function of PCA is simply descriptive – reduction of the dimensionality of a data set to manageable proportions while preserving as much of the original structure as possible. This descriptive function of PCA is served successfully, however, only for data sets whose properties match the PCA model tolerably well (Hotelling 1933; Gauch, Whittaker, & Wentworth 1977). The function and interpretation of PCA results are discussed further by Austin (1968), Isebrands & Crow (1975), Green (1977), and Nichols (1977).

In summary, PCA is an ordination technique for projecting a multidimensional cloud of pints into a space of fewer dimensions, using rigid rotation to derive successive orthogonal axes, which maximize the variance accounted for. This minimizes the distortion of distances between points upon projection into fewer dimensions. When the first few eigenvalues account for much of the total variance, PCA projection allows insight into the structure of the cloud of points. The percentage of variance accounted for and the ecological value of results are not tightly related, however, so the assessment of results must be in terms of ecological insight afforded. Both species and sample ordinations result from a single analysis. Computationally, PCA is essentially an eigenanalysis problem. In comparison with earlier ordination techniques, PCA is relatively objective in that the ordination calculations do not require subjective weights or endpoints. Analysis of community data alone, with complete exclusion of environmental data from the analysis, gives an interesting community-centered viewpoint but at the same time makes environmental interpretation of PCA results a separate step, potentially difficult and rather subjective.

Reciprocal averaging

Reciprocal averaging (RA) is an ordination technique related conceptually to weighted averages but is computationally an eigenanalysis problem similar to that for principal components analysis. It was developed by Hirschfeld (1935) and Fisher (1940); also see Thurstone & Chave (1929), Richardson & Kuder (1933), and Horst (1935). Early ecological applications include those by Goff & Cottam (1967; although applying RA to a secondary matrix, as explained in the previous section on weighted averages), Roux & Roux (1967), Hatheway (1971), and Guinochet (1973). Papers by Hill (1973, 1974) first made RA well known to ecologists, especially after tests by Gauch, Whittaker, & Wentworth (1977) demonstrated its superior performance. Reciprocal averaging is also called *correspondence analysis* (Hill 1974), or *analyse factorielle des correspondances* (Benzécri 1969; Bonin & Roux 1978; Orsay 1979), *reciprocal ordering* (Orlóci 1978*a*:152–68), and *dual scaling* (Nishisato 1980). The name *reciprocal averaging* is fitting because the species ordination scores are averages of the sample ordination scores, and reciprocally, the sample ordination scores are averages of the species ordination scores (Hill 1973, 1974). Benzécri (1973*b*) offers an extensive treatment of reciprocal averaging. Nishisato (1980) discusses reciprocal averaging in a wide context of related methods and statistical test. He also clarifies its assumptions and mathematical properties (Nishisato 1980:14–15, 68–9, 204).

The algorithm of RA involves simple matrix algebra. Earlier it was noted that weighted averages ordination can be used to obtain sample scores from species scores or species scores from sample scores. Using iteration, RA simply does both. For the first iteration, one begins by assigning arbitrary species ordination scores. Then weighted averages are used to obtain sample scores from these species scores. The second iteration produces new species scores by weighted averages of the sample scores; it is convenient to scale the resulting new species scores into a standard range, say, 0 to 100, to keep them from drifting into a very small range. Then new sample scores are produced by weighted averages of the species scores. Iterations are continued until the scores stabilize. The scores converge to a unique solution, and the solution is not affected by the initial choice of arbitrary species scores (except that convergence requires fewer iterations if the initial scores are close to the final scores). Ordinarily, about 20 to 100 iterations result in the convergence of ordination scores with an accuracy of two to four significant digits. When the solution stabilizes, as noted by insignificant changes in species scores from one iteration to the next, the final sample scores may also be scaled into a range of 0 to 100. The contraction in the range of species scores in one iteration (after convergence is reached) is the eigenvalue. Note that RA ordinates samples and species simultaneously. Having extracted one axis, it is possible to extract a second axis by the same iterative method, except correcting for the first axis; a third axis may then be calculated while correcting for the first two, and so on, to as many RA axes as desired. The RA axes so extracted have a sequence of decreasing eigenvalues. Higher axes may be scaled into ranges proportional to the square roots of the ratios of their eigenvalues to the first eigenvalue; for example, given first and second eigenvalues of 0.741 and 0.536, if the first axis is scaled into 0 to 100, the second axis may be scaled into 0 to $100 \cdot (0.536/0.741)^{1/2}$, or 0 to 85. Hill (1973, Appendix 2) gives the details for extracting several RA axes in his example.

The procedure just described, naturally named *reciprocal averaging,* constitutes the direct iteration algorithm for solving RA. Direct iteration is an efficient algorithm for obtaining one to several RA axes and has computer requirements rising only linearly with the amount of data (Hill 1973). Consequently, RA may be applied to very large data sets without difficulty. Also, direct iteration is very simple; indeed, it is hardly twice as complex as weighted averages. Hill (1979*a*) presents an improvement of the basic direct iteration algorithm that is more complex but is even quicker by a factor of two or three. Ordinarily, only one or a few RA axes are desired, but when all axes are required, conventional eigenanalysis may be employed, although computer requirements then increase rapidly and may limit practical problem size (Hill 1973, 1974; Orlóci 1978*a*:152–68). The RA results obtained are not affected by the computational route chosen, however, any more than the square root of seven depends on the method chosen for computation. Only practicalities are affected, namely, the complexity of the computer programming effort required and the speed with which the RA solution is reached.

It is interesting that there are different computational routes for solving RA. One route shows the kinship of RA to simple, traditional,

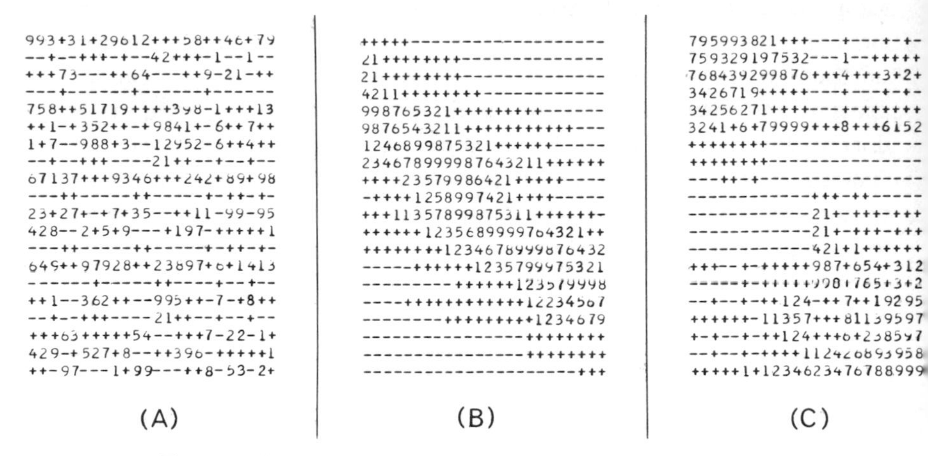

Figure 4.9. A simple, simulated, samples-by-species data matrix with 24 samples (columns) and 20 rows (species), arranged at random (*A*), by reciprocal averaging (RA) scores (*B*), and by centered principal components analysis (PCA) scores (*C*). (From Gauch, Whittaker, & Wentworth 1977:159)

weight averages. Another route shows its kinship to formal, complex, eigenanalysis techniques such as principal components analysis.

The first-axis RA scores have the property of maximizing the correlation of the samples and species. Consequently, if the samples of a data matrix are arranged in rank order of their RA first-axis ordination scores and likewise the species, the resulting RA arranged matrix will have its larger values concentrated along the matrix diagonal. Figure 4.9 shows a data matrix for a simulated one-dimensional community gradient (coenocline) with Gaussian species responses, a moderate gradient length of five half-changes, with 20 species as rows and 24 samples as columns, and datum values rounded into ten equal classes marked "+" and "1" to "9," with "−" for absence (Gauch, Whittaker, & Wentworth 1977). Figure 4.9*A* shows this data set with rows and columns (species and samples) arranged at random. Essentially random order from an ecological perspective is typical for field data when the samples are arranged as encountered and, likewise, when the species are arranged as encountered or alphabetically. Although the data are structured in a coenocline as just described, in the randomly arranged matrix, visual inspection reveals no structure. Figure 4.9*B* shows exactly the same data but the matrix rows and columns have been arranged by RA scores. The structure of this simulated coenocline is then entirely evident, RA arranging the matrix in the desired, natural order. Note that in the RA-arranged matrix, (1) similar species (rows) are brought together and dissimilar species are far apart; (2)

likewise, similar samples (columns) are brought together and dissimilar samples are separated; and (3) the correlation of the samples and species is maximized in that certain samples are seen to contain principally certain species and, likewise, certain species occur in certain samples. The concentration of greater values along the matrix diagonal is a fundamental property and value of RA. Figure 4.9*C* shows this data matrix as arranged by centered principal components analysis. By contrast with the random order, it is evident that the data have some structure, but the simple coenocline structure is not revealed in an entirely satisfactory manner. The RA arrangement is clearly superior and, in fact, could not be improved.

When a community data set is structured by a single or at least a predominating gradient, the RA-arranged matrix will display this gradient effectively (Persson 1981). It must be realized, however, that an arranged data matrix is intrinsically a one-dimensional medium for displaying data structure because the samples are arranged in a single sequence (and likewise the species). Given a data set with two or more important gradients, an arranged matrix cannot possibly be equally satisfactory. In this case the first-axis RA scores may be used to produce an arranged matrix, and likewise, additional matrix arrangements may be based on the second and higher RA axes. Each of these matrix arrangements shows much order, particularly as contrasted to a random arrangement. Each arrangement fails to concentrate larger values along the matrix diagonal terribly well, however, because this one-dimensional medium cannot possibly account for a multidimensional data structure. Note that what is being criticized here is the potential of a one-dimensional display to convey structure in several dimensions; the criticism is not directed to RA or to any other particular means of arranging a data matrix. The effective presentation of multidimensional data structures requires display options other than arranged matrices, such as two- or three-dimensional plots of sample and species ordination scores from RA or other effective ordination techniques.

Reciprocal averaging may be described in geometric terms. Three fundamental options are involved in specifying a given eigenvector ordination technique: the distance measure, the weights attributed to points, and the position of the origin (Chardy, Glemarec, & Laurec 1976; also see Hill 1974; Noy-Meir, Walker, & Williams 1975; and Orlóci 1978*a*:109–37). Centered principal components analysis, as described earlier, uses Euclidean distances, equal weights for points, and location of the origin at the centroid. Like principal components analysis, RA can be viewed geometrically as the derivation of new axes, which maximally account for the structure of the points in a multidimensional cloud of points, making possible the reduction of the dimensionality. Reciprocal averaging, however, uses chi-square distances, weights for sample points proportional to the total for the sample (and likewise species points weighted by species totals), and an origin at the centroid (center of gravity). Geometrically, the general intentions of principal components analysis and RA are identical – a multidimensional cloud of points is to be projected efficiently into fewer dimensions. The details of these two analyses differ, however; hence, the exact meaning of "projected efficiently" differs.

As previously noted, RA is closely related to principal components analysis, both algebraically and geometrically. Furthermore, these two analyses are identical in general function; community data alone are analyzed by a rather objective procedure (not requiring weights or endpoint selection) and environmental interpretation is a separate task. Principal components analysis and RA are also similar in computer requirements. From a practical viewpoint, the most important question in comparing these two ordination techniques is, "how well do RA and principal components analysis perform for analysis of ecological community data?" This question can be answered, in part, from theoretical reasoning (Hill 1973, 1974; Noy-Meir, Walker, & Williams 1975), and, in part, from empirical tests with simulated and field data (Gauch, Whittaker, & Wentworth 1977; Moral 1980).

For most community data sets, RA has been shown to be superior to principal components analysis. Like principal components analysis, RA ordinations have the arch problem, but unlike RA, the additional problem exists for principal components analysis that the arch may be involuted (Gauch, Whittaker, & Wentworth 1977). Consequently, what should be opposite ends of a gradient may be brought into close juxtaposition by principal components analysis, causing serious confusion. For handling long community gradients, RA is superior (Hill 1973; Gauch, Whittaker, & Wentworth 1977). Most community data sets contain one or more gradients with lengths of at least two or three half-changes, and RA results are then ordinarily superior to principal components analysis results. Occasionally, ecologists have relatively homogeneous data sets with shorter gradients, however, and for such data, principal components analysis may be better.

Despite a general and considerable superiority of RA over principal components analysis, RA is not perfect. Two major faults have been recognized ever since RA was introduced to ecologists by Hill (1973).

First, the second axis may be merely a quadratic distortion of the first axis – the arch problem (Hill 1973, 1974; Gauch, Whittaker, & Wentworth 1977; Maarel 1980*a*). This problem persists to many dimensions, the third axis being a cubic distortion, the fourth a quartic, and so on (Gauch, Whittaker, & Wentworth 1977), but as higher axes have decreasing eigenvalues, the distortion in the second RA axis is the most problematic. A particularly unfortunate consequence of this problem is that interesting secondary gradients in the data may be deferred to higher axes because the eigenvalue of the quadratic distortion is greater than the eigenvalue of a secondary gradient when the secondary gradient is less than about half the extent of the primary gradient. A primary and secondary gradient in the data may thus appear in RA axes one and three, rather than the desired one and two. The interpretation of results is thus made more difficult because spurious axes must be distinguished from valid axes, and the search for valid axes may have to be pursued to higher dimensions because of larger eigenvalues given to spurious axes. The second major fault of RA is that the first-axis ends are compressed relative to the axis middle, so that a given distance of separation in the ordination does not carry a consistent meaning in terms of implied differences between the samples (or species).

These two major faults of RA may be illustrated with the simple data set shown in Figure 4.10 (Hill 1979*a*; Hill & Gauch 1980). The rows represent the 18 samples, the columns represent the 23 species, and each dot represents the presence of a species in a sample. Sample 7, for example, contains species 7, 8, 9, 10, 11, and 12. This data set has a regular pattern of species turnover, and an ideal ordination would represent the species (and the samples) as equally spaced points along a one-dimensional gradient, with no extension into a second or higher ordination dimension. In this figure, samples and species are arranged and spaced according to their first-axis RA scores. Note that the axis ends, for both samples and species, are compressed relative to the axis middle. The RA ordination is not quite ideal.

Figure 4.11 shows the RA results for the 18 samples in one and two dimensions (after Hill & Gauch 1980; the results for species, not shown, are analogous). In two dimensions an arch is clearly evident, despite the fact that, ideally, the ordination result for these data should be one-dimensional. The second axis merely separates samples of the gradient middle from samples of the gradient ends; it does not convey new or independent information about the data (because, indeed, the data do not have any structure beyond that expressed by one dimen-

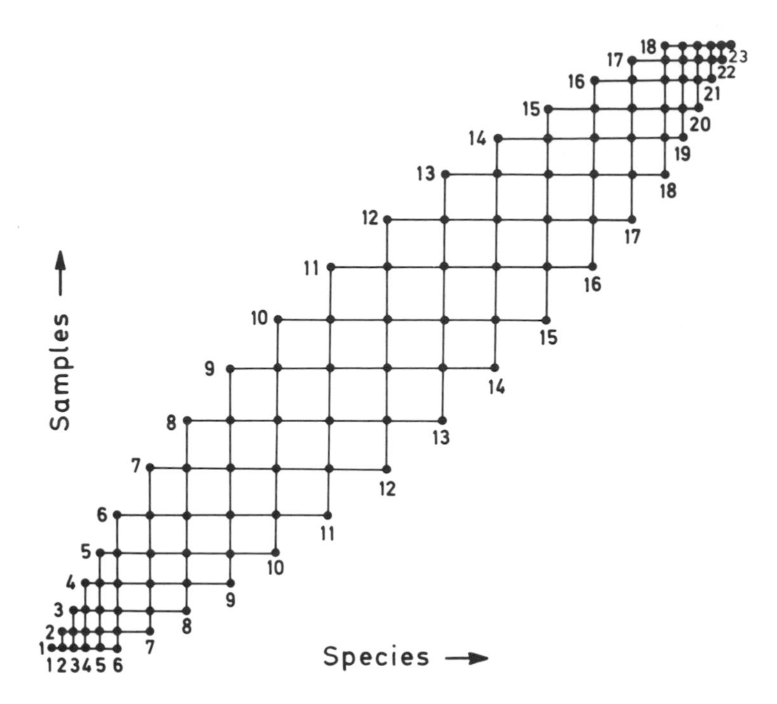

Figure 4.10. Reciprocal averaging arrangement of a regular data structure with 18 samples in rows, spaced according to first-axis reciprocal averaging scores, and, likewise, 23 species in columns. A species's presence in a sample is indicated by a dot. Ideally, these samples would be spaced evenly (and likewise these species), but reciprocal averaging compresses the ends of the gradient relative to the middle. (From Hill & Gauch 1980:48)

sion). Unfortunately, unwary investigators using ordination may be tempted, in a case like that of Figure 4.11, to interpret the first and second axes as being two separate gradients, despite the actual single dimensionality of the data structure (Maarel 1980*a*). The second main fault of RA, compression of the first axis ends, is rendered easier to see by the ordination graph in Figure 4.11*B*, which uses only the first RA axis.

The two major faults of RA are compression of the first-axis ends and extraction frequently of a spurious arch in the second axis. These two problems are intertwined, and both arise from a discrepancy between the underlying model of RA and the mathematical properties of communities as exemplified in the Gaussian model of community structure. Note that the arch on the second axis mainly separates adjacent points located near the axis ends, whereas the first axis mainly separates adjacent points near the axis middle; taken together, the

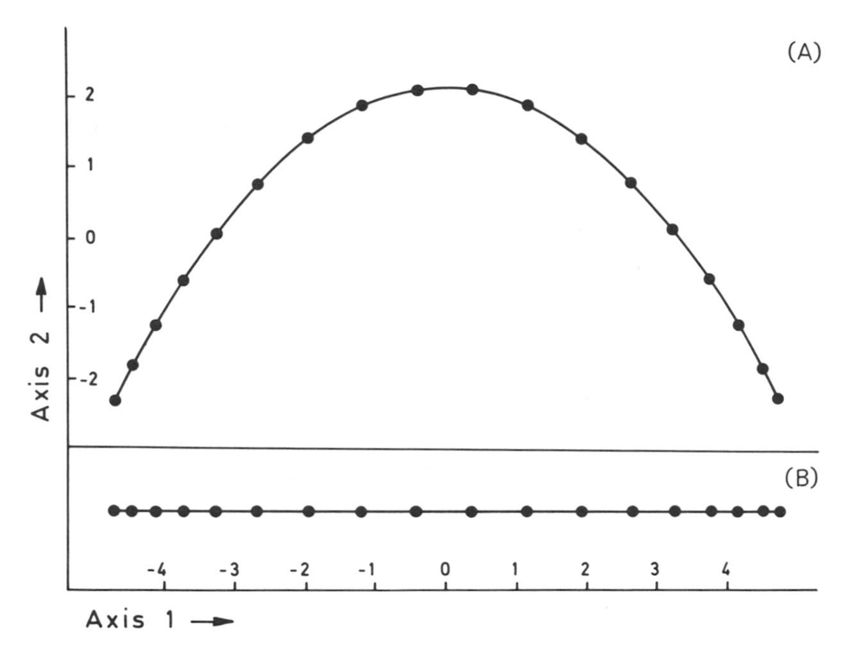

Figure 4.11. Two major faults of reciprocal averaging: the arch distortion of the second axis (*A*) and compression of first-axis ends relative to the middle (*B*). The data ordinated here, given in Figure 4.10, involve 18 regularly placed samples along a one-dimensional community gradient and would be ordinated ideally as equally spaced points in a straight line with no structure extending into higher ordination dimensions. The scaling of the axes is such that $\Sigma\Sigma a_{ij}(x_j - y_i)^2 = \Sigma\Sigma a_{ij}$, where a_{ij} is the abundance of species i in sample j, x_j the score of sample j in the ordination, y_i the score of species i, and $x_j = \Sigma_i a_{ij}y_i/\Sigma_i a_{ij}$. (After Hill & Gauch 1980:48)

two-dimensional ordination gives fairly even spacings between successive points (Figure 4.11). Unfortunately, because the RA model is not entirely appropriate for these data, the ideal ordination solution is not achieved perfectly in the first RA axis, and the discrepancies then become residual data structure expressed in second and higher axes. An ideal ordination would recover data structure entirely in the proper dimensionality and no discrepancies would linger to give energy to higher spurious dimensions.

The RA arch is undesirable, but one small, useful matter associated with the arch deserves mention. Often, points are distributed in two RA dimensions along an arch, but some points are somewhat interior or even central to the arch and other points are on the periphery of the arch. Interior points indicate species with especially broad or undiscriminating (weedy) distributions and, likewise, samples with a broad or mixed collection of species (perhaps due to the sample site being heterogeneous or disturbed). Peripheral points indicate species with especially narrow distributions, having strong indicator potential, and, likewise, indicate samples with a simple collection of similar species.

A third and lesser difficulty of RA may be mentioned. As contrasted with other eigenvector ordinations, RA is a compromise between an emphasis on samples and an emphasis on species (Noy-Meir, Walker, & Williams 1975). Ecologists frequently find this balance natural and desirable. An unfortunate implication of this balance, however, is that species that are both rare and occur in samples with low total abundances are treated as being extremely distinctive. Reciprocal averaging places such rare species (and the samples containing these species) at the extreme ends of ordination axes. This difficulty should not be overstated, however, because it occurs only with data sets having such rare species, and this problem is easily avoided by deleting rare species from a data set (because this deletion removes very little information from the data set). On the whole, the balanced emphasis of RA seems desirable, and the problems generated by this one unfortunate case are easily remedied.

In summary, RA is related both to simple weighted averages and to eigenvector ordinations. Species RA scores are averages of the sample RA scores, and reciprocally, the sample RA scores are averages of the species RA scores; hence the name, reciprocal averaging. An RA-arranged matrix concentrates the larger matrix entries along the matrix diagonal. The RA method serves for relatively objective analysis of community data, requiring no weights or endpoint selections; the environmental interpretation of results is a separate and subsequent step. The computational task for obtaining one or a few RA ordination axes rises only linearly with the amount of data, so large data sets may be analyzed readily. The RA results are generally superior to the results from principal components analysis, especially if the samples are very heterogeneous. Reciprocal averaging has, however, two main faults: The first RA axis ends are compressed relative to the middle and the second RA axis is often a spurious, quadratic (arch) distortion of the first axis.

Detrended correspondence analysis

Detrended correspondence analysis (DCA) is an improved eigenvector ordination technique based on reciprocal averaging but correcting its

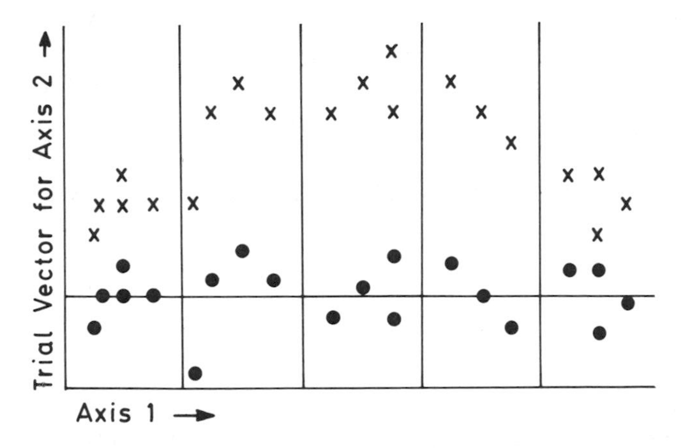

Figure 4.12. Method of detrending used in detrended correspondence analysis. The (×) indicate sample scores before detrending; (•), sample scores after detrending. The first axis is divided into five segments and a local mean is subtracted in each. Note that after detrending there is virtually no systematic relation between the trial vector for axis 2 scores and the axis 1 scores. The computer program for detrended correspondence analysis uses a larger number of segments and a sophisticated running averages procedure, but the essential concept of detrending is displayed by this simplified diagram. (From Hill & Gauch 1980:48)

two main faults (Hill 1979*a*; Hill & Gauch 1980). As reciprocal averaging and correspondence analysis are synonyms, this technique's name indicates a detrended form of reciprocal averaging. Technical details are given by Hill (1979*a*; Hill & Gauch 1980). Here the purpose and general nature of the computations will be described.

The arch distortion of reciprocal averaging arises because when second and higher axes are derived, they are only constrained to be uncorrelated (orthogonal) with lower axes. The arch is indeed uncorrelated (because the positive correlation on one side of the arch and the negative correlation of equal magnitude on the other side, as in Figure 4.11, yield a net correlation of zero). Nevertheless, the arch causes a strong, undesired, systematic relation of the second axis to the first. The orthogonality criterion for second and higher axes of reciprocal averaging is replaced in DCA with the stronger criterion that second and higher axes have no systematic relations of any kind to lower axes. This stronger criterion is implemented by detrending, as will be explained next.

The method of detrending in *detrended correspondence analysis* is division of axis 1 into a number of segments, and within each segment, the axis 2 scores are adjusted to have an average of zero (Figure 4.12).

Detrending is applied to the sample scores at each iteration, except that, once convergence is reached, the final sample scores are derived by weighted averages of the species scores without detrending. This procedure results in a DCA eigenvector ordination of the species with no arch problem and a corresponding set of sample scores, which are simply weighted averages of the species scores (as in reciprocal averaging).

To calculate a third DCA axis, sample scores are detrended with respect to the second axis as well as the first, and so on for higher axes. Because detrending is applied to lower axes individually, it is possible to obtain distortion axes that are interactions of lower axes (linear for one axis for a constant value on another axis and vice versa). Interaction axes can be demonstrated for DCA using appropriate simulated data (Hill & Gauch 1980). They have small eigenvalues, however, and are not likely to be problematic with field data because significant axes would ordinarily predominate small interaction axes.

The other fault of reciprocal averaging is compression of the first-axis ends relative to the axis middle. For example, in Figure 4.10 successive samples (horizontal lines) differ by the same amount (namely, always having two species different out of seven total) and ideally should be equidistant in the first ordination axis. Successive samples in the axis middle, however, are separated several times as much as samples near the axis ends in this reciprocal averaging ordination. In order for distances in the ordination space to have a consistent meaning in terms of the compositional difference of samples (or distributional difference of species), the axis ends need to be stretched out somewhat. Note in Figure 4.10 that both sample and species ordinations suffer from the same problem.

The fundamental intention of the rescaling in DCA is to expand or contract small segments along the species ordination axis such that species turnover occurs at a uniform rate along the species ordination axis and, consequently, that equal distances in the ordination correspond to equal differences in species composition. For technical reasons, this goal is best achieved by an indirect, somewhat complex procedure. Rescaling is achieved by expanding or contracting small segments of the species ordination while trying to equalize the average within-sample dispersion of the species scores at all points along the sample ordination axis (Figure 4.13). This adjustment has the desired effect because demanding uniformity of within-sample dispersion of species scores is tantamount to demanding uniformity of within-species dispersion of sample scores (as is evident in Figure 4.10, in which, at

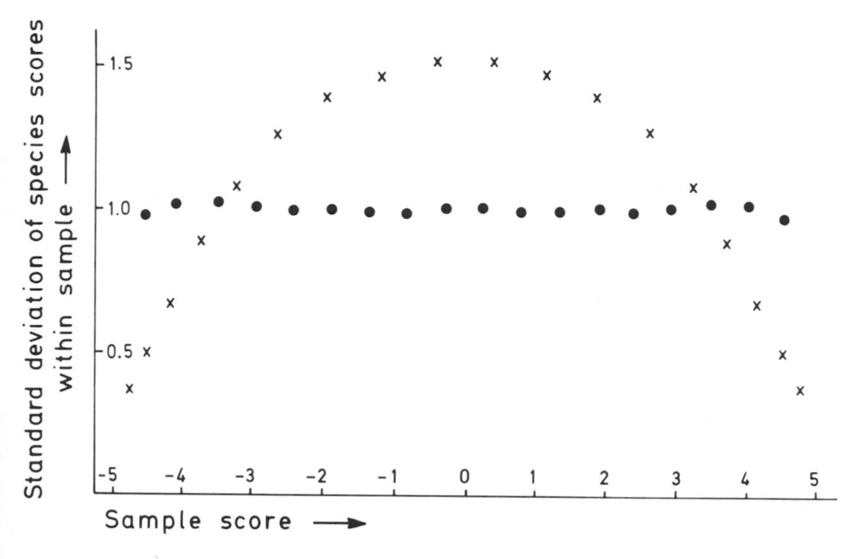

Figure 4.13. Within-sample standard deviation of species scores in relation to position along the first sample ordination axis in detrended correspondence analysis (•), as contrasted with reciprocal averaging (×). With detrended correspondence analysis, the standard deviations are effectively constant as desired, but with reciprocal averaging, compression of axis ends leads to values lower than in the axis middle. Uniformity of within-sample standard deviation of species scores approximately assures uniformity of within-species standard deviation of sample scores, so the axis rescaling of detrended correspondence analysis gives distances in ordination space a consistent meaning. (From Hill & Gauch 1980:49)

any position along the gradient, the length of vertical lines for species is roughly equal to the length of horizontal lines for samples). Adjusting samples is, however, more robust than adjusting species, for technical reasons detailed by Hill (1979*a*; Hill & Gauch 1980).

Given that the within-sample standard deviation is to be constant, as in the DCA standard deviations in Figure 4.13, it is natural to set this constant to 1 to achieve a standard scaling. This value causes the average (more precisely, the root mean square) species abundance profile to have a standard deviation of 1. The resulting unit of ordination length may be called an *average standard deviation of species turnover,* or SD (identical to the *Z* of Gauch & Whittaker 1972*a*). A species appears, rises to its mode, and disappears over a span of about 4 SD, and a full turnover in species composition of samples likewise occurs in about 4 SD. A 50% change in sample composition (a half-change) occurs in about 1 SD or somewhat more (up to about 1.39 SD for data entirely free of noise). As noted earlier, reciprocal averaging

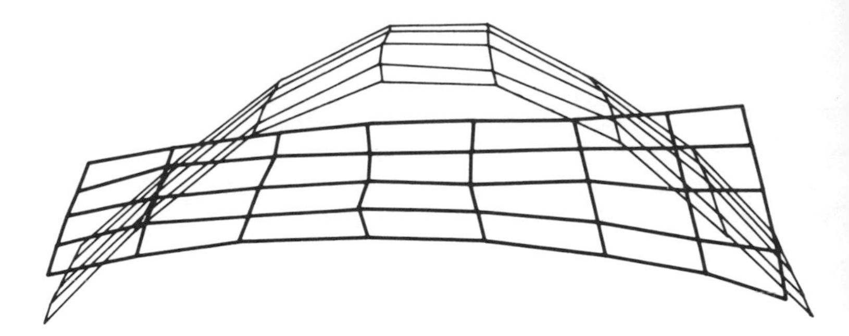

Figure 4.14. Two ordinations of a simulated two-dimensional community pattern with samples in a regular grid three times as long as wide (4.5 × 1.5 half-changes). The detrended correspondence analysis ordination (thick lines) is nearly ideal. The reciprocal averaging ordination (thin lines) shows the characteristic arch distortion and fails to reveal the minor community gradient effectively. (From Hill 1979*a*:2)

axes may be scaled into an arbitrary 0 to 100 range (for the first and higher axes according to eigenvalues) according to Hill's (1973) original suggestion. By contrast, DCA scaling is in natural units. Consequently, if one ordination axis is 6.5 SD long, and another axis (for the same or a different data set) is 4.3 SD long, the first axis represents a longer community gradient than the other. A natural, meaningful, axis scaling is of obvious value for comparing ordination results from different data sets.

The adjustments just outlined, reciprocal averaging with detrending in place of orthogonalization, followed by axis rescaling based on standardization to unit within-sample variance, combine to characterize DCA. Detrending is applied to the second and higher ordination axes and rescaling is applied to all axes. As the first DCA axis differs from the first reciprocal averaging axis only by rescaling, these axes are identical to rank order of samples (or species) along the first ordination axis. The calculations required for DCA rise only in proportion to the number of nonzero items in the samples-by-species data matrix, not the square or cube of the number of samples or species, and hence, there is no difficulty in analyzing large data sets.

The benefits of the DCA refinements, as determined in tests with simulated and field data, are exactly what one would expect – elimination of the arch and uniform axis scaling (Hill & Gauch 1980). Figure 4.14 shows DCA and RA ordinations of a simulated 4.5 × 1.5 half-change two-dimensional community pattern. The DCA result is nearly ideal; the RA second axis is merely a quadratic distortion (arch) of the first axis, with proper expression of the smaller coenoplane axis deferred to the third axis (not shown). For DCA, the first four eigenvalues are 0.400, 0.037, 0.009, and 0.003, so the structure in the spurious third and fourth axes is very small. For reciprocal averaging, the first four eigenvalues are 0.400, 0.089, 0.039, and 0.019, with the rather large spurious axes being the second and fourth. Hence, DCA summarizes this two-dimensional data structure efficiently and cleanly, implying that its underlying model matches the data structure to a considerable degree.

The importance of DCA's ability to correct reciprocal averaging's main two faults is better appreciated when it is emphasized that the two faults are not found in reciprocal averaging only. Rather, these faults also affect most ordination techniques, including polar ordination (especially when using Euclidean distances instead of percentage dissimilarities), principal components analysis, and others yet to be discussed, such as nonmetric multidimensional scaling, principal coordinates analysis, factor analysis, and canonical correlation analysis. These faults are nearly ubiquitous because they originate in the curved locus of points from a simple community gradient viewed in species space (Figure 4.7*B*) or samples space; any plausible dimensionality reduction (ordination) algorithm beginning with species space is bound to encounter these faults. Likewise, because point configurations in species dissimilarity space (or samples dissimilarity space) approximate those in samples (or species) spaces, curved point loci cannot be avoided by shifting to a dissimilarity space. Detrended correspondence analysis is an improvement on reciprocal averaging, but more broadly, it constitutes a significant improvement in ordination methodology.

Figure 4.15 shows DCA ordination for the southern Wisconsin upland forest data (given in Table 4.4). A climax-to-pioneer gradient is evident on the first DCA axis, as in preceding weighted averages and polar ordinations (Figures 4.4 and 4.8). The second axis mainly distinguishes *Juglans cinerea* and *Carya ovata* and does not appear to have ecological significance. (The reciprocal averaging result for these data, not shown, is broadly similar, except that the distinctiveness of *Juglans cinerea* on the second axis is exaggerated even more.) Although secondary gradients have been defined for these southern Wisconsin upland forests using larger data sets (Bray & Curtis 1957; Peet & Loucks 1977), the small data set used here appears to convey enough information to define only a single ecologically meaningful community gradient. The first-axis sample ordination has a length of 2.4 SD.

The real value of DCA is revealed in the analysis of difficult data sets. Hill & Gauch (1980) report DCA results superior to reciprocal

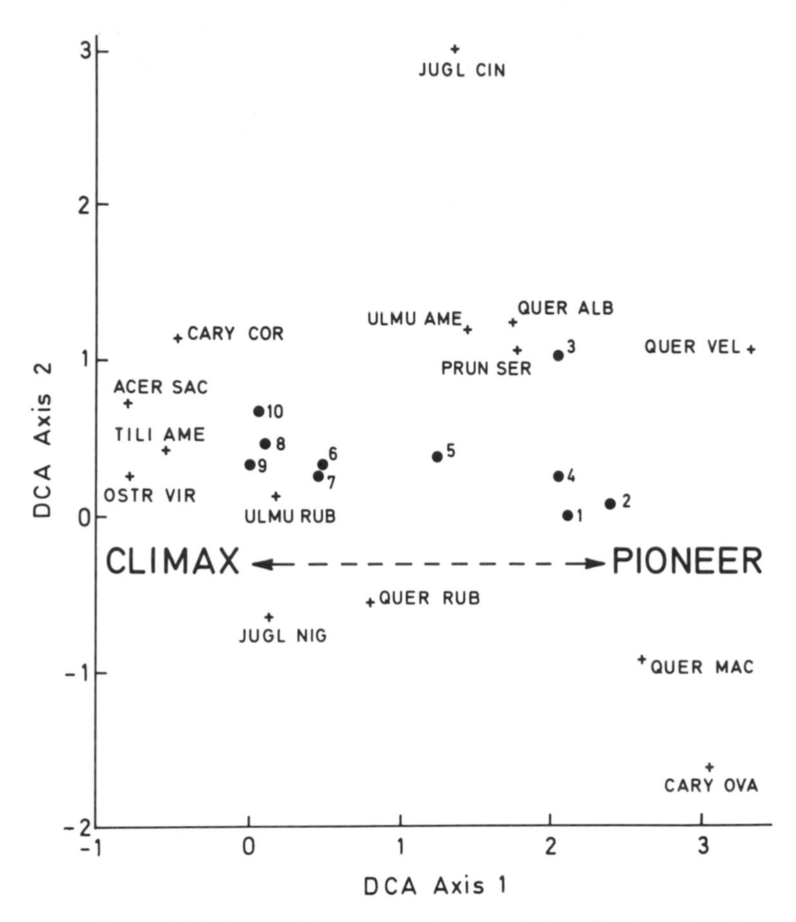

Figure 4.15. Detrended correspondence analysis ordination of southern Wisconsin upland forests. The data and species names are given in Table 4.4. Only the first axis is ecologically meaningful, expressing a climax-to-pioneer gradient (indicated by the dashed line). Axis scales are in units of average standard deviations of species turnover.

averaging and nonmetric multidimensional scaling for a variety of complex field data sets and simulated data sets having one to four community gradients. Figure 4.16, for example, shows a successful DCA ordination of composite samples from a vegetation survey of southeast England (Hill & Gauch 1980; also see Wheeler 1980*a, b*). These extremely diverse samples lead to first and second DCA axes of lengths 6.5 and 4.3 SD, respectively.

The DCA axis lengths for sample ordinations were found to be generally accurate within 10 to 20% (Hill & Gauch 1980). Given a gap

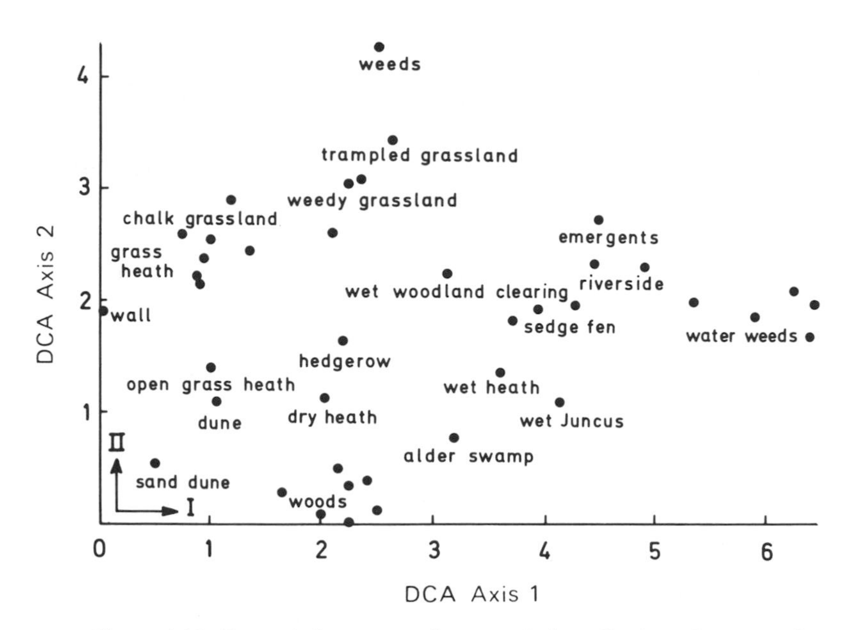

Figure 4.16. Detrended correspondence analysis ordination of a vegetation survey of southeast England (data of J. P. Huntley and H. J. B. Birks, 3270 relevés here clustered into 40 composite samples for ordination). The first axis goes from dry to wet conditions, and the second axis from woodland to weed communities. Axis scales are in units of average standard deviations of species turnover. (From Hill & Gauch 1980:55)

in the sampling design, gap lengths were also estimated accurately provided that a gap was no greater than 3 SD in length (as the problem becomes indeterminate beyond this). Extreme outliers are best removed, and entirely (or nearly entirely) disjunct data subsets are best separated into coherent data subsets prior to DCA ordination. Species ordination lengths are less reliable because of problems with species whose distributions are truncated by the particular sample set chosen (as in Figure 4.10, where species distributions at the axis ends are truncated by the sampling process, resulting in species with only one or two occurrences in the sample set).

Extensive comparative tests with reciprocal averaging and nonmetric multidimensional scaling have shown DCA results at least as good as, and usually superior to, other ordination techniques (Hill & Gauch 1980). Because only community data are analyzed, environmental interpretation is a subsequent task, which is aided greatly by the robustness, freedom from distortion, and meaningful axis units of DCA. In addition, DCA has been found effective for niche ordination of birds by foraging position and behavior (Sabo 1980). Because DCA models variables (species) as occurring along a limited extent of a gradient (as in the Gaussian community model with each species beginning at some point, rising to a mode, and thereafter declining until absent again), DCA is not appropriate for the analysis of a matrix of similarity (or distance) values between community samples, unlike secondary reciprocal averaging or nonmetric multidimensional scaling (Hill 1979*a*:10). For analysis of community data, DCA ordinates samples and species simultaneously, objectively (without endpoints or weights), efficiently (with only linear computer requirements), and effectively. Of the ordinations tested so far, DCA appears to be most appropriate to the Gaussian community model and most successful in applications to community analysis.

Other ordination techniques

Five important ordination techniques have been described in detail. Other ordination techniques exist without number (Orlóci 1978*a*:102–85; Whittaker 1978*c*). Of these, five merit brief mention here: nonmetric multidimensional scaling, principal coordinates analysis, factor analysis, canonical correlation analysis, and Gaussian ordination.

Nonmetric multidimensional scaling is actually a family of related ordination techniques. The central theme is the use of only rank order information in a dissimilarities matrix, rather than its metric information. The intention behind nonmetric methods is to replace the strong and problematic assumption of linearity (of species responses curves to underlying community gradients) made by many ordination techniques, including principal components analysis, with a weaker and, it is hoped, less problematic assumption of monotonicity.

Two forms of nonmetric multidimensional scaling have received substantial attention from ecologists. The earliest and most commonly used method derives from Shepard (1962) and Kruskal (1964*a, b*) and may be referred to as MDS. The goal of MDS is to locate samples (or species) in a low-dimensional ordination space in such a manner that the interpoint distances in the ordination have the same rank order as do the interpoint dissimilarities in the secondary matrix (of samples-by-samples dissimilarities or species-by-species dissimilarities) to as great a degree as possible. The other method, continuity analysis, derives from Shepard & Carroll (1966). Its goal is to find a low-dimensional ordination space in which species response curves are as smooth or continuous as possible, without any other constraints or assumptions as to the form of these curves.

Computationally, nonmetric ordinations are complex procedures. The amount of computer time rises with the square, cube, or more of the number of samples (or species), so data set size is limited even with a computer. Because commonly used nonmetric methods analyze samples or species, not both, development of an integrated perspective may be difficult. The computational procedures are iterative and convergence to the best solution is not guaranteed. In fact, sometimes, several solutions are obtained using different initial trial configurations. The results are then compared in order to assess the likelihood that a satisfactory result has been obtained.

The MDS method has been tested by ecologists for ordination of community data (Anderson 1971; Austin 1976*b*; Fasham 1977; Prentice 1977, 1980*b*), as has continuity analysis (Noy-Meir 1974*a, b*). Systematic tests of these and other nonmetric multidimensional scaling techniques in comparison to reciprocal averaging and detrended correspondence analysis, using a variety of simulated and field data sets, lead to several conclusions (Gauch, Whittaker, & Singer 1981). First, all nonmetric multidimensional scaling techniques give similar results, much like results from principal components analysis or reciprocal averaging. As Gower notes (Sibson 1972:342), ranked dissimilarities actually contain "much metric information, [and] if this were not so, non-metric scaling would not work." Second, individual cases can be constructed for which nonmetric multidimensional scaling or reciprocal averaging gives better results, either marginally or occasionally decidedly so. Third, if two methods (nonmetric ordinations and metric ordinations like reciprocal averaging) give much the same results, the practical view is to choose the analysis requiring the least computation. Unless the results from nonmetric analyses are substantially different, fewer or weaker assumptions are of limited value. Fourth, detrended correspondence analysis usually gives results superior to reciprocal averaging and nonmetric multidimensional scaling and is exceptionally sparing of computer time and memory.

The general conclusion is that, ordinarily, "non-metric methods may not be worth the extra computation" (Gower in Sibson 1972:342). The results from nonmetric ordinations have theoretical interest, however, in confirming the Gaussian model of community structure (Noy-Meir 1974*a*; Gauch, Whittaker, & Singer 1981). Ironically, this confirmation of the Gaussian model provides the basis for explaining the lack of luster of nonmetric ordinations. Nonmetric ordinations assume monotonicity, which is a weaker and better assumption than linearity but is still unrealistic for handling the Gaussian curve, which is ditonic. Consequently, it is no surprise that nonmetric ordination results are much like principal components analysis results and suffer from the same arch distortion. Also, like principal components analysis, nonmetric multidimensional scaling gives effective results only for easy data sets with low diversity, for which species response curves are short segments of a Gaussian curve and, consequently, mostly monotonic (or even linear). Success with easy data sets is of little interest, however, since practically any ordination technique is successful with such data.

Principal coordinates analysis is a generalization of principal components analysis, devised by Gower (1966; also see Pielou 1977:340–5 and Everitt 1978:16–22, 38–9). In principal components analysis, the distances between points in species space (or samples space) are measured as Euclidean distances. Principal coordinates analysis allows the use of distance measures other than Euclidean distance.

Principal coordinates analysis using percentage difference may be expected to perform somewhat better than principal components analysis because percentage difference is usually better than Euclidean distance in polar ordination (Gauch 1973*a*) and in nonmetric multidimensional scaling (Gauch, Whittaker, & Singer 1981). Our tests with simulated and field data confirm a marginal improvement, but the arch problem remains and results are inferior to detrended correspondence analysis (unpublished data). Clymo (1980) finds principal coordinates analysis effective with a peat bog data set, and Williams (1976) uses it extensively. In cases where one wants to ordinate a published secondary samples-by-samples (or species-by-species) matrix, however, principal coordinates analysis may be useful, particularly if the distance measure used for the matrix does not have Euclidean properties.

Factor analysis is similar to principal components analysis, except that instead of trying to account for as much of the total variance as possible, only correlations between variables are of interest as reflecting putative underlying causes or factors (Williams 1976:59–61). There are, however, a number of variants (Jöreskog, Klovan, & Reyment 1976:53–67). Computationally, factor analysis is an eigenanalysis problem. Dagnelie (1978) provides a lucid exposition of factor analysis for community ecology applications (also see Diaz, Novo, & Merino 1976 and Williams 1976:59–61). Because variables (species abundances) are assumed to respond linearly to the underlying factors, effective results are expected only for data sets encompassing a small range of community variation (Whittaker 1967; Dagnelie 1978).

Canonical correlation analysis is another eigenanalysis method applicable to a matrix of vegetational data (samples-by-species) and, simultaneously, a matrix of environmental data (samples-by-environmental measurements). The objective of the analysis is to find ordination axes that maximally reveal the joint, or common, structure of the two matrices (Austin 1968; Williams 1976:66–7; Gittins 1979). This goal is very close to what ecologists should like, an objective mathematical tool to make the basic relationships between vegetation and environment emerge. The assumptions of canonical correlation analysis, however, include linearity of both species responses and environmental effects and orthogonality (lack of correlation) of the underlying gradients or factors. Gauch & Wentworth (1976) review the ecological applications of canonical correlation analysis and present test results with simulated and field data. They find it generally ineffective because its linearity requirements are too stringent to accommodate most ecological data (also see Huntley & Birks 1979*b*).

Gaussian ordination arranges samples along a single ordination axis, which maximizes the least-squares fit of each species's abundance values to a Gaussian curve (Gauch, Chase, & Whittaker 1974). In other words, Gaussian ordination fits community data as well as possible to the Gaussian model of community structure. The result is an ordination value for each sample and a mode, maximum, and standard deviation defining the fitted Gaussian curve for each species, as well as the percentage of total variance accounted for and other performance statistics. A variation of parameters algorithm is used, requiring a rather large amount of computation. Ihm & Groenewoud (1975) and Johnson & Goodall (1979) present similar ordination techniques, and Orlóci (1980) allows for species response forms in addition to Gaussian.

For community data having a single predominant gradient and low noise level, Gaussian ordination is nearly ideal in that its underlying model is exactly the Gaussian model of community structure. For such data, results have been effective. Gaussian ordination and Johnson & Goodall's (1979) algorithm produce residual matrices that may be examined for additional data structure, but production of effective, higher ordination axes is ordinarily unlikely. The method of Ihm & Groenewoud (1975) is intended to yield several ordination axes, but the higher axes are likely to be ineffective because their method makes a simplifying assumption that all species distributions have the same standard deviation. It may be that fitting an exact Gaussian model cannot be extended readily to several dimensions, both because of technical problems and because the implied data base for fitting so many parameters implies far larger data sets than are customary. Because of its computational complexity, Gaussian ordination is of limited utility for routine ordination work (Robertson 1978), but it can be useful in special research applications requiring model parameters and sufficing with one dimension.

Evaluation of ordination techniques

As Dale (1975) observes, the first requirement for evaluating ordination techniques is the selection of goals or criteria that define the desired results (also see Orlóci 1968). Apart from defined criteria, no evaluation is possible. Once criteria are chosen, the methods for making appropriate tests can then be developed. Finally, the application of tests provides results for the evaluation of ordination techniques.

Criteria

Three criteria are basic for ordination techniques.

(1) Effective. An ordination technique should be effective in summarizing community variation. Ordination serves as a tool by which ecologists grasp community variation. Effective ordination techniques must therefore (a) be realistic in their assumptions about the structure of community data (as revealed by direct gradient analysis) and provide parameters or results that are intrinsically powerful for summarizing data structure and (b) be suitable for conveying information in a format agreeable to the habits of human reasoning and communicating.

(2) Robust. An ordination technique should handle a wide range of data sets; that is, it should work frequently. Ideally, good results will be obtained whether a data set involves long or short community gradients, one or many gradients, high or low noise, large or small numbers of samples and species, relatively continuous or discontinuous sample variations, and so on. A robust technique has the further advantage that it obviates the need to quantify various properties of a data set prior to ordinating in order to evaluate the data set's suitability for analysis.

(3) Practical. The implied labor and computer expense of an ordination technique should be reasonable. Ideally, computing expenses should rise only linearly with the amount of data, in which case the analysis cost per sample is a constant. Also, it is desirable that the technique make only ordinary demands on the investigator's time and expertise.

Other criteria for evaluation of ordinations may be listed, but they are subsumed or implied by the preceding three criteria. (For further detail, see the sections on evaluation and purposes of multivariate methods in the Introduction and the references therein.)

Ordination techniques vary in their relative degree of objectivity or subjectivity. Objective ordination techniques (like reciprocal averaging) analyze community data alone, without the investigator supplying any choices or parameters, but leave environmental interpretation to a subsequent, relatively subjective step. Subjective ordination techniques (like weighted averages) require choices or parameters that reflect the investigator's perception of environmental or other factors, but environmental interpretation is, consequently, relatively straightforward. No ordination technique is entirely objective, however, because the selection of an ordination technique is itself a subjective choice; likewise, no technique is entirely subjective because the data have at least partial control over the results and the investigator's role has at least some basis in correctly perceived relationships.

The range in relative objectivity of ordination techniques as such is great, but the range considering the community study process as a whole is not as great. The relatively objective ordinations do not make a community study vastly more objective because they tend to just defer the subjectivity to the next step, environmental interpretation. In other words, the inherent complexity and nature of community studies are such that some subjectivity is inherent to the work. The degree of subjectivity appearing in various steps of a community study (such as the ordination step) may be altered greatly, but the study's overall subjectivity may be kept relatively constant. These observations argue against infatuation with objectivity, but it would be equally naive to consider all community ecology approaches to have equivalent objectivity. In any case, whether relative objectivity or maximal use of ecologists' insights is more valuable depends on the purposes of a given study. Therefore, when listing criteria for evaluating ordination techniques, objectivity is best omitted because it is not always desirable nor always undesirable. It is worthwhile, however, to assess realistically the objectivity of ordination techniques so that this property can be considered in relation to research projects with particular requirements.

Methods

Ordination evaluation proceeds by two principal methods. The first method of ordination evaluation is direct comparison of an ordination technique with models of community structure by means of theoretical, mathematical reasoning. Effectiveness, robustness, and practicality of ordination techniques can be investigated. Because of mathematical intractability, some questions cannot be answered by this approach, however. Consequently, an additional method is needed.

The second method of ordination evaluation employs trials with both field and simulated data having known structure. Ordination results can then be compared against expected results for data of known structure. As emphasized earlier, both field and simulated data are desirable for ordination tests because they are complementary (field data being realistic but expensive and only partially understood and simulated data being inexpensive and exactly known). As data sets vary in several important properties, it is imperative that a large number of data sets be used in order to obtain a balanced evaluation.

It is preferable to apply the same theoretical evaluations and the same trials with data sets to several ordination techniques so that their comparative strengths may be evaluated. As no ordination technique is completely defunct or completely perfect, the final issue is the relative merit of a given ordination technique in comparison with other available ordination techniques.

Results

When an investigator desires to use ecological knowledge of samples to derive an ordination of species (or knowledge of species to ordinate samples), weighted averages may serve admirably (Whittaker 1967; Hill 1973). Polar ordination with subjectively selected endpoints may serve a similar role, and selection of only two endpoint samples (or species) is required instead of a weight for each and every sample (or species). These procedures incorporate environmental information in the ordination step and, consequently, lead to rather straightforward environmental interpretation of results.

When an investigator desires a relatively objective community-centered ordination, detrended correspondence analysis has been found most effective and robust, especially with very heterogeneous and difficult data sets. In the infrequent case of a data set with very little sample variability (about one half-change or less), however, principal components analysis may be best. Reciprocal averaging is effective for data sets of intermediate heterogeneity but does not appear to be superior either to detrended correspondence analysis for most data sets or to principal components analysis for very homogeneous data sets. Polar ordination using automatically selected endpoints is not particularly robust; this function of polar ordination is ordinarily served better by an eigenanalysis ordination such as reciprocal averaging or detrended correspondence analysis.

Much can be said for applying a few good ordination techniques to the same data set (Gauch, Whittaker, & Wentworth 1977; Whittaker & Gauch 1978). The results from various ordinations may be complementary. Provided that several ordination techniques are available in one's computer system, little bother or expense is involved in running several analyses instead of just one; so, to say the least, little is to be lost by trying several. It is reassuring if the same general ordination pattern emerges from different ordination techniques, as ordinarily occurs with simple data sets. In southern Wisconsin upland forests, for example, a pioneer-to-climax gradient clearly predominates. The 20 or so ordinations of these forests available in this chapter and its references all recover the same primary gradient, despite differences in field sampling procedure, inclusion of shrubs or herbs, data transformation, and ordination technique. Difficult data sets, on the other hand, may yield to few or even one ordination technique. Particularly for complex, very heterogeneous data sets, detrended correspondence analysis is distinctive in its effectiveness and robustness.

It may be observed that the preceding results on ordination evaluation are reported in relation to just two principal gradients, a gradient of relatively objective-to-subjective ordination techniques and a gradient of low-to-high sample heterogeneity. Given a modest number of significantly different levels along these two gradients, this observation implies that only several ordination techniques are required. It is suggested that the five ordination techniques emphasized in this chapter (weighted averages, polar ordination, principal components analysis, reciprocal averaging, and detrended correspondence analysis) will suffice for the great majority of ordination applications. Significantly better results from some other ordination technique may be expected to be rare, although not entirely absent. This conclusion, that only several good ordination techniques are sufficient for most purposes, is fortunate because computer implementation of each ordination technique is costly, familiarization with each technique takes time, and communication of ordination results is aided by emphasis on well-known and commonly applied ordination techniques, which of necessity must be few.

Interpretation and presentation of results

Especially when relatively objective eigenanalysis ordinations are used, environmental interpretation of ordination results can be challenging. In this case, interpretation consists of relating two kinds of information (Williams 1962), community gradients (as summarized from community data by ordination) and environmental gradients. (As usual, environmental factors are here construed rather loosely to include what might otherwise be termed disturbance, or historical, factors.) Environmental interpretation involves the perception of common patterns of distribution of community and environmental variables. Such perceptions may help to generate hypotheses about causality or interaction of variables, but these hypotheses must be tested by other means.

A nearly ubiquitous problem is that relationships between community and environmental variables are complex, nonlinear, and numerous. Consequently, only occasionally do community ecologists benefit from customary statistical techniques for analysis of variance and for regression. Instead, community ecologists have mainly relied on informal, visual recognition of pattern in variables. Informal pattern recognition has one exacting requirement, that the investigator be presented as much information as possible in a readily assimilated format. This requires effective options for displaying ordination results (Everitt 1978; Noy-Meir 1979; Green 1980). Because display options are such an integral aspect of environmental interpretation, the interpretation and presentation of results are best discussed together (Scott 1974; Gauch 1977:34–8).

Effective displays are desired for final results, obviously. They are equally important, however, for preliminary scanning because often many ordinations and many variables are scanned in the early stages of successive refinement, and poor results and lack of pattern need to be identified rapidly without wasting time.

Six display options are widely useful.

(1) Samples or species lists. Samples or species may be listed in rank order of ordination scores, and patterns may be sought in terms of known environmental gradients of the samples or known habitat preferences of the species. For example, when samples are listed in order of their position in an ordination, the list may show a clear trend in moisture status; likewise, one end of the ordination axis may group pioneer species and the other end climax species. Ordinations involving related or simultaneous sample and species scores (like weighted

averages and detrended correspondence analysis) require a unified explanation of both sets of scores, but one set may be easier than the other for beginning interpretation. Several lists may be made to display several different gradients.

(2) Arranged matrix. When both samples and species are listed in order, the two lists together may be used to arrange the data matrix. Patterns may become evident that are entirely obscure in a random matrix ordering (Tables 1.3 and 1.4, and Figure 4.9; also see Persson 1981). If there is one main community gradient in the data, the arranged matrix will show a concentration of larger values along the matrix diagonal. If there are two or more gradients, however, especially if they are of similar importance, there is no possible matrix arrangement that will show neat banding.

An arranged matrix makes available at a glance both the raw data in its entirety and the overall pattern of species occurrences. Consequently, the reader can at once gain both detailed (analytic) and general (synthetic) information. This observation may help to explain the effectiveness and popularity of arranged matrices.

(3) Sample and species ordination graphs. Sample and species relationships may be shown by graphs of ordination scores (as in Figures 1.5, 4.4, 4.5, 4.8, 4.15, and 4.16). Commonly, two axes are shown, but sometimes three or more by using several panels. In ordination space, similar entities are nearby and dissimilar entities far apart. An enormous amount of information on sample and species relationships may thereby be conveyed at a glance. The processing of visual information in our minds seems to allow rapid assimilation of more information about relationships than text or tables of numbers.

(4) Graphing of environmental parameters on the sample ordination. Instead of graphing samples on the sample ordination as mere points, the value of an environmental measurement may be shown for each sample. When patterns are evident, isolines may be drawn to clarify the figure. The resulting figures may be compared to see whether environmental variables have common, opposing, or related distributions. Likewise, species properties (such as family membership, geographic origin, and life-form) may be noted in addition to species points in the species ordination and trends observed (Persson 1981).

(5) Hybrid ordination. It is desirable to ordinate samples, species, and environmental factors together, but ordination of a two-way data matrix ordinates only two kinds of entities. Environmental data may be analyzed separately (Bauzon, Ponge, & Dommergues 1974), but

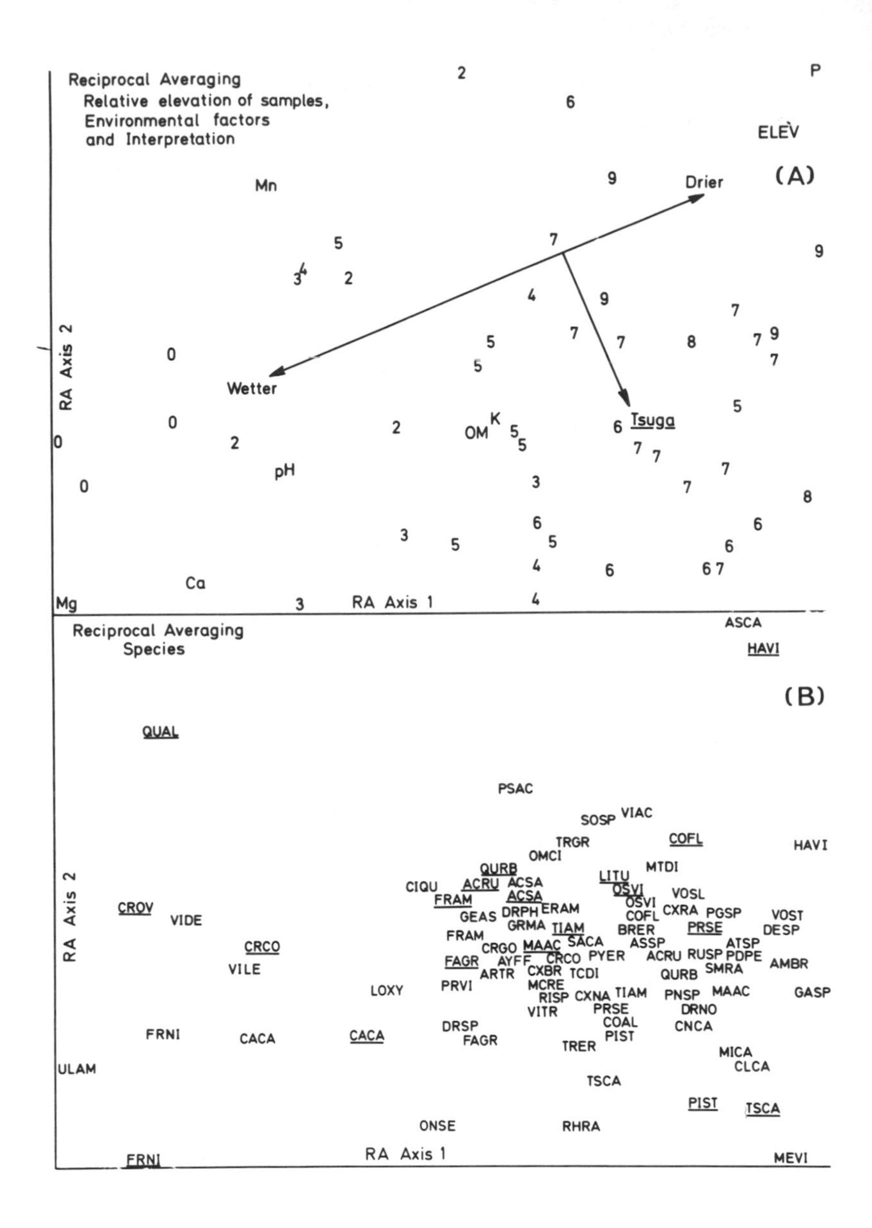

Figure 4.17. Hybrid ordination of samples, species, and environmental factors in a mesophytic forest at Ithaca, New York. Reciprocal averaging (RA) scores are shown for samples (*A*, with relative elevations in deciles from 0 for low to 9 for high) and for species (*B*, with trees underlined and species codes expanded by Gauch & Stone 1979). Both panels are in the same coordinate system; two panels are used merely to avoid clutter. Environmental factors are

the task then remains of relating these results to the community analysis. A simple method of integrating several kinds of information suggested by Gauch & Stone (1979) involves an eigenanalysis community ordination followed by weighted averages to incorporate different data (also see Anderson 1979). First, the samples-by-species community data matrix is analyzed by reciprocal averaging or detrended correspondence analysis. Then, sample ordination scores for as many axes as desired are used as weights for weighted averages ordination of the samples-by-environment matrix to obtain ordination scores for environmental factors. The resulting three sets of scores (for samples, species, and environmental factors) may be scaled into the same range and graphed together. Figure 4.17 shows an example (Gauch & Stone 1979).

The advantage of this hybrid ordination method is that several kinds of entities are placed into the same coordinate system but only the community data are used to determine the configuration, the further weighted averages ordinations being dependent. Additional data matrices can be incorporated, provided that one of their dimensions is samples or species. For example, samples-by-birds data can be used to obtain weighted average ordination scores for bird species. Furthermore, once this is done, bird ordination scores can then be applied to a birds-by-foods matrix to also incorporate the birds' food preferences into the scheme. Hybrid ordination may thus serve to incorporate many kinds of information into a single ordination system.

The quantity and coherence of information made available by hybrid ordination appear to be very helpful in the generally difficult task of environmental interpretation.

(6) Community model. The ultimate and most direct representation of ordination results is a display of species abundances along ordination axes that have been interpreted environmentally. Figure 3.8

Caption to Figure 4.17 (*cont.*)
added to (*A*) by using RA sample scores and the samples-by-environment data matrix to obtain weighted averages ordination scores for the environmental factors. The upper right portion of the ordination space features high, drier sites, with abundant *Hamamelis virginiana* (HAVI), in contrast to the lower left portion with low, wet sites with high pH and abundant Ca and Mg and abundant *Fraxinus nigra* (FRNI), *Ulmus americana* (ULAM) saplings, and *Carpinus caroliniana* (CACA). The lower right portion features *Tsuga canadensis* (TSCA). The interpretation is offered of a major wet-to-dry gradient caused by topographic variation and affecting numerous species distributions and a secondary competition gradient caused by variation in *Tsuga canadensis* abundance. (From Gauch & Stone 1979:335)

showed the distribution of several tree species along a compositional gradient from climax-to-pioneer conditions in southern Wisconsin upland forests. Results using ordination are analogous to those shown in Chapter 3 on direct gradient analysis, except that the axis coordinates are determined from ordination of community data instead of direct physical measurement. These ordination results also substantiate the Gaussian model of community structure (Noy-Meir 1974*a*; Gauch, Chase, & Whittaker 1974).

In considering the preceding six display options, the intrinsic dimensionality of a data set should be considered (Peet & Loucks 1977). For example, if there are two important gradients, like moisture and nutrients, any attempt to reduce the data's structure to one dimension may be unsatisfactory. With more than three important environmental gradients, an initial subdivision of the data into simpler subsets according to suitable geographic or physical criteria may be necessary in order to achieve results that are adaptable to the intrinsic limitations of paper and pen. Classification techniques that will be considered in the next chapter may also be useful for subdividing complex data sets.

Environmental factors may interact in a complex way in their effects on plants and animals. It may be necessary, consequently, to combine several measurements into a single index before environmental patterns are clarified (Loucks 1962; Peet & Loucks 1977).

Environmental interpretation is usually provisional to some degree. Verification of the interpretation by experimental tests may not be feasible or possible. Three things can be readily done, however, to increase confidence in environmental interpretations. (1) The basic plausibility of an environmental interpretation may be considered in terms of the known autecology of the species and the known interrelations of the physical parameters. The greater the volume of information accommodated successfully by an interpretation, the greater is the likelihood that the interpretation is valid. (2) Several random subsets of the community data may be ordinated to see whether the basic patterns remain stable (Wilson 1981). (3) New samples not included in the original community ordination may be used to see whether community composition can be predicted successfully from environmental data, whether the reverse holds, or both. Predictive ability is the most direct test of the generality of ordination results. Samples may be withheld at random from the original ordination, or alternatively, from the outset, a portion of the study area may be selected at random and reserved for later sampling and analysis.

5

Classification

Classification is the assignment of entities to classes or groups. In community ecology, the customary input data are species abundances in a two-way samples-by-species data matrix. The resultant output from classification is either a nonhierarchical or hierarchical arrangement of samples, species, or both, as illustrated earlier in the introduction. An arranged data matrix may be viewed formally as merely a particular option for presenting the results of a hierarchical classification, but this option merits special consideration because it is quite useful and has had an important role from the beginnings of community classification. There are uncountable individual techniques for classification, but the central purpose of community classification is to summarize large community data sets.

Classification is a fundamental, ubiquitous mental activity (Goodall 1953; Sokal 1974; Blashfield & Aldenderfer 1978). Even simple animals must classify past sensory experiences as a prerequisite for future avoidance of deleterious actions and places and for seeking useful foods and habitats. The existence of language is predicated upon the classification of similar entities or concepts under words serving as class labels. It is not surprising that the first conceptual framework used to organize quantitative community data was classification. Becking (1957), Whittaker (1962), and Westhoff & Maarel (1978) trace the earliest origins of relevant concepts to about 1800 and the earliest substantial community classification to about 1900.

Early classification techniques were informal and subjective (Goodall 1953; Whittaker 1962; Sokal 1974). Several limitations were felt. First, it was difficult to assess the consistency of an individual ecologist's work and its agreement with the work of others. Second, the art of classification required a long apprenticeship, and classification of the communities of a particular area placed heavy demands on the time of expert ecologists. Third, the laboriousness of the classification methods made their application to data sets of over about 100 samples tedious or impractical.

Computers became widely available to ecologists around 1960. This timing corresponds with acceleration in work on classification theory and techniques (Sokal 1974; Blashfield & Aldenderfer 1978). Ecologists developed and applied computerized classification techniques extensively in order to reduce the three limitations just mentioned. The computer became a tool to increase the objectivity of classification, to reduce the demands on the time experts, and to enable the analysis of large data sets. Community classification, like most sciences, has a history of progression into the development of increasingly quantitative, objective methods. This progression is linked intimately with the availability of computers.

Extensive treatments of community classification are offered by Shimwell (1971:42–62), Mueller-Dombois & Ellenberg (1974:139–302), Orlóci (1978*a*:186–283), Whittaker (1978*a*), and Maarel (1979*a*). Its early history is covered by Whittaker (1962). Additional treatments of computer classification techniques are offered by Cormack (1971), Sneath & Sokal (1973) and Hartigan (1975) and an extensive bibliography by Blashfield & Aldenderfer (1978). As in the preceding chapter on ordination, the intention of this chapter on classification is to treat principally those moderately few techniques that have been found to be most effective with community data. The usefulness of classification increases with data set size, so the examples in this chapter, which necessarily involve only dozens of samples, can only hint at the value of classification in large projects.

Classification theory

Community classification involves an interaction between ecologists and communities (Whittaker 1962). Consequently, the properties of community classification partly reflect community structure and partly reflect the thought patterns of ecologists.

The classification of communities is useful and natural because humans comfortably think and communicate in terms of classes. Note, however, that this is an observation about human thinking, not about plant and animal communities. It is a separate question whether ecological communities actually occur in discrete, discontinuous classes or types (rather than the alternative of continuous community variation). If community variation is discontinuous, classification is a natural framework for conceptualizing communities; if community variation is continuous, ordination is more natural. Early work in community ecology emphasized the debate over whether community variation is continuous or discontinuous and considered the verdict to imply a preference for ordination or classification, respectively. Current thinking emphasizes the complementary use of ordination and classification and recognizes the utility of classification for many practical purposes even when rather arbitrary dissections must be imposed on essentially continuous community variation (Goodall 1954*a*; Whittaker 1962, 1978*a*:4–6).

Bases and kinds of input data

Ecologists have classified vegetation on many bases, including physiognomy (general physical characteristics of organisms such as height and growth form), characteristics of the environment, and community composition using species, species groups, or other taxonomic levels (Whittaker 1962; Aleksandrova 1978; Bridgewater 1978; Orlóci 1978*a*:186; Dale 1980). Data collection for classifying communities may be limited to certain taxa (such as vascular plants) or to certain strata (such as tree or herb strata). Although the taxonomic unit usually employed is species, one may use broader units (Dale & Clifford 1976; Moral & Denton 1977) or finer units (Baum & Lefkovitch 1972). Likewise, animal communities may be classified on various bases, sometimes with the additional constraint of trying to have results correspond to a relevant classification of plant communities (Whittaker 1962). Nevertheless, in the great majority of classification work, the descriptors (variables) used are species abundances. Analysis of samples-by-species data matrices is thus the primary focus of this chapter, although nothing about the classification techniques discussed here mitigates against application to other kinds of data.

Even if species abundances in samples are accepted as the primary kind of data for community classification, difficult issues remain as to the particular emphases and calculations to be used in classifying such data. The process of classification is essentially the summarization for each sample of the information in many numbers (dozens to hundreds of species abundances) into a single number (a cluster assignment). Clearly, there are countless ways in which many numbers can be summarized into one number. Implicit in each of these ways are various emphases as to the most important information in the data. Various schools of classification have emphasized dominant species, minor species, individual species, groups of species, characteristic species (occurring only in a single community type), or overall species composition (Whittaker 1962). The differences among schools can be understood in part by considering the particular characteristics of their local vegetation and the differences in research purposes. Nevertheless, considerable subjectivity remains.

Of these choices in emphasis, one choice is of particular interest, equal emphasis on all species. Equal emphasis on all the data is characteristic of most multivariate methods (Goodall 1953). This choice has three implications. First, equal emphasis results in relatively objective analyses. The reason for this is simply that the alternative, unequal emphasis, is implemented by making additional choices (sometimes numerous) as to which species or kinds of species to weight more heavily, and these choices are inescapably of a rather subjective nature since other research purposes or other investigators could favor other choices. Second, equal emphasis results in general-purpose classifications rather than special-purpose classifications. Third, equal emphasis (general purpose) is basically a single perspective (despite admitted variations in details), whereas unequal emphasis (special purpose) leads to uncountable variations in perspective.

Even if equal emphasis on all species is adopted, there are still numerous different, particular classification algorithms. The situation is similar to treating all the numbers of a set equally to derive an average, because one might compute an average by any of several specific parameters – mean, mode, median, geometric mean, and so on. This chapter is principally concerned with classification techniques that are applied to samples-by-species data matrices and that treat all species equally. These two decisions, however, still leave room for a great diversity of classification techniques.

Properties of classification techniques

There are many classification techniques. The following nine properties are useful for describing individual techniques succinctly and for developing a classification of classifications (also see Williams 1971*a*; Sneath & Sokal 1973:202–14; Orlóci 1978*a*:186–99; Maarel 1979*a*; Greig-Smith 1980).

(1) Formal or informal. Some community classification techniques are applied by expert ecologists who mentally balance a large number of criteria in a complex manner that cannot be defined precisely. Several ecologists applying the same informal classification technique to

the same data set are likely to obtain somewhat differing results. Informal techniques are problematic for additional reasons, including the excessive time required for learning an art. Also, because an informal algorithm is not well defined, computer implementation of informal classification techniques is partial at best. Formal techniques, on the other hand, can be specified precisely. Consequently, every investigator gets the same answers, the formal techniques can be learned simply and quickly, and the computerization is straightforward.

(2) Nonhierarchical or hierarchical. Nonhierarchical classifications merely assign each entity to a cluster. Hierarchical classifications additionally arrange the clusters into a hierarchy.

(3) Quantitative or qualitative data. Some classification algorithms can accept quantitative values for species abundances; others can only treat species as being present or absent. A third alternative, used by some algorithms, is to generate pseudospecies by noting whether a species is present at several levels of abundance, rather than whether it is merely present (Hill, Bunce, & Shaw 1975; Hill 1977, 1979*b*). For example, mere presence could define one pseudospecies; at least 10 or greater in abundance, a second pseudospecies; 25 or greater abundance, a third pseudospecies; and so on. Then a species abundance value of 12 would count for the first two pseudospecies (because 12 is equal to or greater than both 0 and 10) but not for the third pseudospecies (because 12 is less than 25). By using pseudospecies, an algorithm that, by nature, can only handle qualitative (presence or absence) information can be coaxed into processing semiquantitative information.

(4) General or special purpose. A classification may serve general purposes or a special purpose (Sokal 1974). All species are treated equally in a general-purpose classification, and cluster assignments are in some sense optimal for the communities considered as a whole. If the researcher's purposes are specific, however, such as evaluating the wildflower display of sites, certain species require emphasis and others are less important or even irrelevant. There is basically just one general-purpose perspective but many special-purpose perspectives. General purposes are characteristic of multivariate methods.

(5) Divisive or agglomerative. Divisive algorithms begin with all entities in a single class and divide this class into progressively smaller classes (stopping when each class contains a single member or when the predetermined limit of some stopping rule is reached). Agglomerative algorithms begin with each entity in a class of its own and fuse (agglomerate) the classes into larger classes.

(6) Polythetic or monothetic. Monothetic classification techniques divide sets of samples according to the presence or absence of a single species. Polythetic techniques consider the entire species composition of samples in the process of deriving cluster assignments. Polythetic techniques may be agglomerative or divisive; monothetic techniques can only be divisive (Gower 1967*a*).

(7) Dual or single. Some techniques classify samples and species simultaneously in an integrated manner, yielding a dual analysis. Other classifications analyze only samples (or species). In this case, single analyses of samples and of species can ordinarily both be performed, but there is then no direct mathematical relationship between the two analyses. They have to be interpreted separately, and it may be difficult to relate the two analyses to each other, particularly in any straightforward or simple manner.

(8) Linear or rapidly rising computer requirements. The computing time for some classification techniques rises only linearly with the amount of data. Consequently, the cost per sample is a constant and large data sets present no difficulty. Likewise, the computer memory requirement of some techniques is such that little needs to be stored other than the samples-by-species data matrix itself and, in most cases, only one sample of this matrix needs to be in high-speed memory at one time and the bulk of the data may reside in slower disk or tape storage. Consequently, a modest computer can process enormous data sets. On the other hand, other classification techniques have requirements for computer time rising as a square, cube, or exponential of the amount of data, so even with a computer a practical limit may be reached at only dozens to hundreds of samples. Likewise, computer memory may be required for large matrices in addition to the samples-by-species data matrix, and the process of calculations involved may not lend itself well to the use of anything other than high-speed memory. Consequently, a fast computer with a large high-speed memory is required for less efficient classification algorithms and even then acceptable data set size is limited. Although computer requirements are a practical, rather than conceptual, consideration, they are important for research projects involving large data sets.

(9) Robustness. Robustness of a classification technique means that effective results are produced in most applications of the technique. Robustness has two components: (a) the results for a given data set are stable despite minor perturbations of the data and (b) effective results are produced for a wide variety of data sets. Minor perturbations of a

data set include the alteration of species abundance values by small random fluctuations, the deletion of a few samples, the addition of some samples not unlike others already present, the deletion or addition of rare species, and for a large data set, subdivision of the data set into several replicate sample subsets (see Jardine & Sibson 1971:86–91). Stability against minor perturbations does not mean that classification results are entirely unaltered but that significant features of the classification remain – a small change in the data causes only a small change in the classification. The second component of robustness, effectiveness for a wide variety of data sets, means that the classification technique performs well despite variations in the number and length of community gradients, the degree of continuity or discontinuity in community variation, the noise level, the numbers of samples and species, and the other data set properties. An implication of this breadth of applicability is that it is not necessary to quantify a number of data set properties in advance in order to assess whether or not a robust classification technique is likely to work well. The value of a robust classification technique is that it gives stable, effective results for most data sets.

Cluster properties

The term *cluster* should be recognized as having a vague meaning. The reason for this vagueness is precisely that there are numerous specific, definable parameters that are all facets of what is generally recognized as a cluster. These many parameters intermingle in a complex manner defying exact definition. Sneath & Sokal (1973:194–200) define eight cluster properties: the cluster center, the density, the variance, the dimension, the number of members, the connectivity, the straggliness, and the separateness. They note that there are many additional properties; these are just eight of the more obvious ones.

When the sample points (in species space or samples dissimilarity space) are naturally clustered, the intuitively reasonable clustering criterion is simply that the natural clusters be recovered (Kruskal 1977; Gauch 1979:6).

When the sample points are distributed more or less continuously, cluster boundaries must be imposed, and clustering criteria are more complex. Consider a configuration of points to be partitioned into two clusters. Four cluster properties are especially relevant and lead correspondingly to four clustering criteria.

(1) Equal space. The objective could be to draw a partition that divides the space into two equal volumes.

(2) Compactness. The objective could be to divide the space into two equal volumes along that particular surface perpendicular to the longest extent of the configuration of points, so that the resulting clusters be as compact as possible. (Admittedly, compactness could be defined in various ways, involving various weightings and regular or squared distances; some of these definitions would not imply equal volumes.)

(3) Equal numbers. The objective could be to divide the points into two clusters with equal numbers of points.

(4) Divide at sparse regions. Assuming the usual case in which the points are dispersed throughout the space but in some regions are relatively dense and in others relatively sparse, the objective could be to divide the configuration through its sparsest region.

These four criteria are independent, and cases can be constructed in which these four criteria would lead to four very different partitions. Consequently, it is not possible to optimize these four criteria simultaneously, so a clustering technique must make a choice among these (or other) criteria or else use an explicitly weighted combination of criteria, which then serves as a complex clustering criterion although it no longer optimizes any single criterion. (Incidentally, in the preceding list, only the third criterion is defined exactly; the others have a degree of vagueness. The purpose of the present discussion is not to define an exact, computable clustering algorithm but rather to draw attention to basic design problems for any clustering algorithm. For this purpose, the preceding list is adequately precise.)

Different ecologists give these four criteria different weights and vary these weights for particular community data sets and purposes. A frequently used, perhaps ideal weighting (1) emphasizes compactness and equal space, (2) gives less weight to equal numbers of samples, and (3) given the partition indicated thus far, allows small, but not drastic, adjustments to place partitions in sparser regions. This weighting originates from the objective that resulting clusters represent equal and minimal heterogeneities of community variation, while considering the numbers of samples to be incidental and easily swayed by accidentals of sampling, and considers division through a sparse region desirable but not at the cost of marked departure from division into equal, compact spaces. Usually community ecologists want the samples of a cluster to be similar or, to say the same thing viewed another way, they want the (approximate) species composition of a sample to be predictable from its cluster assignment. This stance implies that the primary clustering criterion adopted be compactness.

Little imagination is required to realize that the richness and complexity of the cluster concept are themselves a fundamental cause of the great diversity in clustering (classification) techniques. Classification techniques differ greatly in their relative emphases on these four and other division criteria. Even if a given classification technique is known to maximize a given criterion (cluster property), the larger question is still the merit of this criterion for community classification in comparison with other laudable criteria.

Classification purposes

Although permutations of the preceding and additional properties of classification techniques (and clusters) lead to numerous classification techniques, their purposes are comparatively uniform. As noted earlier in the Introduction, the basic purposes of multivariate analyses, including classification, are (1) summarizing large, complex data sets, (2) aiding the environmental interpretation of and hypothesis generation about community variation, and (3) refining models of community structure. The fundamental criterion for evaluating the accomplishment of these purposes is that of community understanding returned for a given time input, relative to the productivity of alternative research methods (Moore, Fitzsimons, Lambe, & White 1970). For reasons that will become evident later, nonhierarchical classification is especially useful in the preliminary summary of large data sets, whereas table arrangement and hierarchical classification are especially useful in later stages of analysis for showing relationships among samples.

Classification techniques

Classification techniques used in community ecology may be considered in three groups: table arrangement, nonhierarchical classification, and hierarchical classification. The techniques chosen for presentation in this section are primarily the more effective techniques, but a few are included because of their historical interest or commonness.

Table arrangement

The unique advantage of an arranged table, that is, an arranged samples-by-species data matrix, is that it displays at once both the general features and the full detail of the data set. Table arrangement is the earliest classification technique in community ecology. Braun-Blanquet's original tablework procedure of 1921 and subsequent computerized implementation will be considered here.

Braun-Blanquet tablework. Braun-Blanquet's important paper of 1921 synthesized an approach to vegetation classification that has become the most frequently used method for analyzing plant community data encompassing many thousands of studies (Maarel, Tüxen, & Westhoff 1970; Maarel 1975). His text of 1928 (also translated from German into English in 1932) encouraged widespread adoption of the Braun-Blanquet method of vegetation classification. Becking (1957) provides a review particularly for American readers, recognizing that Americans are less familiar with this approach than Europeans. Whittaker (1962), Mueller-Dombois & Ellenberg (1974:45–66, 177–210), and Westhoff & Maarel (1978) present additional reviews; also see McIntosh (1978).

Braun-Blanquet tablework is an informal, fairly subjective method. It is a polythetic, divisive, hierarchical, dual classification (Moore, Fitzsimons, Lambe, & White 1970; Moore & O'Sullivan 1970).

Three principal ideas are the essence of the Braun-Blanquet approach (Maarel 1975; also see Pignatti 1980). (1) Plant communities are conceived as vegetation types recognized by floristic composition, apart from environmental information. This conception is fundamentally pragmatic; continuity in community composition is allowed and sites affected by succession or disturbance are acceptable. (2) Among the species occurring in a community, some are more sensitive expressions of a given environmental or competitive gradient than others. These diagnostic, indicator or faithful species, as well as characteristic combinations of such species, are emphasized in Braun-Blanquet analysis. (3) Communities are organized into a hierarchical classification on the basis of diagnostic species. The basic unit in this hierarchy is the association, just as the basic unit in taxonomy is the species. Becking (1957) has interesting perceptions on the response of plants to environmental and competitive gradients. These perceptions effectively constitute a model of vegetation and provide a theoretical basis underlying these three principles.

Braun-Blanquet research proceeds in three phases: analytic, synthetic, and syntaxonomical (Westhoff & Maarel 1978).

The *analytic* research phase consists of reconnaissance and data collection. After reconnaissance of an area, sample sites are chosen subjectively to be representative. The relevé sampling procedure is used, requiring only about 30 to 60 minutes for each community sample (Becking 1957). Species abundances are estimated visually with about one-digit accuracy (Maarel 1975). Proponents argue that subjective sample placement is most efficient (characterizes the range of vegetation better with a limited amount of time or limited number of samples than by any alternative procedure such as systematic or random sample placement) and that greater accuracy than visual estimates is fatuous given the inherent spatial and temporal fluctuations in species abundances.

The *synthetic* research phase involves arranging the samples and species (matrix columns and rows) to best show the inherent structure of the data by an arranged table. Mueller-Dombois & Ellenberg (1974:177–210) give a worked example with German meadow samples, the result of which was shown in the Introduction (Table 1.3). Another example is shown in Figure 5.1 (Moore, Fitzsimons, Lambe, & White 1970). Details of the procedure will not be given here but may be found in the work of Becking (1957), Mueller-Dombois & Ellenberg (1974:177–210, with a summary on pages 192–3), and Westhoff & Maarel (1978).

The goals in tablework are to arrange the species in a sequence that brings together species similar in their distributions in the samples and, likewise, to arrange the samples in a sequence that brings together samples similar in species composition (Češka & Roemer 1971; Holzner, Werger, & Ellenbroek 1978; Feoli & Orlóci 1979). A little reflection shows that simultaneous pursuit of these two goals implies that positive data matrix entries will be concentrated into blocks within the data matrix, especially along the matrix diagonal. The procedure for finding species–sample groups has two processes that are alternated iteratively: selection of differential species typifying a given group of samples and selection of those samples typified by a given group of species. This procedure is iterated until a satisfactory, stable result is reached. Recognition of differential species and sample groups is partially an art and cannot be specified exactly, particularly because human judgment must balance conflicting requirements for problematic species or samples; different experts applying Braun-Blanquet analysis to the same data set may obtain somewhat different results.

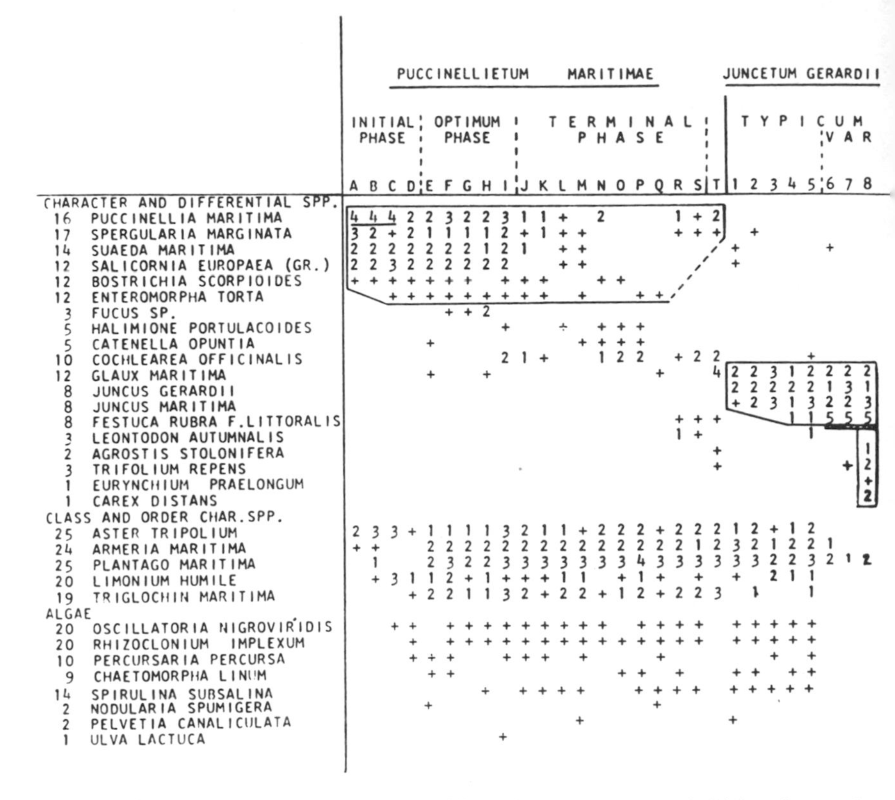

		PUCCINELLIETUM MARITIMAE																				JUNCETUM GERARDII							
		INITIAL PHASE				OPTIMUM PHASE					TERMINAL PHASE											TYPICUM					VAR		
		A	B	C	D	E	F	G	H	I	J	K	L	M	N	O	P	Q	R	S	T	1	2	3	4	5	6	7	8
	CHARACTER AND DIFFERENTIAL SPP.																												
16	PUCCINELLIA MARITIMA	4	4	4	2	2	3	2	2	3	1	1	+		2				1	+	2								
17	SPERGULARIA MARGINATA	3	2	+	2	1	1	1	1	2	+	1	+	+					+	+	+		+						
14	SUAEDA MARITIMA	2	2	2	2	2	2	2	1	2	1		+	+								+					+		
12	SALICORNIA EUROPAEA (GR.)	2	2	3	2	2	2	2	2	2			+	+								+							
12	BOSTRICHIA SCORPIOIDES	+	+	+	+	+	+	+		+	+	+			+	+													
12	ENTEROMORPHA TORTA			+	+	+	+	+	+	+	+	+		+			+	+											
3	FUCUS SP.						+	+	2																				
5	HALIMIONE PORTULACOIDES									+			÷		+	+	+												
5	CATENELLA OPUNTIA					+								+	+	+	+												
10	COCHLEAREA OFFICINALIS									2	1	+			1	2	2		+	2	2					+			
12	GLAUX MARITIMA					+			+									+			4	2	2	3	1	2	2	2	2
8	JUNCUS GERARDII																					2	2	2	2	2	1	3	1
8	JUNCUS MARITIMA																					+	2	3	1	3	2	2	3
8	FESTUCA RUBRA F.LITTORALIS																		+	+	+				1	1	5	5	5
3	LEONTODON AUTUMNALIS																		1	+						1			
2	AGROSTIS STOLONIFERA																				+								1
3	TRIFOLIUM REPENS																				+							+	2
1	EURYNCHIUM PRAELONGUM																												+
1	CAREX DISTANS																												2
	CLASS AND ORDER CHAR.SPP.																												
25	ASTER TRIPOLIUM	2	3	3	+	1	1	1	1	3	2	1	1	+	2	2	2	+	2	2	2	1	2	+	1	2			
24	ARMERIA MARITIMA	+	+			2	2	2	2	2	2	2	2	2	2	2	2	2	2	1	2	3	2	1	2	2	1		
25	PLANTAGO MARITIMA		1			2	3	2	2	3	3	3	3	3	3	3	4	3	3	3	3	3	3	2	2	3	2	1	2
20	LIMONIUM HUMILE		+	3	1	1	2	+	1	+	+	+	1	1		+	1	+		+		+		2	1	1			
19	TRIGLOCHIN MARITIMA				+	2	2	1	1	3	2	+	2	2	+	1	2	+	2	2	3		1			1			
	ALGAE																												
20	OSCILLATORIA NIGROVIRIDIS			+	+		+	+	+	+	+	+	+	+	+		+	+	+	+		+	+	+	+	+			
20	RHIZOCLONIUM IMPLEXUM				+		+	+	+	+	+	+	+	+	+	+	+	+	+	+		+	+	+	+	+			
10	PERCURSARIA PERCURSA				+	+	+			+	+	+		+				+						+		+			
9	CHAETOMORPHA LINUM					+	+									+	+		+			+	+		+	+			
14	SPIRULINA SUBSALINA								+		+	+	+	+			+	+	+	+		+	+	+	+	+			
2	NODULARIA SPUMIGERA					+												+											
2	PELVETIA CANALICULATA													+								+							
1	ULVA LACTUCA									+																			

Figure 5.1. Braun-Blanquet tablework arrangement of Irish salt marsh samples. The table is arranged to concentrate species occurrences into blocks. The character and differential species serve to distinguish samples from these two associations named for their main species *Puccinellia maritima* and *Juncus gerardii*; the class and order character species are common to all salt marsh samples. The sample sequence from left to right corresponds to a gradient of increasing distance from the shore and, consequently, of decreasing frequency of flooding. (From Moore, Fitzsimons, Lambe, & White 1970:5)

The *syntaxonomical* research phase involves (1) the assignment of samples to previously recognized associations or the establishment of new associations, (2) the hierarchical arrangement of associations into higher units (alliances, orders, classes, and finally divisions) or into lower units (subassociations, variants, and facies), and (3) the development of formal, standardized nomenclature.

These three phases intermingle in a complex manner. The last products of Braun-Blanquet analysis are the associations and their hierarchy, but awareness of established associations may affect a researcher in the first step of the analysis, which is subjective placement of the samples in representative portions of the vegetation. Likewise, the table arrangement may necessitate revision of earlier syntaxonomical results.

In comparative tests with other methods, many authors favor the Braun-Blanquet method. "The approach of Braun-Blanquet is, whatever its limitations, the best means of classifying vegetation yet developed" (Becking 1957:474). Moore, Fitzsimons, Lambe, & White (1970) find it most efficient in returning understanding of vegetation per unit of time invested. They consider the Braun-Blanquet approach particularly suitable for preliminary descriptive and classificatory work; it may not be equally suitable for other purposes. Westhoff & Maarel (1978:374) judge the Braun-Blanquet method to be "the most fully developed and most widely useful approach to the classification and interpretation of vegetation." Moore (1972) considers the human judgment integral to Braun-Blanquet analysis useful in dealing with the ecologist's noisy data.

The limitations of the Braun-Blanquet approach have been noted; they explain, in part, the lack of general usage by American ecologists (Lieth & Moore 1971). (1) Its use requires ecologists trained in the Braun-Blanquet method. (2) Application to a new area, where the species are not well known and associations are not yet defined, is difficult (Westhoff & Maarel 1978:372). In some cases, such as alpine vegetation, however, many species have wide geographical distributions, expediting application of Braun-Blanquet methods to American vegetation (Komárkova 1980). (3) Although moderate agreement among experts is typical, Braun-Blanquet analysis is not rigorously standardized and its relatively great subjectivity is problematic for some purposes. (4) Table arrangement is a slow, tedious process, not well suited to large data sets.

The advent of the computer has affected the lines of development for the Braun-Blanquet school in two ways. First, it has made numerical classification techniques and ordination alternative methodologies for organizing community data. Westhoff & Maarel (1978:371) respond to this by viewing Braun-Blanquet analysis as the central research approach to plant communities, with numerical classification and ordination "ancillaries." In their view, "numerical techniques serve to reduce labour, and ordinations to enhance interpretation." Second, the computer can be a labor-saving device within the traditional Braun-Blanquet approach by satisfying the routine aspects of the analysis.

Computer implementations. The computer implementation of Braun-Blanquet tablework has been seen as a means of overcoming or reducing the limitations of the Braun-Blanquet method, specifically, by increasing objectivity, reducing tedious labor, and reducing the requirements of expertise. In manual Braun-Blanquet tablework, progress involves three stages: (1) species–sample groups are recognized, which lead to blocks of positive entries in the data matrix; (2) the blocks are rearranged into the best sequence to show an orderly progression from one block to the next, especially along the matrix diagonal; and (3) a final touch-up is applied by moving occasional species or samples to different groups where they fit better, emphasizing overall species composition and recognizing that with noisy data an expected species may be absent occasionally or a peculiar species may be present. These three stages have proven progressively difficult to formulate precisely for computer implementation.

Benninghoff & Southwood (1964) produced the first computer program for Braun-Blanquet tablework. The "Working Group for Data-Processing in Phytosociology," formed in 1969, had computerization of Braun-Blanquet analysis as one of its main objectives (Moore 1972; Maarel 1974, 1979*a*; Maarel, Orlóci, & Pignatti 1976, 1980). Early programs by Benninghoff & Southwood (1964), Češka & Roemer (1971), Moore (1971, 1972; Moore, Fitzsimons, Lambe, & White 1970; Moore & O'Sullivan 1978), and Lieth & Moore (1971) saved much labor, but final touch-up by hand was required for fully satisfactory results. In fact, the Braun-Blanquet tablework shown in Figure 5.1 was developed with computer assistance (Moore, Fitzsimons, Lambe, & White 1970). Dale & Quadraccia (1973) took a different approach, using interactive computing and graphic display. This program left selection of the species order and sample order entirely to the human operator; the computer was merely a tool to copy the matrix quickly and easily into the order specified by the operator. A sophisticated table arrangement program, *TABORD,* first developed in 1972 but thereafter improved upon, has found wide usage (Maarel, Janssen, & Louppen 1978). Additional references to computer programs or to evaluation of results include those by Spatz (1969, 1972), Dale & Anderson (1973), Stanek (1973; and appendix by Orlóci 1973), Holzner, Werger, & Ellenbroek (1978), Meulen, Morris, & Westfall (1978), Moravec (1978), David, Lepart, & Romane (1979), Feoli & Feoli Chiapella (1979), Feoli & Lagonegro (1979), and Feoli & Orlóci (1979). The literature is reviewed by Westhoff & Maarel (1978).

The concept of reciprocating adjustments to bring similar species together and similar samples together permeates literature on Braun-Blanquet analysis, especially computer-oriented literature (Češka & Roemer 1971; Lieth & Moore 1971; Maarel, Janssen, & Louppen 1978; Moravec 1978). The reader may notice that this concept is close to the concept of the ordination algorithm reciprocal averaging, because reciprocal averaging was shown earlier to concentrate larger matrix values along the matrix diagonal (Figure 4.9). Moore (1972) first commented on the relevance of reciprocal averaging to table arrangement, and Hill (1974), Gauch, Whittaker, & Wentworth (1977), and David, Lepart, & Romane (1979) actually use reciprocal averaging for table arrangement, finding results generally concordant with Braun-Blanquet tablework.

Hill (1979*b*) has developed two-way indicator species analysis (*TWINSPAN*) for table arrangement much along the lines of Braun-Blanquet analysis. The algorithm is sophisticated enough to produce a final product in many cases. The TWINSPAN program also serves for hierarchical classification and will be treated further in the section on hierarchical classification later in this chapter.

Nonhierarchical classification

Nonhierarchical classification merely assigns each sample (or species) to a cluster, placing similar samples (or species) together. Relationships among the clusters are not characterized. Nonhierarchical classification is, of all multivariate analyses, conceptually the simplest.

Purpose and role. The purpose and role of nonhierarchical classification may be appreciated by considering the four aspects of community data described in the introductory chapter: noise, redundancy, relationships, and outliers. Nonhierarchical classification is an excellent means for handling redundancy and outliers, is useful for mitigating problems of noise, but is not appropriate for analyzing relationships.

Because community samples are noisy, samples having a similarity of something under 100% may be considered replicates (typically 60 to 90% similarity; Bray & Curtis 1957; Moore 1972; Janssen 1975). The justification for clustering (grouping) nearly similar samples without requiring exact similarity is that the samples are the same within the limits of resolution imposed by noise. Large data sets may have dozens to hundreds of samples for each discernibly different community type, and these redundant sample groups must be summarized in order to describe communities economically. The samples within a cluster may be averaged to produce one composite sample as a means of summarizing redundancy and of reducing noise.

The sequence of steps desirable in multivariate analysis of community data is determined mostly by the data structure. Much can be said for addressing redundancy first (unless the entire data set is so small that it can be studied intact without problems), because redundant sample groups are uninformative about broader relationships but cause problems of bulk and tediousness. Also, many multivariate analyses are not possible or practical for large numbers of samples. Outliers, likewise, should be addressed at an early point, because outliers may dominate multivariate analyses and obscure other features of the data. Fortunately, redundancy and outliers can be addressed simultaneously by nonhierarchical clustering, because it both clusters redundant samples and leaves outliers solitary or in small clusters, which may then be omitted from further analyses. A natural sequence for community analysis is, therefore, nonhierarchical clustering first to cluster similar samples and to identify outliers, followed by analysis of relationships in the data by other multivariate techniques. Maarel, Janssen, & Louppen (1978) reach a similar conclusion, saying that the first step of the Braun-Blanquet method is to cluster similar samples.

The basic role of nonhierarchical clustering is initial clustering of large data sets. Because of the potential for raw data sets to be large, it is especially important that nonhierarchical clustering programs be rapid.

Specifications. Four specifications for initial nonhierarchical clustering of large data sets require discussion.

(1) Within-cluster homogeneity is the basic cluster property to be achieved in nonhierarchical classification. Within-cluster homogeneity makes possible inference about a sample's properties based on its cluster membership, and such inference is a basic aim of general-purpose classification (Sneath 1969; Sneath & Sokal 1973:25–7; Gower 1974). This one property makes nonhierarchical classification useful for mitigating noise, summarizing redundancy, and identifying outliers, although not for elucidating relationships (because there is no interesting structure within clusters and no definition of relationships among clusters).

(2) Algorithmic uniqueness (the ability to give a unique answer for

each data set) is not required (Goodall 1953). If the data are naturally strongly clustered (having groups of similar samples and great differences between samples from different groups), most clustering techniques readily recover natural clusters correctly. If the data are continuous (having continuous community variation), as is usually the case, there is no natural number of clusters, and there are no natural boundaries to the clusters. Because these features are lacking in continuous data, boundaries must be imposed on the data by the clustering technique.

Algorithmic uniqueness may sound attractive, and some classification techniques are intended to give unique results even for continuous data. In this particular case, however, algorithmic uniqueness is unimportant or illusory for five reasons. (a) Classification algorithms having algorithmic uniqueness compute and rank many numbers. The cost of these extensive calculations is great. Ties can easily occur in the ranking process, however, at which point they must be broken by an arbitrary procedure, causing loss of algorithmic uniqueness. (b) Even if an exact numerical tie does not arise during computations, many close values will arise, and their exact ranking is fatuous given the noise in the data. (c) Limitations in field data collection introduce a subjective element in community ecology that is carried through, rather than eliminated, in subsequent analyses (Moore & O'Sullivan 1970; Moore 1972). There is little point in giving classification an exactitude not borne up by other steps in the overall study. (d) Algorithmic uniqueness is usually achieved by concentrating first on the smallest, finest features of the data. This stance, as will be discussed further, is unfortunate because it appears generally preferable to begin with the larger features and work down to details. (e) An undesirable consequence of the previous points is that a tiny change in the data, such as a slight alteration in a single abundance value, may greatly change cluster assignments. In other words, a classification technique may have algorithmic uniqueness, but if the results are unstable, given vanishingly small perturbations of the data, then this uniqueness is fatuous. The fundamental problem is not in the classification technique but rather in the data; the data, given their noise and, usually, their continuity, do not support a unique classification. No amount of computational effort can grant unique classes to the data when such features are simply lacking.

If algorithmic uniqueness did not involve additional computational load, no real advantage would accrue from excising it. In reality, algorithmic uniqueness is costly, and because a corresponding benefit is lacking, the best decision in nonhierarchical classification is to dispense with it.

The rejection of algorithmic uniqueness for continuous data sets does not mean that any cluster assignments will do nor that the resulting clusters are devoid of specified desirable properties. This rejection merely means that (a) for a given data set, there are a large number of alternative, equally satisfactory possibilities for cluster assignments and (b) the task in nonhierarchical clustering is to find one of these satisfactory classifications. In particular, nonhierarchical clustering techniques should achieve within-cluster homogeneity.

A related, but different, question is whether clustering results should be biased systematically by the sequence in which samples happen to be supplied (Louppen & Maarel 1979). It seems best to avoid such a bias.

(3) A hierarchy is not needed in initial clustering of large data sets. It is better to begin with a nonhierarchical classification for two reasons. First, the resulting homogeneous clusters of essentially replicate samples do not contain enough sample variation to merit analysis of within-cluster variation (Williams 1971*a*). Second, initial nonhierarchical clusters may be represented by a composite (average) sample for each cluster, and these far fewer composite samples are then practical for submitting subsequently to a hierarchical clustering in order to elucidate relationships on a higher level. The alternative, initial hierarchical classification of a large data set, is prohibitively costly for most techniques (although Jardine & Sibson 1971 and Bruynooghe 1978 present relatively rapid hierarchical techniques, and Hill 1979*b* a very rapid technique) and, in any case, generates a greater volume of information than is ecologically meaningful or humanly comprehensible.

(4) Computer requirements must be minimal. Because initial clustering may involve large data sets, both computer time and memory must be conserved. This specification deserves detailed consideration (Gauch 1980).

Computer time and memory requirements for various nonhierarchical classification techniques are functions of five numbers: I, the number of species; J, the number of samples; K, the number of clusters; N, the average number of species per sample; and A, the amount of stored data, $2NJ + J$.

Minimization of computer memory is relatively straightforward. The samples-by-species data matrix is usually sparse (mostly zeros), especially for large data sets. Rather than store the entire data matrix with IJ abundance values, a condensed storage is preferable, storing only positive data matrix entries. One vector of length NJ stores the positive data entries and another vector of the same length stores the species identification number of each entry. The data are stored sequentially by samples, and a third vector of length J indexes the beginning location of each sample's data. Hence the total amount of memory required to store the samples-by-species data matrix is only $2NJ + J$, which is the amount of data A. For data sets 5 to 20% dense (95 to 80% sparse), condensed storage is about 10 to 2.5 times as efficient in the use of computer memory.

A clustering algorithm requires minimal computer memory if it involves storage of A numbers and comparatively little else. The computation and storage of a secondary samples-by-samples dissimilarities matrix are especially to be avoided. If the sequence of computations is such that the samples are accessed sequentially, it is possible to store the bulk of the data in slow memory (disk or tape), and consequently, even a small computer may suffice to tackle large data sets.

Minimization of computer time for nonhierarchical classification involves two devices.

First, condensed storage saves memory, but it also yields a gain in computing speed because only positive data are consulted and the data are accessed in singly subscripted arrays. This gain is on the order of I/N times a factor for using single instead of double subscripts (say, 2). For example, given 1000 species with an average of 40 species per sample, the gain in computing speed is a factor of about $(1000/40) \cdot 2 = 50$. Furthermore, as community samples are added, the number of species I usually increases but the average number of species per sample N is relatively constant. Consequently, the ratio I/N tends to increase with data set size, being largest when efficiency is needed most.

Second, the computation of fewer dissimilarity calculations saves computer time. Most clustering techniques compute a complete samples-by-samples dissimilarities matrix (Sneath & Sokal 1973). This computation is on the order of J^2 and becomes prohibitive with large J. Other clustering algorithms use a samples-by-clusters dissimilarities matrix of order JK instead. This change increases computing speed by a factor of $J^2/(JK) = J/K$. For example, given 2000 samples and 50 clusters, this is a factor of 40. The latter, smaller computation is closer to the usual mental process in classifying objects because the tendency is to compare objects with classes into which they may be placed, rather than to compare each object against all others (especially for large data sets). Furthermore, the number of samples J may increase indefinitely, but for reasons of human comprehensibility, the number of clusters K is rarely more than 50 or at most 100. Consequently, K may be viewed as a constant with a value of 100 (with deviations from this ordinarily being on the smaller side). Hence the ratio J/K also tends to increase with data set size, which again is convenient because efficiency is needed most with large data sets.

The preceding two devices for saving computer time are independent, and the effects of using both are multiplicative. The first device replaces a dependency on I for one on N, and the second replaces one on J for K; the joint effect is to replace a dependency on IJ for one on NK. In the examples given, the first device increases computing speed by a factor of 50 and the second by 40; the joint effect is a factor of 2000. In this example, an algorithm including these two devices accomplishes in 15 seconds what one without them accomplishes in 8.3 hours.

When discussing computing speed, it helps to know the theoretical limit for any algorithm that could be considered a nonhierarchical clustering algorithm. Comparison of a given algorithm with the theoretical limit then indicates what room is or is not left for improvement.

A theoretical minimal computational load for nonhierarchical classification may be derived as follows. (a) Classification may be viewed as having two components: cluster definition and assignment of each sample to a cluster. (b) Assignment of samples to previously defined clusters is termed *identification* (Sneath & Sokal 1973:3; Orlóci 1978*a*: 284–308). The usual method of identification is simply the comparison of a sample with all the clusters and the assignment to that cluster with which it has greatest similarity. Given J samples and K clusters, this requires computation of JK dissimilarities. Assuming that all the I species are to be used (a polythetic rather than monothetic method), each dissimilarity computation requires comparison of I numbers or, with efficient programming and condensed storage, only N numbers (the average number of species per sample). Hence the computational task of identification is on the order of NJK. Because NJ is the same order as the amount of data A, the task is of order KA. (c) There is no way to specify a minimal algorithm for cluster definition and, hence, to declare its computational requirements. The strictest assumption that can be made is to require little computation in comparison with that required for identification. (d) Combining the preceding three points, the lower limit for computational load of nonhierarchical classification is on the order of $KA + 0 = KA$, where K is the number of clusters (which, as noted earlier, is effectively a constant of 100 at most) and A the amount of data.

This argument leads to the conclusion that rapid classification should involve a computational load that is merely a linear function of the amount of data. Also note that for a given community, with a given average number of species per sample, this implies a linear dependency on the number of samples, and consequently, the computational cost per sample is a constant regardless of the number of samples.

Wishart (1978:43) describes a nonhierarchical classification algorithm that involves iterative reallocations of samples and an enormous amount of computation (also see Hartigan 1975:84–112, Gordon & Henderson 1977, and Frieze 1980). Sneath & Sokal (1973:483) also cite several papers pertaining to rapid clustering, but these techniques are either monothetic or not particularly fast (not linear with respect to the number of samples). These algorithms are far from the theoretical limit in speed and they do not confer significant advantages over much faster techniques. Interest here centers on techniques for nonhierarchical classification that are at the theoretical limit in speed.

Composite clustering. Rapid initial nonhierarchical classification may be done by a technique termed *composite clustering* (Gauch 1980), implemented in a computer program *COMPCLUS* (Gauch 1979), which uses the preceding two devices for computational speed and has computer requirements rising only linearly with the amount of data. Hence the computational speed of *COMPCLUS* is at the theoretical limit.

The algorithm of composite clustering is explained most readily by conceiving samples as points in samples dissimilarity space. Several dissimilarity measures may be used (ED, PD, or CD, as defined in the ordination chapter), with percentage dissimilarity PD generally recommended. Composite clustering involves two phases.

Phase 1 picks a sample at random and clusters samples within a specified radius to that sample. This process is repeated until finished, ignoring any sample previously clustered. Because center samples are selected at random, the order in which the samples of the data set are given introduces no systematic bias in the results.

The first phase has an undesirable property: as spherical clusters are placed, small regions can be left behind in the interstices, resulting in clusters with only one or a few members. This is undesirable because these extra, small clusters increase the number of clusters required to account for the data and they obscure the identification of outliers (because outliers are detected as members of very small clusters). This problem can be ameliorated rather simply, as will be explained next.

Phase 2 reassigns samples from small clusters (defined as having fewer than a specified number of members) into the nearest large cluster, provided that the sample is within a specified radius. The radius used in phase 2 is usually somewhat (but not greatly) larger than that used in phase 1.

For each resulting cluster, a composite sample is produced by averaging the sample it contains. Each composite sample is also characterized briefly by listing the main species it contains. By controlling the radius used in phases 1 and 2, the investigator can control the number of clusters obtained; a small radius leads to numerous clusters and a large radius to few.

Essentially the same algorithm as used in *COMPCLUS,* but differing in various details, has been presented by Baum & Lefkovitch (1972), Diday (1971), Benzécri (1973*a*), Hartigan (1975:74–83), Janssen (1975), Briane, Lazare, & Salanon (1977), Maarel, Janssen, & Louppen (1978), Salton & Wong (1978), Swain (1978), and Louppen & Maarel (1979). In all cases, the single criterion achieved is within-cluster homogeneity, and the results are, in general, similar. All these algorithms have a linear computing dependency on the number of samples, but *COMPCLUS* is fastest, apparently being the only program with condensed data storage.

In summary, nonhierarchical classification is useful for rapid initial clustering of large data sets. Such clustering mitigates noise, identifies outliers, and summarizes redundancy; it does not elucidate relationships, but it can produce far fewer composite samples, which then make other multivariate analyses feasible and effective for analyzing relationships. Effective initial clustering must achieve within-cluster homogeneity and must be rapid; algorithmic uniqueness and a hierarchy are not needed. Because of the simple requirements, nonhierarchical classification is essentially a single approach, unlike ordination or hierarchical classification for which numerous, quite distinctive techniques exist. A computer program for nonhierarchical classification, *COMPCLUS,* operates at the theoretical limit in computing speed and requires storage of little more than the positive entries in the samples-by-species data matrix. Computing time rises only linearly with the amount of data, so the cost per sample is a constant, and there is no problem with analysis of large data sets. As there is only one goal (within-cluster homogeneity) in nonhierarchical classification and it is achieved at the theoretical limit in speed, there does not appear to be any possibility of significant further advances in nonhierarchical classification.

Hierarchical classification

Hierarchical classification groups similar entities together into classes as does nonhierarchical classification but, additionally, arranges these classes into a hierarchy. Figure 1.4 in the Introduction presents a hierarchical classification in the form of a dendrogram.

Purpose and role. The advantages of a hierarchy are that a single analysis may be viewed on several levels, variously general or detailed, and that relationships are expressed among the entities classified. Hierarchies are problematic, however, with large data sets because most hierarchical classification algorithms have nonlinear computer requirements and, hence, become costly for large data sets; in any case, a hierarchy with more than 50 or, perhaps, 100 entities is difficult to display or absorb. In applications involving the computerized inventory of a large data base, a hierarchical arrangement of thousands of samples may be useful, particularly if only small, manageable portions of the hierarchy need be recalled at a time. The computer is a medium that, unlike paper of ordinary size and human comprehension, can contain a large hierarchy. Most research results are printed on paper and absorbed visually, however, so hierarchical classification is limited to 50 to 100 samples for the most part.

Hierarchical classification is complementary to nonhierarchical classification because (a) nonhierarchical classification is ideal with large data sets and hierarchical with small and (b) hierarchical classification has the purpose of revealing relationships in the data and nonhierarchical does not. Hence nonhierarchical classification is useful in the early stages of data analysis and hierarchical classification is useful in later stages. Often a nonhierarchical classification can be used initially to summarize a large data set (with hundreds to thousands of samples) by 20 to 100 composite samples, and these fewer composite samples can then be analyzed by hierarchical classification (and, perhaps, also ordination) in order to elucidate relationships in the data.

Hierarchical classification is treated at length by Sneath & Sokal

(1973). Applications in community ecology are presented by Goodall (1978*a*), Robertson (1978, 1979), and Orlóci (1978*a*:186–283). Gauch & Whittaker (1981) discuss the theory of hierarchical classification in application to ecological community data and compare several techniques using numerous simulated and field data sets. Simon (1962) provides a philosophical basis for viewing communities hierarchically, showing that a hierarchical structure is fundamental to almost all complex systems and, likewise, that our concepts about complex systems are structured hierarchically. Blashfield & Aldenderfer (1978) review the literature.

The numerous hierarchical classification techniques may be approached in three groups as follows: monothetic divisive, polythetic agglomerative, and polythetic divisive.

Monothetic divisive. Monothetic divisive classification techniques begin with all the samples in a single cluster and then divide them hierarchically into progressively smaller clusters on the basis of presence or absence of a single species.

Williams & Lambert (1959) invented a monothetic divisive technique, termed *association-analysis,* which has found wide usage in plant community ecology and is based on similar, earlier work by Goodall (1953). Community samples are subdivided into two groups according to the presence or absence of a single species, and this process is repeated for a number of cycles in order to yield a hierarchy. The division species chosen is that species having the maximum ability to separate one group or association of species from another association; more specifically, the maximum sum of chi-squared values with all other species. Only presence/absence information enters into the calculations, not quantitative information. They tested association-analysis with field data sets and found it effective, especially for a "primary survey," in contrast with "later more detailed investigation" (Williams & Lambert 1959:100). Williams & Lambert (1960) then programmed association-analysis for a digital computer, which could handle thousands of samples. The computing time rose linearly with the number of samples but rose with the square of the number of species. Williams & Lambert (1961) further described inverse association-analysis, classifying species instead of samples (also see Lance & Williams 1965). Both sample and species analyses could be applied to the same data set, but the two analyses had no direct or simple relationship. Subsequently, Lambert & Williams (1962) integrated the sample and species classifications into a single analysis suitable for producing arranged data matrices (also see Ivimey-Cook & Proctor 1966 and Dale & Anderson 1973). Lambert & Williams (1966), Williams, Lambert, & Lance (1966), and Pritchard & Anderson (1971) compared association-analysis with polythetic agglomerative classification techniques, and Ivimey-Cook & Proctor (1966) compared it with traditional Braun-Blanquet analysis. Also see Crawford & Wishart (1967, 1968).

Orlóci (1978*a*:225–33) and Podani (1979) describe variations of association-analysis, particularly those involving different criteria for selecting dividing species. Sometimes, association-analysis produces a satisfactory result (as in Kachi & Hirose 1979*a*). A consistent problem, however, has been the high misclassification rate due to samples lacking a single species ordinarily present or, conversely, having a single species ordinarily absent, given the others present (Coetzee & Werger 1975; Hill, Bunce, & Shaw 1975; Orlóci 1978*a*:229; Ladd 1979). Community data are noisy, so classification on the basis of a single species, even the very best species for this purpose, is bound to misclassify many samples, as judged by overall species composition. Another problem is that data sets with a small range of community variation may necessitate the analysis of quantitative differences in species abundances (Hill 1977).

A major motivation for association-analysis in its original presentation (Williams & Lambert 1959) was its computational feasibility. This motivation is now gone, given advances in computers and, more importantly, in polythetic classification algorithms. The misclassification problem with any monothetic technique is so serious that at present such techniques are essentially relegated to historical interest. In contrast, polythetic techniques use the full compositional information of community samples (Sneath & Sokal 1973:20–3; Szőcs 1973).

Polythetic agglomerative. Polythetic agglomerative classification techniques use information on all the species; they begin with each sample assigned to a cluster with a single member and agglomerate these in a hierarchy of larger and larger clusters until finally a single cluster contains all the samples. Classification techniques of this kind are employed several times as frequently as all other kinds combined (Sneath & Sokal 1973:214; Blashfield & Aldenderfer 1978); a profusion of polythetic agglomerative techniques exists (Sneath & Sokal 1973:214–45; Williams 1976:84–90; Blashfield & Aldenderfer 1978; Goodall 1978*a*; Orlóci 1978*a*:199–219).

Agglomerative classification has two steps. First, from the samples-by-species data matrix, one computes a samples-by-samples dissimilarities matrix, using any of many dissimilarity (distance) measures (such as percentage dissimilarity or Euclidean distance). The samples may then be conceived geometrically as points in samples dissimilarity space, having the dissimilarity with each sample as an axis of the space (analogous to the species dissimilarity space shown earlier in Figure 4.2). Second, some agglomeration procedure is applied successively to build up a hierarchy of increasingly large clusters.

These two steps imply a rather heavy computational load for agglomerative classification. Concerning the first step, given J samples, there are $(J^2 - J)/2$ dissimilarity values to compute and to store in memory, and each computation involves as many comparisons as there are species I (or, with efficient computer programming, as the average number of species per sample N). At best, computation of the samples dissimilarities matrix is an operation of order NJ^2. This computing requirement is nonlinear and costly for large J. The second step, the actual agglomerating, has even more severe requirements for most techniques, rising faster than NJ^2. Consequently, no more than about 50–500 samples are typically practical from a computational viewpoint. In practice, this limitation is generally inconsequential, however, because one is limited first by the number of samples (perhaps 50 to 100) that can be displayed readily by a dendrogram printed on a sheet of paper of ordinary size.

Complete-linkage clustering, also called the *maximum* or *furthest-neighbor* method, is a polythetic agglomerative clustering technique noted for producing tight clusters of similar samples (Lance & Williams 1967; Sneath & Sokal 1973:222–5). The samples are conceived as points in samples dissimilarity space. A sample's dissimilarity (distance) to a cluster is defined to be equal to its dissimilarity to the furthest sample in that cluster; when two clusters agglomerate, their dissimilarity is equal to the greatest dissimilarity for any pair of samples with one in each cluster. During clustering, the most similar sample pair is joined first, and the process continues until a single cluster contains all the samples.

Average-linkage clustering is similar to complete linkage, except that the dissimilarity between clusters is equal to the average dissimilarity rather than the maximum dissimilarity (Sokal & Michener 1958; Lance & Williams 1967; Sneath & Sokal 1973:228–40; Orlóci 1978*a*:199–204). There are several ways to compute an average and, consequently, there are several, somewhat different average-linkage techniques, but the most common is the unweighted pair-groups method using arithmetic averages (UPGMA), which uses the simple, unweighted, arithmetic average. Sneath & Sokal (1973:230) consider UPGMA the most frequently used classification technique and recommend its use for hierarchical classification when there is no specific reason for choosing some other technique. The UPGMA algorithm is unique in that it appears to maximize the correlation between input dissimilarities (in the samples-by-samples dissimilarities matrix) and the output dissimilarities implied by the resulting dendrogram (using the lowest level required to join any given sample pair in the dendrogram). This is termed the *cophenetic correlation,* and its maximization implies a laudable match between distances in the input data and output results (Sneath & Sokal 1973:278).

Single-linkage clustering, also called the *minimum* or *nearest-neighbor* method, is the opposite of complete linkage (Sneath & Sokal 1973:216–22; Orlóci 1978*a*:214–9). The dissimilarity between clusters is equal to the minimum dissimilarity (rather than the maximum or the average as in the two preceding clustering techniques). Single-linkage clustering tends to produce straggly clusters, which quickly agglomerate very dissimilar samples. It is not appropriate for the usual research purposes of community ecology (Hill 1977) but is mentioned here because of its relationship to complete and average linkage and because it is frequently used in taxonomy research. Lance & Williams (1966, 1967) produced a general scheme encompassing complete-, average-, and single-linkage clustering, as well as additional related techniques (also see Jardine & Sibson 1971:45–58, Pritchard & Anderson 1971, Sneath & Sokal 1973:214–45, Pakarinen 1976, and Everitt 1978:43).

Minimum-variance clustering, also called *minimization of within-group dispersion,* agglomerates clusters, provided that the increase in within-group dispersion is less than it would be if either of the two clusters were joined with any other cluster (Orlóci 1967, 1978*a*:204–14). It is similar to average-linkage clustering, except that instead of minimizing an average distance one minimizes a squared distance weighted by cluster size. Penalty by squared distance makes minimum-variance clusters tighter than average-linkage clusters.

Other polythetic agglomerative classification techniques are discussed by Crawford & Wishart (1968), Sneath & Sokal (1973:240–5), Blashfield & Aldenderfer (1978), and Orlóci (1978*a*:220–4). Information theory is used to provide dissimilarity measures for some (Robertson 1979).

Although technical details differ, the general goal is ordinarily the same as usual: to agglomerate similar samples together hierarchically into larger and larger clusters. None of these other techniques have found common usage in community ecology.

Although previously discussed in terms of classifying samples, species may be classified instead by the preceding technique by substituting a species-by-species dissimilarities matrix. Orlóci (1978*a*:224–33) further discusses species classification. Complete-, average-, and single-linkage clustering and minimum-variance clustering may be applied to both the samples and the species of a data matrix by analyzing both the samples-by-samples and species-by-species dissimilarities matrices. Unfortunately, however, such sample and species classifications are separate, unintegrated analyses.

Polythetic divisive. Polythetic divisive classification techniques use information on all the species. They begin with all samples together in a single cluster and successively divide the samples into a hierarchy of smaller and smaller clusters until, finally, each cluster contains only one sample or some specified small number of samples. Although the theoretical merit of this approach has long been appreciated, technical challenges delayed substantial development until the past few years.

The simplest polythetic divisive classification is *ordination space partitioning* (Roux & Roux 1967; Noy-Meir 1973*b*; Hall & Swaine 1976; Swaine & Hall 1976; Verneaux 1976*a,b*; Hill 1977; Peet 1980; Prentice 1980*b*). Because detrended correspondence analysis has been found to be an especially robust and effective ordination technique, Gauch & Whittaker (1981) recommend this ordination for the positioning of sample points in a low-dimensional space. Successive partitions are then drawn in the ordination to generate a divisive, hierarchical classification. The partitions may be placed subjectively by drawing boundaries in ordination graphs by hand. Subjective partitions can be particularly useful when (1) divisions through sparse regions of the cloud of sample points are desired, because none of the other clustering techniques considered here can take sparse regions into consideration, (2) field experience or previous analyses have provided a general understanding of the data that the investigator wants to incorporate into the analysis but cannot specify precisely or supply to a computer (Williams 1971*a*), and (3) subjective clustering is sufficient for the purposes of a given study. Of course, the partitioning procedure can be made automatic and objective in a variety of ways (for example, see Wildi

1979, 1980), and this might sometimes be preferable, but a subjective choice of which procedure to use still has to be made.

Two-way indicator species analysis, TWINSPAN, is another polythetic divisive technique (Hill 1979*b*; Gauch & Whittaker 1981; compare AXOR in Lambert, Meacock, Barrs, & Smartt 1973, POLYDIV and REMUL in Williams 1976:92–4, 118, an iterative technique in Jancey 1980, and the precursor indicator species analysis that only classified samples in Hill, Bunce, & Shaw 1975). The data are first ordinated by reciprocal averaging. Then those species that characterize the reciprocal averaging axis extremes are emphasized in order to polarize the samples, and the samples are divided into two clusters by breaking the ordination axis near its middle. The sample division is refined by a reclassification using species with maximum value for indicating the poles of the ordination axis. The division process is then repeated on the two sample subsets to give four clusters, and so on, until each cluster has no more than a chosen minimum number of members. A corresponding species classification is produced, and the sample and species hierarchical classifications are used together to produce an arranged data matrix. In its emphasis on indicator species and production of an arranged matrix, TWINSPAN has similarities to the approach of Braun-Blanquet. Its resultant sample hierarchy (and species hierarchy) may also be displayed as a dendrogram, using the sequences of divisions as integral levels or computing the levels as the average distances between samples in ordination space (Gauch & Whittaker 1981). Figure 5.2 shows the TWINSPAN hierarchy for the southeast England vegetation samples, which were ordinated earlier by detrended correspondence analysis (Figure 4.16). The sample sequence in this dendrogram effectively reflects moisture status from xeric sand dune and wall communities (left) to aquatic water weed communities (right). The computer program for TWINSPAN (Hill 1979*b*), unlike any other hierarchical classification program, deliberately arranges the two clusters at each node in the way that results in placing the most similar samples together in the dendrogram's sample sequence. This makes the information in the dendrograms more lucid. Because of the intrinsic one-dimensionality of a sample sequence, however, the additional weediness gradient revealed in the ordination of these data (Figure 4.16) cannot be expressed in this dendrogram (Everitt 1978:46).

These two polythetic divisive classification techniques, ordination space partitioning using detrended correspondence analysis and TWINSPAN, are similar. They differ, however, in that (1) ordination space

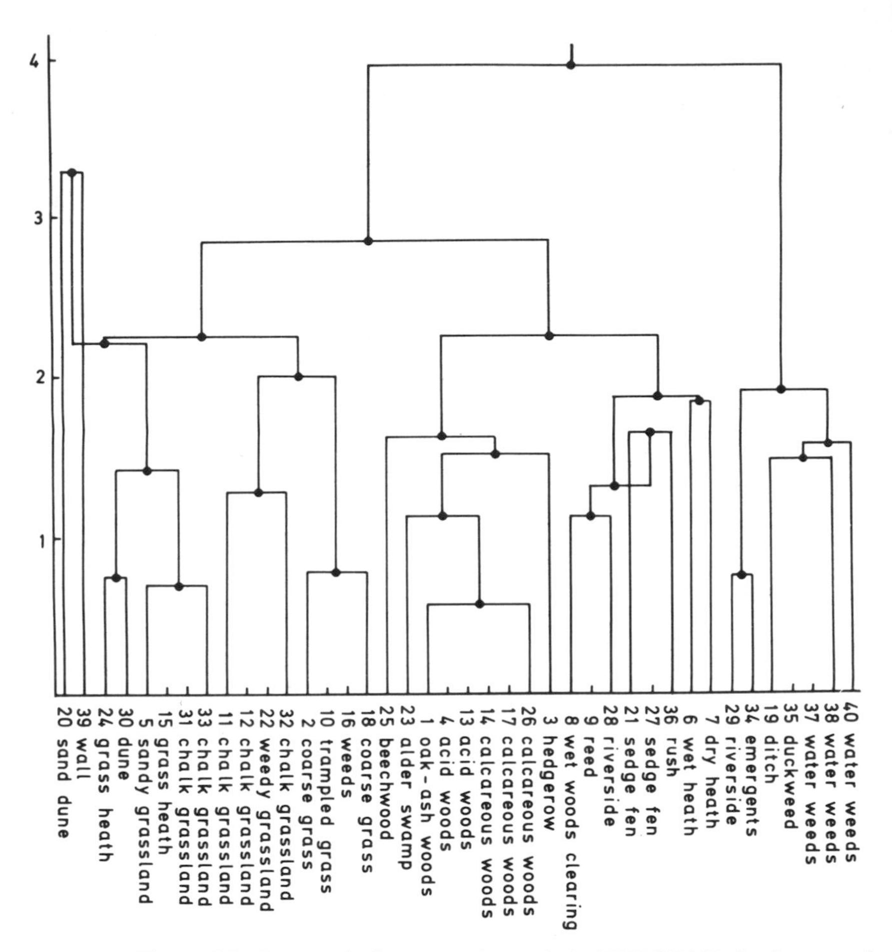

Figure 5.2. Two-way indicator species analysis, TWINSPAN, dendrogram of 40 composite samples from a vegetation survey of southeast England (the data are part of the National Vegetation Classification program, supported by the Nature Conservancy Council, made available by Mrs. J. P. Huntley and Dr. H. J. B. Birks). These data are also ordinated by detrended correspondence analysis in Figure 4.16. The sample sequence reflects moisture status from xeric (sand dune and wall) to aquatic (water weeds). The ordinate is the average Euclidean distance in the detrended correspondence analysis sample ordination (within a cluster using all pairs of members and between clusters using all pairs with one member from each cluster). Similar samples are joined at a low level in the dendrogram, whereas dissimilar samples are not joined until high levels. (From Gauch & Whittaker 1981)

partitioning uses detrended correspondence analysis, but TWINSPAN uses reciprocal averaging, (2) for ordination space partitioning, a single ordination is performed having one unified coordinate system, whereas TWINSPAN uses separate ordinations for each data subset it produces, and consequently, the various divisions at a given level may be focusing on different community gradients that are important for the individual data subsets, and (3) ordination space partitions are usually imposed subjectively, whereas TWINSPAN partitions are objective and automatic. These classification techniques are complementary.

Computer requirements for these two polythetic divisive classification techniques are minimal. The amount of calculation required rises only linearly with the data. No dissimilarity matrix is stored or computed, and little needs to be stored in memory besides the positive data matrix entries (and the bulk of these may reside in slow memory if need be). Consequently, there is no problem in analyzing large data sets. It is ironic that for many years polythetic divisive techniques were lacking because of excessive computational demands of the suggested approaches, but now polythetic divisive techniques are far more rapid than agglomerative techniques. The key to this development has been the employment of rapid ordination techniques to arrange the samples in a manner suitable for straightforward divisive classification. A corresponding increase in the speed of agglomerative techniques is not possible because even if the agglomerating task is minimized somehow, the initial production of a samples-by-samples dissimilarities matrix still involves a computational task rising with the square of the number of samples. Consequently, divisive hierarchical classification presently poses no computing problem. Limitations arise, however, in the communication and assimilation of the results of hierarchical classification, at least within the usual context of ordinary paper or human contemplation.

Evaluation of classification techniques

The evaluation of classification techniques involves principles already emphasized for the evaluation of ordination techniques. Evaluation proceeds from mathematical reasoning and from empirical tests with data sets. Because performance varies with variation in several data set properties, balanced assessment requires tests with a number of data sets, including both simulated and field data. As no classification technique is perfect, the performance of a given technique is judged in relation to others, which means that there is need to apply several classification techniques to the same data sets and compare results.

Criteria

Apart from explicit performance criteria, evaluation of classification techniques is undefined and impossible (Blashfield & Aldenderfer

1978). The fundamental criterion for community classification is understanding in return for a given time input, relative to alternative methodologies (Moore, Fitzsimons, Lambe, & White 1970). This general criterion may be discussed in greater detail using four components: computational feasibility, robustness, effectiveness, and extent of results (compare Maarel 1979*a*).

(1) *Computational feasibility* may be considered on two scales: a larger scale with the original samples that may range into the thousands and a smaller scale for small data sets or composite samples limited to 50 or 100 samples (Gauch & Whittaker 1981).

(2) *Robustness,* as defined earlier in this chapter, means that the important features of a classification remain stable despite minor perturbations of the data set and that the classification technique works well on a wide variety of data sets.

(3) The *effectiveness of results* means that a classification technique summarizes the data well and aids understanding. Effectiveness can be assessed by empirical tests using data sets for which proper results (a null hypothesis) are known in advance. Hence tests must employ well-known field data sets or else simulated data for which expected results are known exactly. Once the effectiveness of a classification technique is established with known data sets, the technique may then be applied profitably for the analysis of less familiar data sets. An alternative possibility for evaluating effectiveness is mathematical reasoning to establish desirable properties of a given classification technique. Effectiveness is evaluated by these two complementary approaches, empirical tests with known data sets and mathematical deduction (Mannetje 1967; Jackson 1969; Sneath 1969; Jardine & Sibson 1971:102–12; Rand 1971; Rohlf 1974; Robertson 1979). The evaluation of effectiveness is partially a subjective art because one classification may display one feature of a data set best and another classification another feature; not every feature's display can be optimized at once (Hill 1979*b*). Optimal weighting of various data features is partially a subjective matter of the researcher's purposes and taste. Although the selection of the very best results may be somewhat subjective, the discrimination of good from bad results is less delicate and usually falls within the range of simple, objective criteria of judgment.

(4) The *extent of results* concerns whether a classification computer program produces various useful outputs. For nonhierarchical classification, useful features include a list of major species in each cluster, composite samples, and alphameric names as well as numbers for identification of samples and species. For hierarchical classification, useful

features include the computation of dendrogram levels, the ordering of the dendrogram to bring similar samples in proximity, and the corresponding production of a species classification and arranged data matrix. The extent of results is mostly an incidental feature of particular computer programs but is considered here because, in practice, these features affect research effectiveness and efficiency.

Because both classification techniques and clusters have numerous properties, one could choose criteria for evaluating classification techniques additional to the four just mentioned. A little reflection, however, indicates that most additional criteria are largely subsumed by these. Little more could be desired of a classification technique if it is computationally feasible, robust, effective, and produces extensive results.

Evaluations of classification techniques will be considered next on two levels. The first level concerns the choice among the basic classification strategies: table arrangement, nonhierarchical classification, and hierarchical classification. The second level is the choice within one of these strategies, say, hierarchical classification, of the better techniques. In the following evaluations, the preceding four classification criteria will be applied.

Choice of strategy

The choice among table arrangement, nonhierarchical classification, and hierarchical classification is largely a matter of research purpose and data set size and complexity. Table arrangement is just a particular display option for a hierarchical classification, so the choice is essentially between nonhierarchical or hierarchical classification.

If the data set is large, initial nonhierarchical classification is indicated, and ordinarily, 20 to 100 composite samples produced thereby can represent the variation in the sample set adequately. (For special research purposes with computer inventories of large data sets, however, hierarchical classification may be useful.) Subsequent hierarchical classification of the composite samples can then reveal sample relationships, and resultant table arrangements or dendrograms are of practical proportions.

If the data set is small, hierarchical classification is ordinarily preferred because it shows sample relationships. If research purposes clearly call for classes but not for a hierarchy (for example, assignment of each sample to one of five community types for the purpose of producing a vegetation map of a park), nonhierarchical classification may be better. Hierarchical classifications at any given level are compromised somewhat by demands at other levels in the hierarchy, so a simple division of samples into a given number of groups is accomplished most cleanly by nonhierarchical classification (Williams 1971*a*, 1976:80). Application of both nonhierarchical and hierarchical classifications to the same data set is sometimes useful, as these strategies are complementary.

The intrinsic dimensionality or complexity of a data set affects the suitability of various analyses or presentation options. Dendrograms are intrinsically appropriate for high-dimensional data structures, or alternatively, the sample sequence in a dendrogram can convey a single dimension or gradient effectively (although it is not possible for the indicated levels to faithfully reflect the actual dissimilarity values in the data; Holman 1972; Gauch & Whittaker 1981). Likewise, table arrangements are effective with very complex, heterogeneous data sets having numerous underlying gradients in community variation because the larger matrix entries may then be concentrated into a number of blocks in the data matrix. Alternatively, table arrangements are suitable, given a single dimension of community variation, because this leads to a neat banding of larger matrix entries along the matrix diagonal. In short, hierarchical classification, expressed by either table arrangement or a dendrogram, is effective, given data with numerous gradients or one gradient. Given an intermediate number of community gradients, two to, perhaps, four, hierarchical classification is intrinsically awkward.

Nonhierarchical classification is suitable for data of any intrinsic dimensionality because it does not express or depend on relationships in the data.

Probably most community data sets are of intermediate intrinsic dimensionality, having two to four important gradients. Plant communities often have main gradients of moisture and nutrient status, for example. It is unfortunate that hierarchical classification is especially weak in this dimensionality range. Ordination, however, is especially strong in this range and may especially complement classification for such data.

Table arrangement

Braun-Blanquet tablework is the indicated procedure for table arrangement when plant community results are to be incorporated within the Braun-Blanquet system. European plant ecologists have done thousands of studies with this technique, which grants to each individual study a broad context and, usually, numerous closely related studies. Unless an investigator specifically intends to express results within the Braun-Blanquet framework, however, several problems make the choice of this technique unlikely: its use requires extensive apprenticeship in the Braun-Blanquet methodology, the results are relatively subjective, the large sample sets are problematic, and the application to unfamiliar species and unfamiliar vegetation is difficult.

Reciprocal averaging is a simple, objective algorithm that generates ordination scores for species and for samples. Arranging the data matrix according to ranked reciprocal averaging scores maximizes the concentration of larger matrix entries along the matrix diagonal. An advantage of this approach is that a single analysis provides both an arranged table and an ordination. Probably the best computerized technique for table arrangement, however, is two-way indicator species analysis, TWINSPAN.

Nonhierarchical classification

Because nonhierarchical classification involves a single, readily achieved goal (namely within-cluster homogeneity), essentially ideal effectiveness is attainable. For nonhierarchical classification, composite clustering or any similar technique appears quite satisfactory.

A related question is how one can compare several nonhierarchical classifications of the same data set (as contrasted with the preceding discussion about comparing nonhierarchical classification techniques). Gauch (1980) defines an index, *percent mutual matches* (PMM), for comparing two nonhierarchical clusterings of a given data set, where PMM equals 200 times the number of mutual matches divided by the total number of matches, where a *match* is a sample pair assigned to the same cluster in either clustering and a *mutual match* is a sample pair assigned together in both clusterings (a mutual match thus counts twice among the matches). The values of PMM range from 100 for identical clusterings to 0 for completely different clusterings. Given two random clusterings with K clusters of approximately even sizes, PMM has an expected value of about $100/K$. For comparing a number of clusterings simultaneously, a clustering-by-clustering matrix is computed of PMM values subtracted from 100 to yield distances. This matrix is then ordinated using reciprocal averaging. In the ordination, similar clusterings are placed in proximity and dissimilar clusterings far apart (an example is given by Gauch 1980).

Hierarchical classification

For hierarchical classification, there is generally a clear preference for polythetic over monothetic techniques. Because ecological community data are noisy, classification on the basis of the presence or absence of a single species leads to an unacceptable rate of misclassification, even when the best possible species is selected for division.

There is also a marked preference for divisive over agglomerative techniques (Hill, Bunce, & Shaw 1975; Hill 1977; Gauch & Whittaker 1981). Agglomerative techniques begin by examining small distances between similar samples. In community data, these small distances are more a reflection of noise than anything else. It is not surprising that agglomerative techniques tend not to be robust. Polythetic techniques, however, begin by examining overall, major gradients in the data. As Lambert, Meacock, Barrs, & Smartt (1973:173) observe, "polythetic-divisive methods have theoretical advantages in that all the available information is used to make the critical topmost divisions." The larger, significant differences in community composition, but not the tiny differences, are related to differences in environment and history, which an ecologist wants to express in a classification. Hence two-way indicator species analysis (TWINSPAN), being polythetic and divisive, is recommended for hierarchical classification because of its effectiveness and robustness.

The further advantages of TWINSPAN are (1) its use of the original vegetation data, rather than a secondary dissimilarities matrix with information only on samples (or species), (2) integrated classifications of both samples and species, and, consequently, (3) production of an arranged data matrix. As implemented by Hill's (1979*b*) program, additional advantages include (4) ordering of the sample sequence to place most similar samples together, making dendrograms clearer and (5) minimal computer requirements. Although TWINSPAN could be improved somewhat along the lines suggested by Hill (1979*b*:46), this would involve a difficult programming task. In many cases, however, TWINSPAN already provides a satisfactory hierarchical classification of a data set.

As alternative or complementary approaches to hierarchical classification using TWINSPAN, one may consider either (1) initial nonhierarchical clustering to produce composite samples, followed by TWINSPAN analysis of the composites (and perhaps ordination) or (2) ordination space partitioning by subjective or dichotomous division. These two approaches are hybrid ordination–classification techniques; they combine the power of classification for summarization with the effectiveness of ordination in revealing directions of relationship.

Because of the general superiority of divisive algorithms over agglomerative ones for community classification, the evaluation of finer differences among agglomerative techniques takes on secondary interest. Complete linkage produces tight clusters, average linkage maximizes the cophenetic correlation, and minimum variance minimizes within/between cluster variance. As these are all laudable goals, slight differences in performance can only be evaluated by empirical tests (Williams & Lance 1968; Gauch & Whittaker 1981). Complete linkage and minimum variance appear preferable because tight clusters are a primary consideration (Hill 1977). Differences among these agglomerative techniques are slight, however, in a larger perspective in which they are compared with polythetic techniques.

The comparison of several hierarchical classifications of a given data set is a difficult matter because a hierarchy is a complex entity with numerous properties. Effective indices for evaluating the similarity or effectiveness of entire hierarchies have not yet been devised, although interesting concepts may be found in the work of Lance & Williams (1967), Jardine & Sibson (1968), Gower (1971), Williams (1971*a*), and Cunningham & Ogilvie (1972). Clusterings at a given level may be compared, however, as if they were nonhierarchical clusterings by using the percent mutual matches index described earlier.

Discussion

Data structures that are naturally strongly clustered are rare in community ecology; ordinarily, community variation is relatively continuous, and consequently, the classification is imposed. The number of clusters desired and the details affecting desirable locations of boundaries are then controlled by the investigator, not by structure internal to the data (Goodall 1953). For this reason, both the process of classification and the choice among classification techniques tend to be more complex and more subjective than those with ordination. Likewise, the burden is greater upon the user to make effective choices of classification technique and to discern the proper results. Classification remains partly an art to which the investigator's experience and understanding may contribute much.

The implications of the Gaussian model of community structure for community classification are analyzed by Hill (1977). The basic implications are that classification involves imposing a dissection upon essentially continuous variation (although this is quite justified for practical purposes) and that the broad view is important, rather than the details (and hence divisive algorithms are better than agglomerative ones). He acknowledges the high noise level typical of field data, and after concluding that abundance values should be taken seriously only to several levels (in effect, one-digit accuracy or somewhat less), he derives the implications of noise limitations for the design of classification algorithms. Hill comments perceptively that the numerical methods of classification should be founded on two considerations: the purpose of the analysis and the underlying model of community structure.

Community classification appears to have progressed to an encouraging point. Rapid initial nonhierarchical clustering by a technique such as composite clustering makes large data sets tractable and has computing requirements linear with the amount of data. Hierarchical classification by two-way indicator species analysis, TWINSPAN, is generally effective and robust and also has linear computing requirements. For certain purposes, Braun-Blanquet tablework and ordination space partitioning may also be useful. Although an enormous number of classification techniques have been developed (Whittaker 1962, 1978*a*; Sneath & Sokal 1973; Blashfield & Aldenderfer 1978; Orlóci 1978*a*:186–286), much can be said in favor of concentrating on the use of relatively few techniques for the sake of comparability of results and facility of communication. In the preceding chapter it was argued that a modest number of ordination techniques effectively cover most cases. The same argument holds for classification, and the several especially effective classification techniques mentioned in this paragraph may be expected to suffice for most applications.

It is encouraging that at present there are effective and robust techniques with computer requirements rising only linearly with the amount of data for each of the three basic classification strategies: table arrangement, nonhierarchical classification, and hierarchical classification.

6

Applications

This chapter will be concerned with the applications of multivariate analysis. First, general recommendations will be given for community ecology applications regarding data editing in preparation for multivariate analysis and selecting multivariate techniques appropriate for a given data set and purpose. These are based on the theory and evaluations presented in earlier chapters on direct gradient analysis, ordination, and classification. Then, representative literature references will be given for applications of multivariate analysis in applied community ecology and in related and distant fields.

General recommendations

First, general recommendations will be given for multivariate analysis in community ecology. Although presented in the context of community ecology, these recommendations, for the most part, concern common features of any kind of two-way individuals-by-attributes matrix and, consequently, apply to many kinds of data.

Especially for applied researchers, multivariate analysis is just one of many tools serving as a means to specific ends. An applied researcher's time for learning multivariate techniques and for analyzing a particular data set is limited, so it is necessary to provide straightforward, general recommendations that usually work. Given the greater robustness of recent multivariate techniques, such recommendations are now possible. As Williams (1976:28) notes, "Where a subject is ancillary to one's own discipline, the important thing is not how much, but how little, one needs to learn."

Data editing

A samples-by-species community data matrix may be edited prior to multivariate analysis (Singer 1980). Editing possibilities include the application of transformations, the deletion of species or samples, the selection of sample subsets, and the formation of composite samples.

There are two basic motivations for data editing. First, editing affects the range of community variation present in a data set and the relative emphasis on various features of the data. Editing allows different facets of complex data sets to be examined and exhibited. Second, some data sets in their original form may be intractable for a given multivariate technique or even for any multivariate technique. Editing is then indispensable for producing satisfactory results.

A distinction is important: All multivariate techniques have more or less the same goal of making multivariate data comprehensible, whereas different editings are intended to feature different sides of the data. The range of results obtainable by different data editings is greater than that obtainable by employing different multivariate techniques because data editing changes the data submitted for analysis. Likewise, certain research problems that cannot be solved by changing or improving multivariate analyses may yield easily upon suitable data editing. The full potential and importance of data editing are rarely appreciated.

Transformations of the abundance values in the samples-by-species matrix are one data-editing option (Bannister 1966; Sneath & Sokal 1973:152–7; Jensen 1978; Maarel 1979*b;* Clymo 1980; also see Armitage 1971:349–59). At one extreme, abundance values may have an enormous range, over many orders of magnitude. At the opposite extreme, mere presence or absence may be noted (Stanek 1973; Williams, Lance, Webb, & Tracey 1973; Frenkel & Harrison 1974; Lausi & Feoli 1979). General experience indicates a preference for abundance values with an intermediate range of approximately 0 to 10 (or somewhat less; Maarel 1979*a*). This range allows both quantitative and qualitative information to be expressed without either dominating the other. Compression from a greater range is usually accomplished by a logarithmic transformation such as the octave scale (given in Chapter 2). Five reasons bear upon this recommendation to use an intermediate range. (1) Biological processes responsible for the abundances of species are of an exponential nature leading to an enormous range of abundances (Maarel 1979*b;* also see Preston 1948, 1980). Consequently, only the few dominant species, rather than the entire species composition, control the results of many multivariate analyses unless logarithmic (or similar) transformation is applied in order to put the species on a more equitable footing. (2) Human perception of species abundances is, like perception of sound and light intensity, essentially logarithmic. Consequently, multivariate analyses based on logarithmic scalings of species abundances usually best match the researcher's perceptions and expectations. (3) Sampling limitations and spatial and temporal fluctuations in species populations imply that for most community data the reliable information can be carried by one digit (as discussed earlier in Chapter 2). (4) One-digit values, using an integer range of 0 to 9, are especially convenient for printing and for computer storage. (5) The results of multivariate analysis are affected little by finer differences in the input data (Hill 1977; Hamer & Soulsby 1980; Gauch 1981).

The general recommendation to use a species abundance range of approximately 0 to 10 has exceptions. (1) If the raw data are collected using a somewhat, but not vastly, greater range, say, 0 to 100 or, perhaps, 0 to 300, multivariate analysis of the raw data may be fine without bothering to apply a transformation. Use of the raw data then confers an advantage of simplicity. (2) If the community samples are rather homogeneous, the sample variation present may reside in small differences in species abundances (hopefully measured accurately), and in this case, compression of abundance values into a smaller range may destroy important information.

The choice of a dissimilarity measure for multivariate analyses employing a secondary matrix of samples-by-samples or species-by-species dissimilarities has similar effects as does the choice of data transformation. The Euclidean distance (ED) emphasizes dominant species, the complemented coefficient of community (CD) emphasizes rare species, and the percentage dissimilarity (PD) emphasizes dominants somewhat but still considers minor species. The intermediate weighting of PD is generally best for the same reasons that a data transformation yielding an intermediate range (of about 0 to 10) is preferred. The effects of different distance measures have been described for ordination (Noy-Meir 1973*a*; Goff 1975; Noy-Meir, Walker, & Williams 1975; Chardy, Glemarec, & Laurec 1976; Prentice 1980*a*), hierarchical classification (Williams, Lambert, & Lance 1966; Moore, Fitzsimons, Lambe, & White 1970; Moore & O'Sullivan 1970; Frenkel & Harrison 1974; Watanabe & Miyai 1978), and nonhierarchical classification (Gauch 1979:9–11, 1980).

Rare species are usually deleted from a data matrix prior to multivariate analysis (Webb, Tracey, Williams, & Lance 1967; Austin & Greig-Smith 1968; Orlóci & Mukkattu 1973; Goff 1975; Majer 1976; Onyekwelu & Okafor 1979; also see Mirkin & Rozenberg 1977). Obviously, *rare* is a relative term, but typical criteria include species occurring in less than about 5% of the samples or in fewer than about 5 to 20 of the samples. The justifications for deleting rare species are partly theoretical and partly pragmatic. (1) The occurrences of rare species are usually more a matter of chance than an indication of ecological conditions. (2) Most multivariate techniques are affected very little by rare species carrying such a small percentage of the overall information or variance. Other multivariate techniques, especially many ordination techniques, perceive rare species as outliers, thus obscuring the analysis of the data set as a whole. In the first case, rare species are unnecessary; in the latter case, they are deleterious. (3) Deletion of rare species reduces the amount of data storage required (especially if one stores the full data matrix including zero values). The force of this observation is recognized when one recalls that most species are rare (Preston 1948, 1980). (4) Fieldwork is expedited if rare species are not even recorded in the first place, provided that it is possible to list common species in advance. Fieldworkers sometimes find the identification of common species challenging enough, and rare species further increase the taxonomic challenge by a factor of two to several times. On the other hand, information on rare species may be valuable for purposes (both foreseen and unforeseen) other than multivariate analysis, so a careful decision should be made.

Computation of sample totals (or, likewise, species averages) at an early point in data analysis may reveal great variability in sample totals. In this case, it is usually best to relativize the abundance values for each sample so that all samples have the same total, in order to treat the samples moderately equitably. Likewise, if the species vary greatly in their average value (or variance or maximum value), they may be standardized to have the same average (or variance or maximum). Species standardization is employed less frequently than sample standardization, however, because species standardization may give too much emphasis to minor species. A traditional standardization, the Wisconsin double standardization, involves standardization of species maxima to 100, followed by relativization of sample totals to 100 (Bray & Curtis 1957). The choice of standardization affects the relative emphasis on different features of the data and is principally a matter of research purpose (Goff 1975; Noy-Meir, Walker, & Williams 1975; Green 1977; Maarel 1979*a*). This choice is also affected, however, by the weightings or standardizations implicit in various ordination and classification techniques and in various dissimilarity measures used by many techniques. It may be suspected, for example, that the standardizations by sample totals and by species totals implicit in reciprocal averaging (Hill 1973, Appendix 2) contribute significantly to the effectiveness of this ordination technique. Because reciprocal averaging applies these standardizations internally, in comparison to algorithms without internal standardization, there is less need to standardize prior to analysis and there is less difference in results caused by prior standardization.

If the sample set contains peculiar samples unlike the others, outliers, it is best to delete these samples prior to data analysis, especially with ordination (Gauch, Whittaker, & Wentworth 1977; Green 1977; Dume 1978; Everitt 1978:11–16; Moral & Watson 1978; Reyment 1980:10–12; also see Armitage 1971:359–61). Outliers may be identified most easily by a preliminary nonhierarchical clustering, distinguishing themselves by remaining in solitary or small clusters. The justification for omitting outliers is that their relationships to other samples in the data set are not expressed by information in the data anyway and they cause problems for many multivariate techniques, particularly ordinations (Figure 6.1). Likewise, if the sample set is disjunct, having several subsets of related samples that have little or nothing in common, it is best to separate the samples into several data subsets suitable for further analysis (Green 1980). Disjunction may be detected easily by a preliminary ordination using reciprocal averaging to produce an arranged data matrix; this also reveals outliers.

In summary, it is generally recommended that outliers be deleted and disjunct sample sets separated, rare species be omitted, sample totals be standardized, and abundance values be expressed logarithmically using an integer range of 0 to 9. Individual cases vary, however, depending on the research purposes, the data set, and the implicit weightings embodied in the calculations of subsequent multivariate analyses. Sometimes it is useful to compare the results from multivariate analyses based on different editings. Simple data sets may not require any editing. Complex data sets, however, may require editing to make subsequent multivariate analysis possible, and informed editing decisions may require preliminary ordination, nonhierarchical classification, and calculation of simple statistics.

The preceding recommendations pertain to editing a data matrix as a whole. Considered next will be the options for summarizing or splitting matrices.

Composite samples may be formed by averaging several samples together. Composite samples can be useful for two purposes: to summarize a large data set by a workable number of composite samples

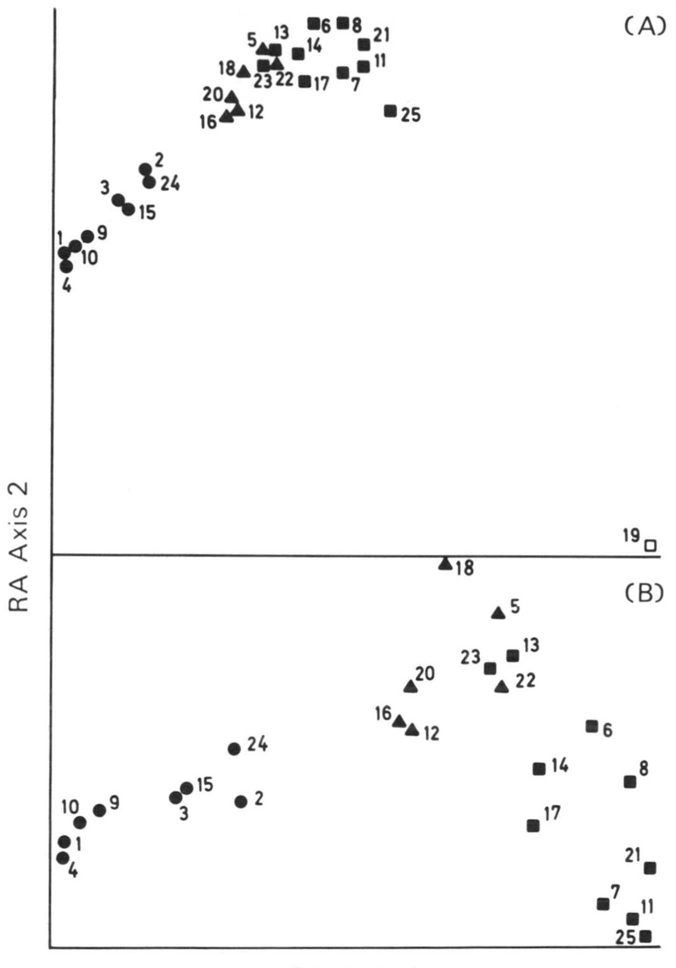

Figure 6.1. Reciprocal averaging (RA) ordination of German meadow samples, showing the improvement obtained by deleting an outlier (data tabulated in Table 1.2; also see Table 1.4). The samples were previously classified by Braun-Blanquet analysis into three community types occurring along a dry-to-wet soil moisture gradient: *Bromus–Arrhenatherum* (●), *Geum–Arrhenatherum* (▲), and *Cirsium–Arrhenatherum* (■, with one outlier from a periodically flooded site, □). In part (*A*), the outlier (sample 19) is so far removed from the other samples that it compresses the other samples into an undesirably small area. In part (*B*), the outlier was removed prior to ordination, and RA then produces good results. In this particular case, a more robust ordination, detrended correspondence analysis, can handle these data even with the outlier present (Figure 1.5) and thus offers an alternative solution to the problem. In other cases, however, data editing may offer the only remedy for making multivariate analysis succeed.

and to reduce noise by averaging together a number of replicate samples. Especially for large data sets, such summarization and noise reduction are often a valuable preliminary to subsequent detailed ordination and classification. By averaging out small differences among samples, the formation of composites tends to raise the level of abstraction somewhat, so that the broader features of the data are emphasized. The samples going into each composite may be chosen by several means, including (1) subjective classification of the samples, (2) stratification of the samples along a measured environmental gradient (such as elevation), (3) dissection of a reciprocal averaging (or other ordination) axis, (4) nonhierarchical classification such as composite clustering, (5) hierarchical classification such as two-way indicator species analysis using a given level, and (6) random partitioning. Compositing of nested samples placed adjacently is a special case used to produce samples of various sizes as a means of studying species contagion and association on various size scales (Goff & Mitchell 1975).

Data subsets may be formed by selecting some samples from the original entire data set (Forgas 1976; Feoli & Feoli Chiapella 1979; Louppen & Maarel 1979; Peet 1980). Selection may be done by the same means as just given for composite samples. Subsets have three principal uses. First, subsets of certain kinds of samples can be selected for further study, for example, just the forest samples from a heterogeneous regional survey of vegetation or just those samples with a certain cluster assignment or with an ordination score within a certain range. This refines the level of abstraction, permitting the analysis of community gradients within a certain subset of the entire data collection. Second, subsets may be chosen to be essentially replicate samples. Such subsets may be analyzed by ordination or classification to confirm that, indeed, they do not still contain significant community variation. Third, a large data set may be subdivided at random into two to several subsets, which are then analyzed separately in order to test whether the results are stable.

These possibilities for data editing may be classified into two groups, which serve two rather different roles. (1) One group is deletion of outliers, separation of disjunctions, deletion of rare species, standardization of sample totals, and expression of species abundances logarithmically (or by some similar transformation resulting in a modest numerical range). These options serve to make the data set amenable to multivariate analysis or, restated in rather different terms, to impose those emphases upon the data set that generally correspond to the investigator's purposes and perspectives. (2) The second group is formation of composites and subsets. These options serve principally to raise or lower the level of abstraction of a data set and, consequently, of its analysis, to emphasize the broader or finer features of the data. Often it is useful to view a data set on several levels of abstraction, for the same reason that it is useful to examine a microscope slide using objectives of various powers or to survey a terrain using maps with different scales. Secondarily, composites and subsets may serve to make data analysis or presentation of results manageable by providing fewer entities.

Data editing is an important preliminary to multivariate analysis. Ineffective ordination and classification analyses, either published or unpublished, suffer from inappropriate data editing fully as frequently as from an inappropriate choice of multivariate analysis technique. From the preceding discussion (and references cited), it is clear that data-editing options are numerous. Fortunately, however, the simple, general recommendations given are frequently adequate. One may debate whether to use a square root or logarithmic transformation, for example, but in practice, the effects of these and similar transformations are usually comparable, as contrasted with using no transformation. The development in recent years of more robust ordination and classification techniques does not obviate the need for preliminary data editing, but it does make acceptable a wider selection of details of editing options. Consequently, it is usually an easy matter to edit a data set (if need be) in preparation for multivariate analysis.

Multivariate analysis

Preferred techniques for direct gradient analysis, ordination, and classification were reviewed in the three preceding chapters. Here these recommendations will be summarized and an integrated research approach will be developed to employ all three methodologies.

The first choice in analyzing community data is whether to emphasize and begin with environmental data or community data. If important environmental factors are obvious and have been measured, direct gradient analysis has the advantages of simplicity and ease of interpretation. Alternatively, if environmental factors are not clear or if community variation is the primary focus and environmental factors are to be assayed or scaled in terms of their impact on plant and animal communities, the appropriate methods include ordination and classification.

This choice between direct gradient analysis with environmental emphasis and ordination and classification with community emphasis is not necessarily exclusive. Some research purposes call for both, and a comparison of the results is often illuminating. Indeed, relative to the time and cost of collecting field data, it may be quite reasonable to analyze the data using several different methods. An additional alternative is to use weighted averages ordination to derive sample ordination scores from species weights reflecting known habitat preferences or to derive species ordination scores from sample weights reflecting the ecologist's assessment or measurement of site conditions. This alternative is intermediate in the contrast between environmental and community emphases because community data are employed but the weights reflect environmental features. Polar ordination with endpoints chosen deliberately to reflect environmental gradients similarly has an intermediate status.

Direct gradient analysis is a relatively simple method not requiring a choice among numerous techniques. Likewise, weighted averages ordination is rather simple. The remainder of this discussion will concern those methods with community emphasis that do have difficult choices among numerous techniques: ordination and classification.

The second data analysis choice, given a community emphasis, is between ordination displaying continuous community variation in a low-dimensional space and classification placing communities into discontinuous types. The Gaussian model of community structure implies continuous community variation, making ordination somewhat more natural. Practical purposes, however, frequently mandate results in terms of community types. Again, this choice is not necessarily exclusive, as these approaches are complementary (Kachi & Hirose 1979*a,b*; Robertson 1978; Green 1980).

Nonhierarchical classification is useful for preliminary summarization of large data sets. Given several hundreds of samples or more, nonhierarchical classification is ordinarily best for starting data analysis. Composite clustering is a rapid, suitable technique. It summarizes redundancy, reduces noise, and identifies outliers. It does not analyze relationships in the data, but its production of fewer composite samples provides appropriate data for elucidating relationships by means of subsequent ordination and hierarchical classification.

Among ordination techniques, general preference is for detrended correspondence analysis. In the rare case of very low sample heterogeneity, however, principal components analysis is better. Reciprocal averaging is frequently effective and involves simple calculations, but it is not as robust as its successor, detrended correspondence analysis.

Among hierarchical classification techniques, general preference is for two-way indicator species analysis, TWINSPAN, a robust and effective polythetic divisive technique. Table arrangement by TWINSPAN is also recommended, with reciprocal averaging arrangement an alternative.

Four general comments should be kept in mind during data analysis. (1) The choice of data analysis techniques depends as much on research purposes as on anything else. Research purposes should be considered explicitly and frequently, or else data analysis may become aimless. (2) Several analysis techniques may be applied to the same data set, as the various results are frequently complementary. Usually it takes less work to try several techniques and compare results than it takes to accurately assess numerous data set properties and deduce the best choice of technique. Given a limited amount of time for data analysis, it is better to try a few substantially different methods (such as an ordination and a hierarchical classification), rather than a few closely related techniques (such as several agglomerative classifications), which are practically replicate techniques within the larger context of available methods. (3) The analysis of complex community data usually progresses by successive refinement. For example, one result may reveal major sample groups in the data, and subsequently, these groups may be analyzed separately. Generally, it is best to progress from broader down to finer features. (4) The communication of results is promoted by employing a moderate number of commonly used, relatively standard techniques (Pielou 1977:331; Romesburg 1979).

The environmental interpretation of community variation is an integral part of most community studies, rather than the mere characterization of community variation. Two principles are important. (1) Direct gradient analysis confers different, complementary advantages to those of ordination and classification. Environmental information is incorporated most straightforwardly in direct gradient analysis, and such analysis is often satisfactory, given a specific environmental gradient of interest such as elevation, precipitation, or distance from a pollution source. Ordination and classification, however, have the advantage of scaling environmental gradients in terms of their impact upon plant and animal communities. (2) Regardless of which methods are used, environmental interpretation is essentially the search for congruent changes in communities and environment. Consequently, effective environmental interpretation requires collection of appropriate environmental data in addition to community data.

The interpretation of results for other than community ecology data involves these same two principles: the employment of direct gradient analysis and complementary ordination and classification techniques and the collection of data on variables likely to figure in the interpretation. The interpretation of a data matrix in a business application, for example, a matrix of customers-by-purchases, involves underlying variables concerning the customers' incomes and attitudes. Both direct gradient analysis along these variables and ordination and classification analyses would be useful and complementary. Data on these underlying variables would have to be collected in order to make interpretation possible. If underlying variables are uncertain, a pilot study for their identification should precede extensive data collection.

Applied community ecology

Multivariate analysis of community and related data is used frequently in applied ecological work. Applications involve land use and management, assessment of pollution, and agriculture and fishing.

Land use and management

Classification of land into categories suitable for different uses requires sophisticated techniques because of numerous, often conflicting purposes and because of the need for repeatable, objective procedures. A multivariate approach is indicated because "individual parameters considered in isolation are rarely adequate and many factors need to be taken into account" (Bunce, Morrell, & Stel 1975:152). The rapidness of a given approach to land classification is also an important consideration, especially for large regions and for fine resolution.

Bunce, Morrell, & Stel (1975) classify land into eight land types, based on data concerning 152 attributes ascertainable from a map of the area for a square region with sides of about 50 km around Ambleside in the Lake District of northern England (Figure 6.2; also see Bunce & Shaw 1973 and Bunce & Smith 1978). Indicator species analysis (Hill, Bunce, & Shaw 1975; the forerunner of two-way indicator species analysis, TWINSPAN, Hill 1979*b*) was used to derive the

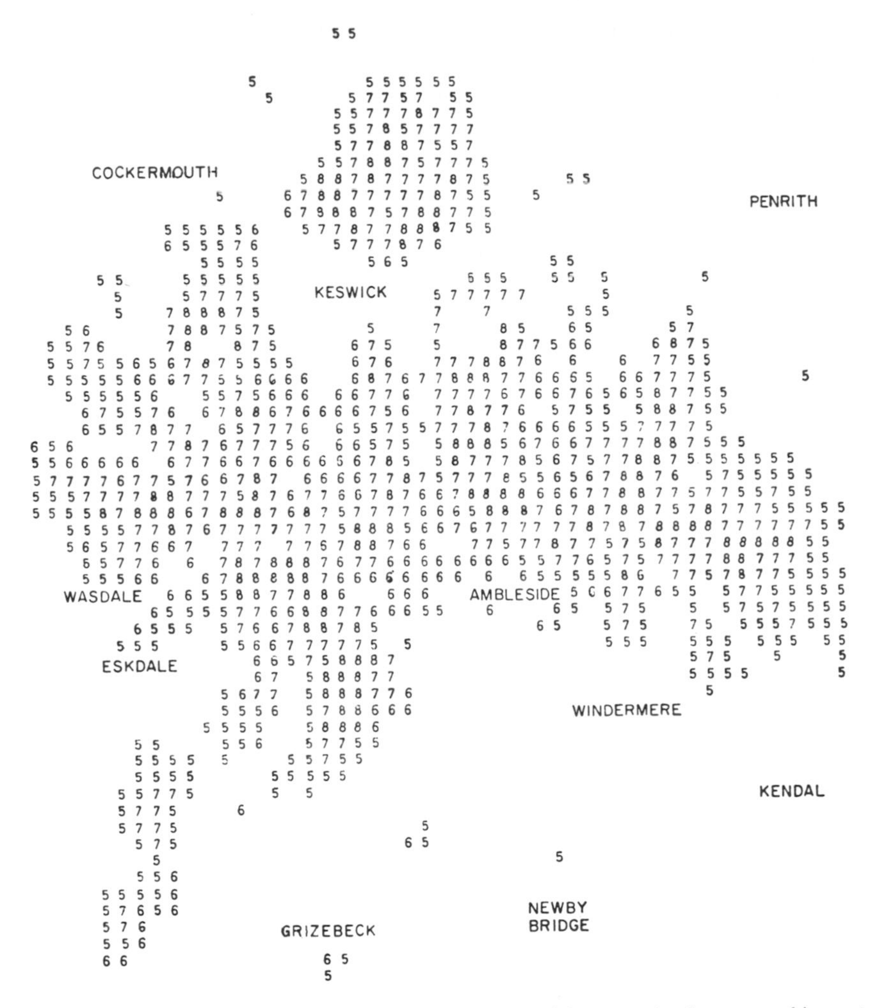

Figure 6.2. Distribution map showing four of the many land types used in a set of figures from Bunce, Morrell, & Stel in a 1-km grid for a square region with sides of about 50 km around Ambleside in the Lake District of northern England. The four types are: *5*, rounded upland; *6*, rocky upland; *7*, steep high fells; and *8*, mountain tops. This classification is based on indicator species analysis (the forerunner of two-way indicator species analysis) of 152 attributes available from a map and accords well with vegetational and environmental data. (From Bunce, Morrell, & Stel 1975:161)

classification for each grid of 1 km^2. This classification was shown to accord well with vegetational and environmental data. Elevation was the dominating environmental gradient, ranging from sea level to about 1000 m. This successful general-purpose classification into eight land types provided a stratification for further sampling of "any data that would be likely to be correlated with the overall environment," for example, data on bird populations (Bunce, Morrell, & Stel 1975:163). This classification was expected to be useful for planning recreation areas and road patterns. The method is repeatable, objective, and rapid. Also see Pritchard & Anderson (1971), Buse (1974), Colquhoun & Watson (1975), Dale (1976*a*), Stocker, Gilbert, & Smith (1977), Ladd (1979), Matthews (1979*a,b*), and Denys (1980) for similar studies. Computer classification is also used for mapping soils (Webster & Burrough 1972*a,b*; Tothill 1976), weather patterns (Russell & Moore 1970, 1976), and snowdrifts (Rawls & Jackson 1979).

Additional applications of multivariate analysis to land classification and land-usage decisions may be cited. Hill, Bunce, & Shaw (1975) use indicator species analysis to classify native pinewoods of Scotland and to identify major environmental gradients of wetness and pH. Pemadasa & Mueller-Dombois (1979) use reciprocal averaging of higher plant and soil fungi data to distinguish five types of montane grasslands in Sri Lanka. Kessell (1979:120–36), Huschle & Hironaka (1980), Maarel (1980*b*), and Onans & Parsons (1980) use multivariate analyses to model succession of disturbed vegetation and relate their models to management and land-use decisions. Crow & Grigal (1979) and Pfister & Arno (1980) use multivariate analysis of vegetation for classifying forest habitat types as a basis for various management purposes. Pakarinen (1976) and Pakarinen & Ruuhijärvi (1978) ordinate and classify Finnish peatlands. Lacoste (1975) classifies subalpine vegetation in the French Alps and Jeník, Bureš, & Burešová (1980) in the Sudeten Mountains. Regnéll (1979) studies grazed and ungrazed meadows in Sweden. Johnson & Rowe (1977) study fire and vegetation change pertinent to land-use regulation and management in the Canadian subarctic (also see Krajina 1975, Black & Bliss 1978, and Johnson 1981). Thompson (1980) classifies the vegetation of Keewatin into types particularly reflecting wildlife habitat use. Longton (1979) classifies Antarctic vegetation types by methods similar to Braun-Blanquet analysis. The impact of people on vegetation has been investigated in Belgium (Hermy & Stieperaere 1981) and South America (Ellenberg 1979). Sobolev (1975) considers classification methodology for agricultural lands. Ellis, Fallat, Reece, & Riordan (1977) review classification systems for land cover and use for the western United States and Canada. Orlóci & Stanek (1979) use multivariate analysis to predict vegetation sensitivity relating to plans to construct a gas pipeline through the Yukon Territory. Austin (1978) uses vegetation classification in an

Australian land-use study. Whitney & Adams (1980) use reciprocal averaging to discern five major tree community types in Ohio urban areas and discuss the role of man as a maker of new plant communities (also see Dume 1978 and Olsson 1978). Likewise, Crowe (1979) classifies urban weed communities in Illinois. Chappell (1976) uses ordination to calculate a socioeconomic index for watershed subareas in the Tennessee River Basin. Hierarchical classification of insect communities is developed by Refseth (1980) as a basis for documenting the conservation value of different habitats and, likewise, vegetation classification (Bakker 1979). Lebrun (1977) uses Braun-Blanquet analysis for planning a new town, a recreation area, and agricultural improvement. Management of wildlands involves assessing environmental sensitivity and suitability for various purposes (such as timber, forage, water yield, scenic beauty, recreation, and firebreak), and Betters & Rubingh (1978) find hierarchical classification a valuable tool for identifying those areas with dominant land-use characteristics and those exhibiting multiple-use potential. Bailey, Pfister, & Henderson (1978) present a brief review of land and resource classification.

In addition to classification and management of land, multivariate analyses have been applied to various problems concerning aquatic or marine environments. Fishery habitat classification has been poorly developed, although there are cogent reasons for developing a workable classification system (Platts 1980). Multivariate methods are used by Jensen (1979) and Jensen & Maarel (1980) to classify Swedish lakes on the basis of their macrophyte composition (also see Wiegleb 1980). Likewise, Green & Vascotto (1978) classify Ontario lakes on the basis of zooplankton composition. Gladfelter, Ogden, & Gladfelter (1980) classify coral reef fish communities. Elemental composition of Norwegian rivers is analyzed in order to elucidate terrestrial and atmospheric sources of elements (Salbu, Pappas, & Steinnes 1979). Classification of data describing the flood hydrology of New Zealand catchments is used to identify regions with similar hydrologic regimes (Mosley 1981). Multivariate analysis elucidates the best terrain variables for locating snowdrifts in Idaho, with the location of these drifts essential to developing and managing this water resource (Rawls & Jackson 1979). Verneaux (1976*a,b*) uses reciprocal averaging to typify invertebrate and fish communities at various distances from the mouths of streams.

Natural revegetation of acid coal spoils in southeast Iowa is analyzed by reciprocal averaging ordination (Glenn-Lewin 1979, 1980). Revegetation is found not to follow a simple successional scheme but, rather, varies according to site conditions such as substrate acidity and moisture. Likewise, natural revegetation of strip-mined land in lignite coalfields of southeastern Saskatchewan is studied using hierarchical classification and ordination in order to determine the order of importance of various environmental factors (Jonescu 1979). Important factors are moisture, time since disturbance, and cattle grazing. Ridge vegetation is highly variable and has poor recovery. Valley vegetation has a more stable environment, and succession to vegetation similar to undisturbed sites is more rapid. Stocum (1980) studies the vegetation of abandoned coal surface mines in Tennessee. These and further results are needed in order to plan reclamation appropriate to particular site conditions. Roberts, Marrs, Skeffington, & Bradshaw (1981) study revegetation on wastes from mining operations for china clay in England.

Several management problems have been studied by multivariate techniques in the Great Smoky Mountains National Park, Tennessee and North Carolina. Ordination of the vegetation elucidates important environmental and historical factors (Bratton 1975*b;* Lindsay & Bratton 1979). Such results provide a general framework for the study of the impact of wild boar (*Sus scrofa*) disturbance in various vegetation types (Bratton 1975*b;* Bratton, Harmon, & White 1981). Management of wild boar populations and their locations is necessary to reduce soil erosion and the destruction of herbaceous communities (including rare wildflowers). Also studied using ordination is the origin and management of grassy balds (Lindsay 1978; Lindsay & Bratton 1979). Continuing disturbance is found to be required to prevent succession to forest. Finally, attributes of trails are ordinated to identify factors associated with soil erosion (Bratton, Hickler, & Graves 1979; also see Liddle & Greig-Smith 1975). This study is used in the environmental analysis of proposed development and in long-range planning for the park trail system.

Resource and fire management, particularly in Glacier National Park, Montana (Kessell 1976, 1979), was discussed earlier (Table 3.1 and Figure 3.13). Multivariate data on plant and animal communities, forest succession, fuel loadings, and weather were integrated in a computer system used to make complex and necessarily rapid decisions for fire management (Kessell 1976, 1979; Kessell & Cattelino 1978). These decisions also incorporated a diversity of long-range management objectives.

The management of bird communities was studied by multivariate analysis in order to cluster similar bird species, relate bird distribution to vegetation and other habitat factors, and predict consequences of

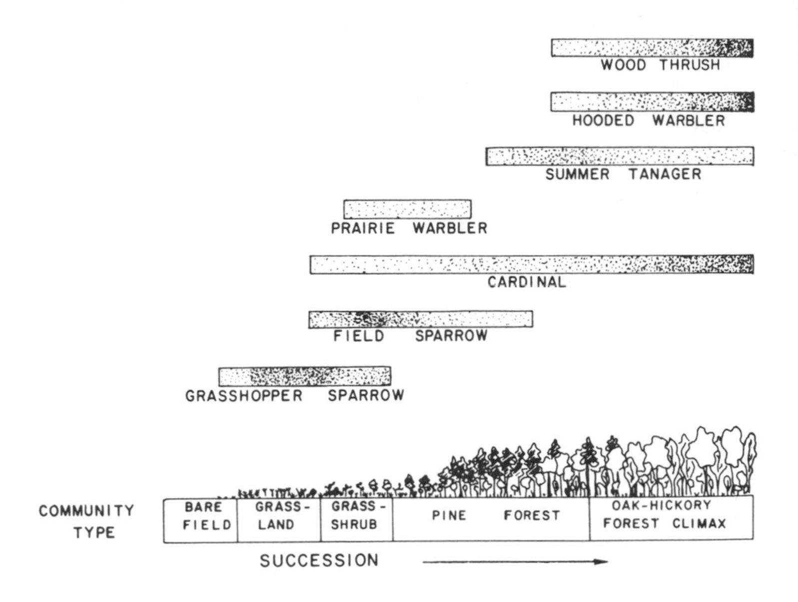

Figure 6.3. Direct gradient analysis of bird species distributions along a plant succession gradient in the Piedmont region of Georgia. Differential stippling in the occurrence bars for each bird species indicates relative abundance. The time sequence is grassland for 2 years following bare field, grass–shrub for years 3 to 25, pine forest for years 25 to 100, and finally a gradual transition to oak–hickory forest climax beginning about 150 years after the initial clearing. Bird succession strongly relates to plant succession, so the management of plant communities is a major factor in the management of bird communities. (From Gauthreaux 1978: 21; based on Johnston & Odum 1956)

habitat alteration or management (Gauthreaux 1978; Meyers & Johnson 1978; Anderson 1979; Crawford & Titterington 1979; Niemi & Pfannmuller 1979; Capen 1981; Karr 1980; Rotenberry & Wiens 1980, Wiens & Rotenberry 1981). Figure 6.3 shows bird succession as a function of plant succession in the Georgia Piedmont region (Gauthreaux 1978; likewise, see Matveev & Tikhomirova 1975 concerning insect succession tracking forest succession). Species composition and age and structure of the vegetation are major habitat features for birds, and the vegetation in turn is affected by soil and climate, diseases, and disturbances such as lumbering or grazing. Much bird habitat is owned and managed primarily for lumber or as watershed, so preservation of a suitable habitat for birds is often in a complex context of economic factors and multiple land use. Multivariate approaches to bird communities confer two advantages in particular: relative objectivity and clustering of numerous species into fewer groups. There are about 800 bird species in North America and over 100 species in most states, so managers need economy of thought through studying numerous species in terms of a manageable number of bird communities. Parren, Thompson, & Capen (1980) provide a bibliography on the use of multivariate methods in studies of wildlife habitat.

Braun-Blanquet vegetation analysis indicates the suitability of sites for mosquito breeding, and this information helps in directing a larvaciding program in Quebec (Lamoureux & Lacoursière 1976).

Multivariate analysis is used to group recreation participants into a workable number of recreation types with similar activity patterns (Ditton 1975; Avedon 1977; Duncan 1978; Romesburg 1979). The definition of recreation types and their percentages are useful in park planning and management intended to consider equity of opportunities for various activities and in predicting the behavior of park visitors (Beaman & Lindsay 1976). Water recreation types, likewise, are derived by numerical classification (Ditton, Goodale, & Johnsen 1975). Motivations of sportsmen are analyzed by Benzécri (1969), using reciprocal averaging ordination.

Remote sensing using airplanes or satellites readily generates enormous numbers of samples, making classification by computerized multivariate methods necessary (Swain & Davis 1978). The data consist of brightnesses in typically 4 to 20 spectral bands for each picture element (pixel). Often the goal is to define classes by spectral characteristics that correspond to specific crops or land-use types. Standard applications include crop identification and acreage estimation, regional land-use inventory, vegetation and geologic mapping, and water temperature mapping. Remote sensing lacks the detail of ground studies but confers unique advantages for surveying large areas. Laperriere, Lent, Gassaway, & Nodler (1980) use Landsat data to classify 13 000 000 ha of Alaska for moose habitat. An unsuccessful attempt to identify grassland types in Belgium on the basis of spectral brightnesses, however, illustrates the need for continued methodological research in remote sensing to develop more robust methods (Hecke, Impens, Goossens, & Hebrant 1980). One of the main classification techniques used with remote sensing data is much like composite clustering (Swain & Davis 1978:181).

Environmental impact statements frequently involve baseline data on vegetation and animal communities, classification of communities, and prediction of consequences of development. Wikum & Shanholtzer

(1978) recommend one-digit accuracy in field methods, such as the Braun-Blanquet scale, as being sufficiently accurate for environmental impact statements while being rapid and cost effective (also see Hamer & Soulsby 1980). They recommend multivariate analysis as a means of establishing community types relatively objectively and for elucidating relationships to the environment. Myers & Shelton (1980: 332–4) also discuss the use of multivariate community analyses for ecosystem management. Levin & Tedrow (1980) use ordination of vegetation to clarify vegetational and environmental history of forest in New Jersey; their results were used in litigation.

Pollution

Air pollution is a complex problem involving numerous variables. Multivariate analysis helps to identify the major factors affecting air pollution (Peterson 1970; Gether & Seip 1979; Henry & Hidy 1979). Multivariate analysis of element concentrations in aerosols serves to identify the sources of air pollution around Boston (Hopke et al. 1976). Principal components analysis of St. Louis thermal air pollution explains heat patterns as functions primarily of land use and weather parameters (Clarke & Peterson 1973).

Water pollution studies involving multivariate analysis include the classification of waste solids in the coastal ocean around New York (Ali, Gross, & Kishpaugh 1975), the ordination of zooplankton in a marine area exposed to sewage (Arfi, Champalbert, & Patriti 1981), the diet of a gudgeon as affected by mining pollution (Jeffree & Williams 1980), and the classification of macrobenthos communities around a farm sewage drain with the elucidation of important environmental variables (Poore & Mobley 1980). The use of changes in lake zooplankton communities to monitor and assess pollution is suggested by Green & Vascotto (1978); likewise, see the macrobenthos and pollution study of Nalepa & Thomas (1976). Hamer & Soulsby (1980) apply multivariate analysis to biological monitoring of river pollution. Although not applied to pollution problems directly, the relevance is apparent of a study by Foster et al. (1976), which detected four distinct water types in the western Irish Sea by means of multivariate analysis of the physical, chemical, and biological characteristics of the water.

Pollution effects on terrestrial vegetation have been studied by direct gradient analysis of forest communities near a heavy metals smelter at Sudbury, Ontario (Freedman & Hutchinson 1980), and near a natural gas refinery near Fox Creek, Alberta (Winner & Bewley 1978*a*), and by ordination of vegetation following pollution control of a heavy metals smelter at Trail, British Columbia (Archibold 1978; also see Tazaki & Ushijima 1977 and Greszta, Braniewski, Marczyńska-Gałkowska, & Nosek 1979). The results are useful for identifying especially sensitive species and for predicting the consequences of similar industrial efforts elsewhere. Winner & Bewley (1978*b*) and Johnson (1979) calculate an index of atmospheric purity from weighted averages of terrestrial mosses sensitive to sulphur dioxide pollution (compare Marsh & Nash 1979). Will-Wolf (1980) analyzes lichen communities to assess sulphur dioxide pollution from a coal-fired generating station in Wisconsin (also see Mueller-Dombois & Spatz 1975; Grodzińska 1977; Lötschert 1977; Saeki et al. 1977; Schubert 1977; and Steubing 1977). Westman (1980, 1981) uses multivariate analysis to scale the relative importance of 40 environmental factors including 10 indicating pollution on coastal sage scrub of southern California. Bouche & Beugnot (1978) study the effects of pesticides on earthworms.

Agriculture and fisheries

Multivariate analyses have been applied to a variety of data sets concerning agriculture. For example, Webb, Tracey, Williams, & Lance (1971) use ordination to predict the agricultural potential of Australian rain forest sites on the basis of the original natural vegetation. Likewise, Fourt, Donald, Jeffers, & Binns (1971) relate an ordination of climatic and soil variables to success and growth rate of Corsican pine (*Pinus nigra* var. *maritima*) in southern Britain and draw implications for forest managers. Streibig (1979) uses ordination and classification to analyze weed communities in relation to 19 different crops in Denmark; also see Trenbath & Harper (1973), Trenbath (1974), Sharp (1976), Borowiec, Kutyna, & Skrzyczyńska (1977), and Lavrentiades (1980). Majer (1976) characterizes the influence of ant communities on cocoa cultivation (also see Doncaster 1981). Mittelhammer, Young, Tasanasanta, & Donnelly (1980) employ a modified form of principal components analysis to identify the important factors affecting the agricultural production of Thailand. Horticultural applications are discussed by Broschat (1979), particularly the derivation of commodity quality indices by ordination to replace subjective visual quality ratings, and elucidation of relationships among cultivars.

Plant breeding and genetics applications of multivariate analyses serve a variety of purposes, including: (1) comparing or grouping genotypes, lines, plant introductions (even many thousands), hybrids, cytoplasms, and responses to grazing, fertilizers, and symbiotic fungi; (2) selecting traits affecting yield and quality and lines having potential for high yield; (3) comparing test sites regarding soil and climate and selecting sites appropriate for a breeding project; (4) evaluating phenotypic stability (or instability) in the presence of environmental variation, particularly as this affects the design of field-testing procedures and the discrimination of environmental from genetic effects; (5) detecting errors in pedigree records; (6) identifying cultivars more accurately and objectively, even by automated methods; and (7) establishing or studying taxonomic, phylogenetic, and phenetic relationships, including elucidating the wild progenitors of cultivated crops in order to expand gene pool resources (see Sneath & Sokal 1973:371–3). Multivariate analyses are used to study pathogenic organisms of concern to plant breeders (Madden & Pennypacker 1978; Riggs, Hamblen, & Rakes 1981). Also relevant to plant breeding will be citations later in this chapter concerning niche and habitat, taxonomy and genetics, agronomy, and meteorology. Various crops have been studied using multivariate analyses, including: wheat (Walton 1971; Bhatt 1976; Tsunewaki et al. 1976; Campbell & Lafever 1977; Ghaderi, Everson, & Cress 1980; Kosina 1980), rye (Kaltsikes 1973), the wheat and rye hybrid *Triticale* (Kaltsikes 1974), oats (Baum & Lefkovitch 1972, 1973; Baum & Brach 1975; Baum & Thompson 1976), barley (Molina-Cano 1976; Molina-Cano & Elena Rosselló 1978; Baum 1980; Baum, Petruk, & Bailey 1980), the Triticeae (Baum 1978; Tulloch, Baum, & Hoffman 1980), millet (Hussaini, Goodman, & Timothy 1977), rice (Jacquot & Arnaud 1979), corn (Jancey 1975; Stalker, Harlan, & Wet 1977), sorghum and sugar cane (Gupta, Wet, & Harlan 1978), the lawn grass *Poa* (Williamson & Killick 1978), pasture species (Reid 1973), hops (Small 1978*a,* 1980, 1981), mulberry (Hirano, Inokuchi, & Nakajima 1980), tomato (Sachan & Sharma 1971; Cavicchi & Giorgi 1976), potato (Tai & Jong 1980; Tai & Tarn 1980), pepper (Jensen, McLeod, Eshbaugh, & Guttman 1979; Pickersgill, Heiser, & McNeill 1979), eggplant (Pearce & Lester 1979), citrus (Barrett & Rhodes 1976), cucurbits (Rhodes, Bemis, Whitaker, & Carmer 1968), horseradish (Rhodes, Carmer, & Courter 1969), carrot (Small 1978*b*), beets (Ford-Lloyd & Williams 1975), mango (Rhodes, Campbell, Malo, & Carmer 1970), cotton (Abou-El-Fittouh, Rawlings, & Miller 1969), okra (Singh, Singh, & Rai 1980), avocado (Rhodes, Malo, Campbell, & Carmer 1971), roses (Roberts 1977), soybeans

(Edye, Williams, & Pritchard 1970; Mungomery, Shorter, & Byth 1974; Verma, Murty, Jain, & Rao 1974; Shorter, Byth, & Mungomery 1977; Newell & Hymowitz 1978; Broich & Palmer 1980), mung beans (Ghaderi, Shishegar, Rezai, & Ehdaie 1979), dry beans (Denis & Adams 1978), pigeon pea (Akinola & Whiteman 1972), clover (Mannetje 1967), vetch (Couderc 1977, 1978), and *Stylosanthes* (Burt et al. 1971, 1974; Williams, Edye, Burt, & Grof 1973; Edye et al. 1975; Burt 1976*a,b,c;* Burt & Reid 1976; Burt, Reid, & Williams 1976; Edye 1976*a,b;* Jones 1976; Reid, Ryan, & Burt 1976; Robinson 1976; Date, Burt, & Williams 1979; Burt, Isbell, & Williams 1979; Burt & Williams 1979*a,b;* Burt, Pengelly, & Williams 1980; Robinson, Burt, & Williams 1980; Williams, Burt, Pengelly, & Robinson 1980).

Animal husbandry applications of multivariate analysis include the summarization of data on cow size and shape (Carpenter et al. 1978) and the prediction of subsequent productivity from early performance records (Young, Johnson, & Omtvedt 1977; Brown, Frahm, Morrison, & McNew 1979), the analysis of cow conception rate (Williams & Edye 1974), and the study of the effects of steer on a pasture (Williams & Gillard 1971). Analyses concerning sheep include those by Hedges (1976*a,b*), Winter (1976), and Orr (1980). An application in poultry science concerns the texture of cooked chicken meat (Frijters 1976).

Fish, echinoderm, mollusc, and arthropod distributions on the continental slope south of New England have been studied by Haedrich, Rowe, & Polloni (1975). They compare direct gradient analyses using depth and cluster analyses of the animal communities to examine zonation in the epibenthic macrofauna and find rapid changes at depths of about 300 to 400 m and about 1000 to 1100 m. Hughes & Thomas (1971) use direct gradient analysis, ordination, and classification of plant and animal community data from shallow-water benthic samples from Prince Edward Island, Canada, to identify community types and important environmental factors. Boesch (1973) uses ordination and classification of estuarine macrofauna samples from Virginia to define community types and to assess the effects of natural environmental gradients and of pollution. Felley & Avise (1980) classify Florida bluegill populations on the basis of genetic and morphological variation.

Related fields

Ecology is a singularly interdisciplinary field, and most community ecology studies require background information from at least several related fields. Applications of multivariate analysis in other fields related to community ecology will be considered next.

Niche and habitat

The physical and chemical environment of a site may be termed its *habitat,* and each species is found in a distinctive range of habitats (Whittaker, Levin, & Root 1973; Grime 1979). The physical and biotic variables to which species are related adaptively define the species's *niche.* Habitat and niche concepts are basic in theoretical community ecology for understanding community organization and in applied community ecology for predicting the implications of management alternatives. Multivariate analysis of the usual samples-by-species data matrix helps to clarify habitat relationships; analysis of a species-by-attributes matrix reveals niche characteristics. Bird niches have been studied by James (1971), Cody (1974), Holmes, Bonney, & Pacala (1979), Sabo & Whittaker (1979), and Sabo (1980). Many of the references cited earlier in this chapter concerning wildlife management are also pertinent. Niche and habitat were investigated for higher plants (Bratton 1975*a,* 1976; Johnson 1977*a,b;* Garten 1978; Olsvig, Cryan, & Whittaker 1979; Hicks 1980; Pentecost 1980; Beatty 1981; Weimarck 1981), algae (Allen 1971; Allen & Koonce 1973; Allen, Bartell, & Koonce 1977; Bruno & Lowe 1980), fungi (Bissett & Parkinson 1979*a,b,c*), and lichens (Larson 1980). Niches of bivalve molluscs of central Canada were characterized by Green (1971).

Taxonomy and genetics

Because most community data are recorded in terms of species abundances, taxonomy is a necessary prerequisite to collecting community ecology field data. Taxonomists have been using multivariate methods increasingly, especially classification, for grouping specimens into species units (or higher or lower units), arranging species into a hierarchy, and guiding the choice of characters for taxonomic evidence (Sneath & Sokal 1973). Because taxonomic characters are often quite varied, special consideration must be given to the coding and scaling of characters (Sneath & Sokal 1973:147–57) and to multivariate analysis (Smith & Hill 1975; Hill & Smith 1976). Representative studies in taxonomy and genetics include leaf variation in the elms (*Ulmus*) of England and northern France (Richens & Jeffers 1978), range-wide genetic variation in black spruce (*Picea mariana;* Morgenstern 1978), phenotypic variation in clover (*Trifolium repens;* Burdon 1980), and taxonomy of the Melanthioideae (Liliaceae; Ambrose 1980). Because of extensive hybridization, the taxonomy of oaks (*Quercus*) is particularly complex and, consequently, suitable for multivariate analysis (Jensen & Eshbaugh 1976*a,b;* Jensen 1977*a,b*). Phenotypic relationships among lizard populations (*Podarcis sicula* and *P. melisellensis,* Sauria: Lacertidae) from islands in the Adriatic Sea off Yugoslavia have been studied using agglomerative classification and nonmetric multidimensional scaling (Clover 1979). Hierarchical agglomerative classification of genetic variation in biochemical traits in walleye pollock (*Theragra chalcogramma*) in the Bering Sea and Gulf of Alaska suggest that these fish have two breeding stocks corresponding roughly with these two water bodies, with little variation within each stock (Grant & Utter 1980; also see Avise & Smith 1977 and Felley & Avise 1980). Multivariate methods and applications are reviewed for mammalian taxonomy by Neff & Marcus (1980) and for insect taxonomy by Moss & Hendrickson (1973). Multivariate analyses are also used in taxonomy of fossils (Niklas 1976; Niklas & Gensel 1976, 1977, 1978; Hanks & Fryxell 1979; Puckett & Finegan 1980). Studies using multivariate analysis to define and distinguish taxa include those by Hill & Smith (1976), Barkworth, McNeill, & Maze (1979), Clucas & Ladiges (1979), Parker, Bradfield, Maze, & Lin (1979), Baum (1980), Drake (1980), Reynolds & Crawford (1980), and Small (1980). For classification, taxonomists principally use agglomerative techniques, probably because of the influential text by Sneath & Sokal (1973). For ordination, most applications employ principal components analysis or nonmetric multidimensional scaling.

Agronomy

Soil classification is usually performed with the intention of its being usable for multiple purposes, and consequently, most researchers take a general-purpose viewpoint (Webster & Butler 1976). Mapping and other applications require large numbers of samples, making the rapidity of measurements important. The success of soil classification depends on a good correlation between the soil parameters used for classification and many other soil parameters, especially those of particular concern to agronomists, engineers, and other soil map users. Ordination reveals the correlation structure of soil parameters, aiding the choice of parameters for an economical and effective survey procedure. The resultant soil samples may be classified rapidly and objectively by numerical classification techniques. The resultant classification is useful for characterizing spatial scales of variation and, hence, for determining the required spacing of soil samples. Burrough & Webster (1976) use multivariate analysis of soils in east Malaysia to develop successful procedures for soil sampling, classification, and mapping. Cipra, Bidwell, & Rohlf (1970) find ordination and classification valuable for classifying soil types from several orders and, likewise, Muir, Hardie, Inkson, & Anderson (1970) for soils from several series. Teil & Cheminee (1975) use reciprocal averaging to study major and trace elements in Ethiopian soils. Also see Moore, Russell, & Ward (1972), Webster & Burrough (1972*a,b*), Jöreskog, Klovan, & Reyment (1976:165–9), Moore & Russell (1976*a,b*), Thompson, Lloyd, & Moore (1976), and Tothill (1976).

The influence of numerous environmental and soil factors upon 10 biochemical soil properties in 9 soil types in tussock grasslands in New Zealand has been studied by principal components analysis (Ross et al. 1975). Simple correlation coefficients and multiple regression are found not to be suitable analytical tools because of the large number of variables investigated; the problem was distinctively multivariate. Ordination arranges the soils on the basis of biochemical properties in a manner consistent with either the pedological classification or a dominant environmental factor such as moisture supply. Likewise, ordination of chemical and physical properties of soil samples from a beach in British Columbia, Canada, reveals podzolic pedogenic processes and sea spray input (Sondheim, Singleton, & Lavkulich 1981).

Relationships between soil factors and vegetation are analyzed using multivariate techniques by Grigal & Arneman (1970), Fourt, Donald, Jeffers, & Binns (1971), Gauch & Stone (1979), Huntley & Birks (1979*a,b*), and Dye & Walker (1980). Garten (1978) uses principal components analysis to summarize the mineral nutrient requirements of several plant species from South Carolina coastal plain habitats (similarly, see Puckett & Finegan 1980 concerning lichens and Niklas 1976 and Niklas & Gensel 1976 concerning fossils).

Meteorology

Principal components analysis is used extensively to ordinate meteorological data. Kidson (1975) analyzes monthly means worldwide of surface pressure, temperature, and rainfall. Pressure and temperature show coherent patterns on a hemispheric scale, whereas rainfall variations are regional. Ordination produces a comparatively small number of climatological indices, which effectively summarize gross features of circulation. Hardy & Walton (1978) ordinate regional wind data in order to define prototype days, to group similar weather stations, and to predict air pollution episodes. Essenwanger (1976:325–81) describes the eigenanalysis computations necessary for principal components analysis of meteorological data.

The assignment of rainfall stations into homogeneous groups is a necessary prelude to delimiting climate boundaries on a map and to reducing weather records in forecasting models with heavy computational demands (Dyer 1975). Likewise, homogeneous rainfall groups are determined in order to evaluate a meteorological network's representativeness and spatial distribution (Morin, Fortin, Sochanska, & Lardeau 1979).

Climatologically homogeneous areas are used to stratify a region for studies on the ecology of vegetation (Paterson, Goodchild, & Boyd 1978). Climate is a major factor determining the distributions of tree species in British Columbia (Newnham 1968). Evaluation and comparison of climate is important in crop tests (Abou-El-Fittouh, Rawlings, & Miller 1969; Russell & Moore 1970, 1976; Reid 1973, 1976; Mungomery, Shorter, & Byth 1974; Campbell & LaFever 1977; Shorter, Byth, & Mungomery 1977).

Behavior

Animal behaviorists ordinarily study numerous individuals and numerous behaviors; hence behavioral data are commonly multivariate. Multivariate analysis is helpful in summarizing and comprehending such complex data. Representative applications include territorial disputes of spiders (Riechert 1978) and fish (Huntingford 1976), shell selection by hermit crabs (Kuris & Brody 1976; also see Cuadras & Pereira 1977), postures in the Chilean flamingo (Davies 1978), vocalizations of birds (Sparling & Williams 1978) and squirrels (Koeppl, Hoffmann, & Nadler 1978), and chimpanzee social organization (Morgan, Simpson, Hanby, & Hall-Craggs 1976).

Paleoecology

Paleoecologists study fossils in order to reconstruct past communities and climates. Such results also help to explain the species content and distributions of present communities.

Fossil community composition is useful for detecting sea level changes and for dating samples (as in Figure 3.10 shown earlier; Cisne & Rabe 1978) and, hence, contributes to paleoecological methodology. Representative paleoecological studies employing multivariate analyses include those by Gordon & Birks (1972, 1974), Birks (1974, 1977, 1980), Birks, Webb, & Berti (1975), Blanc et al. (1976), Jöreskog, Klovan, & Reyment (1976:169–71), Cisne & Rabe (1978), Ritchie & Yarranton (1978*a,b*), Birks & Berglund (1979), Hayward & Buzas (1979), Birks & Birks (1980), Cisne, Chandlee, Rabe, & Cohen (1980), Cisne, Molenock, & Rabe (1980), Prentice (1980*a*), Rabe & Cisne (1980), and Reyment (1980).

Distant fields

For fundamental reasons having to do with the architecture of complexity and with the organization of human thinking, multivariate data of any sort have much in common, despite a certain degree of concerns unique to each specialty (Simon 1962; Benzécri 1969; Sneath & Sokal 1973:435–50; Blashfield & Aldenderfer 1978). For this reason, it is desirable for ecologists to be aware of the broader context of applications of multivariate techniques. In particular, ecology is enriched by importing multivariate techniques and presentation formats developed in other fields. Conversely, techniques originally developed in ecology may be useful in other disciplines. Applications of multivariate analysis in fields distant to community ecology will be reviewed here briefly.

Business is one of the major fields in which multivariate analysis is used, especially agglomerative classification and ordination by nonmetric multidimensional scaling (Green & Rao 1972). Ordination is used for market segmentation of compact automobiles (Figure 6.4 after Nevers 1972; also see Green & Rao 1972, 1977, and Myers 1977). An ordination of products (as in Figure 6.4) can be used to identify the closest competitors of a given product and to identify market segments that are crowded or open. Multivariate analysis helps to characterize the stimuli affecting purchasing decisions (Rao 1977; Green, Rao, & DeSarbo 1978), to compare products such as television programs (Rao 1975) and commercials (Green 1978), and to evaluate advertising alternatives (Wilkes & Uhr 1978). A review of books on multivariate methods for consumer research is given by Rao (1980). Because divisive classification techniques have been found superior to agglomerative techniques in community ecology, it may be suspected that recently

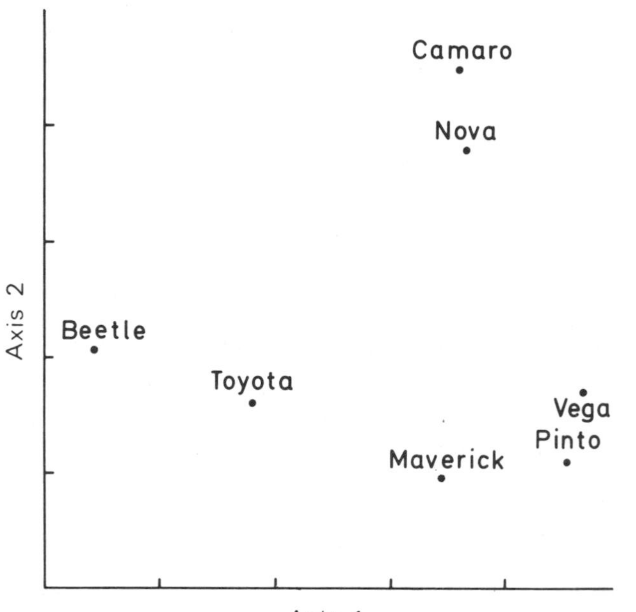

Figure 6.4. Ordination of consumer perception of seven brands of compact cars. Nonmetric multidimensional scaling is applied to pairwise similarity comparisons of 698 consumers. Similar cars ordinate near each other. Axis scales are arbitrary. Such analysis of product similarities is useful in market segmentation. (After Nevers 1972:136)

developed divisive techniques such as two-way indicator species analysis would give superior results in many business applications. Likewise, the very fact that nonmetric ordination is popular implies that heterogeneous, nonlinear data sets are common and, consequently, that exceptionally robust ordination techniques like detrended correspondence analysis are likely to give superior performance. These techniques (two-way indicator species analysis and detrended correspondence analysis), originally developed in community ecology together with composite clustering for nonhierarchical classification, have the further advantage of minimal computer demands.

Economics is related to business and is also studied by multivariate techniques. Barrett, Gerardi, & Hart (1974) use ordination to develop a theory of inflation for the United States economy, considering data on investments, credit, wholesale prices, the labor force and wages, unemployment, and government policies and taxes. They use these results to predict the consequences and successes of alternative economic policies.

Psychology, another major field for application of multivariate analysis, gave birth to principal components analysis (Hotelling 1933) and nonmetric multidimensional scaling (Kruskal 1964*a,b*). Multivariate analysis is used to study perception, including speech tones (Gandour 1978) and normal voice qualities (Singh & Murry 1978), visual processing of numbers and letters (Levine 1977; Hanley & Cox 1979), color perception (White, Lockhead, & Evans 1977; McCullough, Martinsen, & Moinpour 1978), taste differences among sweeteners (Schiffman, Reilly, & Clark 1979; also see McCullough, Martinsen, & Moinpour 1978), and brain activity as a function of selective attention level (Rösler 1978). Benzécri (1969) uses reciprocal averaging to ordinate children's fears. Gillen & Sherman (1980) apply nonmetric multidimensional scaling to data concerning the influence of physical attractiveness upon how people are perceived. Everett & Entrekin (1980) use principal components analysis to study work-related attitudes of academics. Linguistic applications may be found in the work of Benzécri (1969) and Arabie & Carroll (1980). The main multivariate techniques used in theoretical and applied psychology, however, are those invented in that field, principal components analysis and, especially, nonmetric multidimensional scaling.

Educational research employing multivariate analyses includes representative topics such as assessment of teachers' attitudes toward small-group teaching (Darom, Sharan, & Hertz-Lazarowitz 1978), cognitive structures of preschool children (Fenker & Tees 1976), learning success of 16-year-old French high school students (Benzécri 1969), comprehension of prose (LaPorte & Voss 1979), appropriate and inappropriate classroom behavior (Sanson-Fisher & Mulligan 1977), and the relationship between declared major of college freshmen and measures of occupational interests (Knapp, Knapp, & Michael 1979). Nonmetric multidimensional scaling is frequently used.

The social and political sciences use multivariate analyses to summarize data and to help elucidate relationships and causal factors. Agglomerative classification, principal components analysis, factor analysis, and nonmetric multidimensional scaling are employed most commonly. Representative applications of multivariate analysis in the social sciences concern prestige among Costa Rican peasants (Seligson 1977), caste groups in Bengal (Jardine & Sibson 1971:250–8), gene differentiation among Dhangar caste groups in India (Malhotra, Chakraborty, & Chakravarti 1978), the American occupational structure (Vanneman 1977), perceptions of typical social episodes in homemakers and students (Forgas 1976), perceived risk and attractiveness in gambling (Nygren 1977), power strategies (Falbo 1977), juvenile delinquency (Braithwaite & Law 1978), and causes of negative behavior in a marriage (Passer, Kelley, & Michela 1978). Ordination is used to analyze a swarm of economic, political, cultural, and ecological variables in Mexico (Young, Freebairn, & Snipper 1979), Caribbean Islands (Young 1976), Tunisia (Young, Bertoli, & Bertoli 1981), India (Adams & Bumb 1973), and Malaysia (Schlegel 1981). Stopher & Meyburg (1979) survey methods and applications of multivariate analysis in the social sciences. Nishisato (1980) applies reciprocal averaging and related techniques. In the political sciences, Clausen & Horn (1977) use multivariate analysis to assess the policy positions of members of legislative bodies based on voting records regarding economic policy, social welfare, civil rights, and foreign and defense policy. Weisberg (1980) uses factor analysis to conceptualize the party identification and voting decisions of citizens.

Medical applications of multivariate analysis are varied and numerous. White & Lewinson (1977) classify patients in order to identify clusters of patients with similar background, pretreatment symptom ratings, and treatment responses. They also classify pharmaceuticals to elucidate dependencies between chemical structure and biological activity as a means of expediting assessment in the pharmaceutical industry of analog series of chemical compounds. Manion & Hassanein (1980) use classification to evaluate hospital usage and costs and find ecological factors of home and work environments particularly important. Representative applications of multivariate analysis in medicine include those by Benzécri (1969), Jones et al. (1970), Stitt, Frane, & Frane (1977), Zajicek, Maayan, & Rosenmann (1977), Mueller & Reid (1979), and Rohlf, Rodman, & Flehinger (1980). Sherman (1977) presents a nonmetric multidimensional scaling ordination of about 100 medical schools in the United States.

Locating oil and gas deposits draws upon many kinds of geological data, including the age of rock strata and the ocean depth at the time of deposition (Poag 1977). Fossil community types have been used for some time to indicate age and environmental conditions (Ziegler 1965; Ziegler, Cocks, & Bambach 1968; Watts 1980). Ordination of fossil communities has been found to be more accurate, however, by an order of magnitude (providing time correlation of roughly 100 000 year's accuracy, using methods illustrated earlier in Figure 3.10; Cisne & Rabe 1978; Rabe & Cisne 1980). Additional important advantages of multivariate analyses for biostratigraphy are objectivity, repeatability, and facility with large data sets (Hazel 1977; Reyment 1980).

Ordination of plant communities provides information on soil substrates, which aids exploration for minerals (Yates, Brooks, & Boswell 1974). Likewise, ordination of geological data may indicate prospective sites for ores (Jöreskog, Klovan, & Reyment 1976:2–7). Certain plant species or communities are indicative of soils rich in copper (Wild 1968; Shewry, Woolhouse, & Thompson 1979), nickel (Wild 1970), and arsenic (Wild 1974).

Other fields of application of multivariate analysis include library science and document retrieval (Salton 1975; Salton & Wong 1978). Classification and ordination of archeological data serve to define artifact types and organize collections, to identify particular artifact groups in certain site groups, and to sequence and date artifact collections (Hodson 1970; Kendall 1971, further analyzed by Hill 1974; Dumond 1974; Matson & True 1974; Drennan 1976). Applications in military science include classifications of personnel, jobs, and training programs (Ward 1963; Christal 1974; Phalen 1975). In the arts, examples of multivariate analysis include musicology (Filip 1970), analysis of Greek poetry (Wishart & Leach 1970), and comparison of paintings and analysis of characters and vocabulary in a play (Benzécri 1969); also see Sneath & Sokal (1973:449–50).

Discussion

Applications-oriented researchers frequently study numerous attributes of numerous individuals and, consequently, generate multivariate data. The preceding references demonstrate the ubiquity of multivariate data. Fields do differ, however, in the commonness of multivariate data because they differ in the number of variables ordinarily considered simultaneously and the feasibility of direct experimental investigation (yielding data usually studied by statistical tests and analysis of variance, rather than multivariate analysis). Consequently, there are substantial causes for some of the variation among fields in the commonness of multivariate analysis.

In addition to these substantial causes, fields also differ in the frequency and kinds of multivariate analyses for incidental historical reasons. Blashfield & Aldenderfer (1978) document the tendency of researchers within a given field to limit their multivariate analyses to those traditional to their field. A consequence of this is that in fields in which traditional analyses are not able to handle nonlinearity problems well, multivariate analyses have not been as successful as they could be, and consequently, there is limited impetus to apply multivariate analyses. Recently extensive development and evaluation of multivariate techniques have been accomplished, however, leading to more robust techniques and to more straightforward recommendations for their application.

Several principal advantages of multivariate analysis in applied research, given the commonness of multivariate data, may be noted from the preceding examples. (1) Especially when the data base is large, the ability of multivariate analysis to summarize data and reveal relationships is crucial for comprehending the data. (2) Multivariate analysis is cost-effective because computing costs continue to decrease with progress in computer technology, and the fairly routine nature of most analyses places modest demands on the time of experts. (3) Decision making based on multivariate data is basically a process of transforming the numerous data variables into a summarizing parameter or classification assignment to be used for making decisions. Applied problems are frequently characterized by a tension: many variables must be considered, but the ultimate goal is a single score or a cluster assignment. For example, multivariate data concerning the abundances of numerous plant species may be needed to assess site conditions, but the desired management decision may be merely a binary decision to fertilize or not. Consequently, in this example, for each sample site, the input set of possibly dozens of numbers must be transformed somehow into a single binary variable for output. Multivariate analysis is a natural tool in the data reduction required in many procedures for decision making. (4) Multivariate analysis is relatively objective and repeatable.

7

Conclusions

In this book multivariate analysis in community ecology has been reviewed and the theory, methods, and applications of direct gradient analysis, ordination, and classification have been presented. In conclusion, a few comments will be offered concerning inherent limitations, model accuracy, and prospectus.

Several inherent limitations should be acknowledged in community ecology. They may be mitigated in part by strenuous efforts but cannot be eliminated. These limitations must be recognized or the likely result will be unrealistic and ineffective research. (1) Community ecology field data are noisy. (2) Multivariate analysis summarizes community data and functions because numerous species abundances are controlled largely by a relatively small number of environmental and historical factors. The word *largely* is required, however, because each species is unique and individual (Gleason 1926; Whittaker, Levin, & Root 1973; Macfadyen 1975; McIntosh 1975, 1980; Simberloff 1980; Strong 1980), and likewise, each site and each environmental factor are to some degree unique. Consequently, any general community description, even the best possible such description, can capture only a fraction of the total variation. In contrast, the objects of other sciences may be more uniform, such as in chemistry, where trillions of atoms of carbon or iron may be treated as identical for many purposes. (3) Community ecology is partially subjective. The origins of subjectivity include the choices of research purposes and perspectives, the field data collection procedures, the multivariate analyses, and the methods of interpreting the results. Subjectivity exists because the inherent complexity of ecological communities in relation to the investigative power of ecologists is practically infinite; the resulting necessarily partial study requires subjective choices by the investigator of what will be emphasized or done and what omitted. On the other hand, community ecology is not completely subjective because this would imply that available data have no bearing on acceptable viewpoints, and such an implication is patently absurd. A balanced assessment of subjectivity strengthens community ecology by encouraging realistic examination of the accuracy, confidence, and limitations of results. (4) The goals of community ecology are partially indefinite and frequently progress by successive refinement. The general goal is to describe and understand communities, but little insight is required to recognize that this statement is far removed from a workable experimental design or data analysis algorithm. What ecologists should observe and how the data should be analyzed to best reveal inherent structure are questions resolvable only subsequent to some data collection and analysis. Later research, consequently, may be pursued somewhat differently, incorporating new or more precisely defined goals.

Matching the underlying assumptions of multivariate analyses with known community structure is the central requirement for effective data analysis. Numerous community studies, especially by direct gradient analysis, support two fundamental generalizations: (1) the Gaussian model of community structure summarizes much of the data and (2) the data are noisy and depart partially from the idealization of the Gaussian model. Ordination algorithms have been implemented to fit community data directly to the Gaussian model (for example, Gauch, Chase, & Whittaker 1974). These algorithms can only handle one dimension of community variation effectively, however, and even then require data with exceptionally low noise. For special research purposes, such algorithms are valuable. It must be suspected, however, that they are ineffective for general use because their design emphasizes the idealizations of the Gaussian model too much. On the other hand, ordination and classification algorithms have been developed that are too simple or too different from the Gaussian model, and these also prove to be ineffective in practice. Experience indicates that there are penalties for incorporating into the design of a multivariate analysis too much or too little of the Gaussian model (see Dale 1976*b*). The same point may be stated in positive terms by saying that robust multivariate techniques must both incorporate the Gaussian model in general and handle noisy data with partial departure from idealized structure. Detrended correspondence analysis, for example, assumes that species abundances generally have a modal optimum along underlying gradients and decline to either side but does not make more specific demands and does not emphasize those details in the data that are noise. This intermediate stance, embracing the Gaussian model but not too tightly, is the fundamental reason for the superior performance of this ordination technique. In general, the best policy in developing multivariate analysis algorithms is to have an algorithm reflect known data structure faithfully, not to idealize known structure beyond the level conveyed by individual, actual data sets such as are to be submitted for analysis.

Future developments in multivariate analysis of community data are difficult to predict, but a few speculations may be offered (also see Maarel, Orlóci, & Pignatti 1976, Noy-Meir & Whittaker 1977, and Maarel 1979*a*). (1) Perhaps the greatest need is to develop statistical tests of the significance of ordination and classification results. This is not easy because relevant statistical tools are poorly developed and, more importantly, because empirical community data cannot fulfill exactly the ever-present list of mathematical requirements for a given statistical test. Also, much can be said in favor of developing methods of multivariate analysis not incorporating statistical tests, at the same time recognizing the utility of statistical tests applied to results from multivariate analyses (Williams & Gillard 1971; Williams & Edye 1974; Edye 1976*b*; Williams 1976:130–6). The motivations are strong for improving the statistical basis of multivariate analysis, however, and doubtless progress is possible. Just how much progress is possible, especially without new and excessive sampling demands, is not yet clear. (2) It may be anticipated that larger, more integrated ecological studies will become more common. This will require new and better ways to integrate community analyses with additional data bases. (3) A particularly subjective and complex situation in community studies is the final step of environmental interpretation of results. Perhaps, better multivariate methods can be developed to aid in this process of comparing patterns in community and environmental data. (4) Continuing improvements in computer technology may offer new possibilities. Large computers are already widely available to ecologists at low cost and are quite adequate for implementing many multivariate algorithms. The future development of large computers, consequently, may have little practical impact upon community ecologists. Small computers, on the other hand, are becoming increasingly portable and inexpensive. In the future, small computers may be the instrument of choice for entering field data, replacing conventional note pads, and eliminating tedious keypunching or written notes. If sufficiently powerful, it would be possible to obtain an immediate, on-site ordination or classification of each new community sample. Alternatively, occasional telephone communications between a small field computer and a large office computer could serve to add new data to the office data base and to print at the field site the results of analyses performed by the office computer.

Developments in the field of multivariate analysis of community data have been rapid. Quantitative field sampling began around 1900, early direct gradient analysis in 1930, and computer analyses around 1960. Most of the preferred multivariate analysis techniques were developed in just the past few years. Such rapid developments, together with the uncertainties of future developments, make it difficult to assess the present state of the field. An assessment is offered here, although only the test of time will tell for sure. Present algorithms appear to warrant three claims: they suffice in a majority of research applications; they produce general descriptions of communities that are about as accurate as possible, given inherent limitations in the data; and they have modest computer requirements. The use of multivariate analysis in applied ecological research is expected to continue increasing because the methodology has reached a reasonable level of robustness, synthesis, and standardization.

Appendix

Available computer programs

Multivariate analyses routinely require millions of mathematical steps and hence necessitate solution by computers. The algorithms are often complex, however, requiring mathematical and computing expertise for programming them and sometimes many weeks of programming labor. Consequently, the development of computer programs for multivariate analysis has occurred mainly at a relatively small number of laboratories, and the programs are distributed to numerous other laboratories. This Appendix will list several sources of computer programs for multivariate analysis of community data.

The Cornell Ecology Program, series, edited by the author, is perhaps the most widely used package, having been used in hundreds of laboratories in over 30 countries. Information on these programs is available upon request from the author. All programs are written in FORTRAN. Several of the commonly used programs are as follows: ORDIFLEX, a flexible ordination program performing weighted averages, polar ordination, principal components analysis, and reciprocal averaging; DECORANA, providing reciprocal averaging and detrended correspondence analysis ordinations; TWINSPAN, producing hierarchical classifications by two-way indicator species analysis; COMPCLUS, performing nonhierarchical classifications by composite clustering; DATAEDIT, offering a variety of data–matrix-editing options, along the lines discussed in Chapter 6; and CONDENSE, a utility program for reading data matrices keypunched in a wide variety of convenient formats and copying them into a single standardized format efficient for computer processing. There are additional programs, including one for Gaussian ordination and two for simulating community data. These programs, other than ORDIFLEX, require minimal computer memory and time and hence are suitable for large data sets.

Orlóci (1978*a*:348–435) provides programs in BASIC for several ordination and classification algorithms. Wildi (1980) and Wildi & Orlóci (1980) present a larger, integrated, FORTRAN program package for data management, ordination, and classification. (Also see Orlóci, Feoli, & Fewster 1977.) Milne (1976) describes the extensive program library at Canberra.

Volland & Connelly (1978) are foresters whose FORTRAN program package helps with organizing data into a suitable format for computer processing and performs several ordinations and classifications.

Classification program packages, mainly featuring agglomerative polythetic techniques, include CLUSTAN (Wishart 1978), NT-SYS (Rohlf, Kishpaugh, & Kirk 1972), SYN-TAX (Podani 1980), and CLASP (Ross, Lauckner, &

Hawkins 1976). The program PHYTO is a table arrangement with computer requirements rising only linearly with the amount of data (Moore & O'Sullivan 1978). Nonmetric multidimensional scaling packages are available from the Computer Information Service Group of Bell Laboratories (Murray Hill, New Jersey) and from the L. L. Thurstone Psychometric Laboratory (University of North Carolina, Chapel Hill, North Carolina). A general review of statistical computer programs including many multivariate analyses is provided by Francis (1979).

References

Aart, P. J. M. van der (1973). Distribution analysis of wolf spiders (Araneae, Lycosidae) in a dune area by means of principal components analysis. *Netherlands Journal of Zoology,* 23, 266–329.

Abou-El-Fittouh, H. A., Rawlings, J. O., & Miller, P. A. (1969). Classification of environments to control genotype by environment interactions with an application to cotton. *Crop Science,* 9, 135–40.

Adam, F. (1977). Données préliminaires sur l'habitat et la stratification des rongeurs en forêt de Basse Côte-d'Ivoire. *Mammalia,* 41, 283–90.

Adams, J. & Bumb, B. (1973). The economic, political and social dimensions of an Indian state: A factor analysis of district data for Rajasthan. *Journal of Asian Studies,* 33, 5–23.

Akinola, J. O. & Whiteman, P. C. (1972). A numerical classification of *Cajanus cajan* (L.) Millsp. accessions based on morphological and agronomic attributes. *Australian Journal of Agricultural Research,* 23, 995–1005.

Aleksandrova, V. D. (1978). Russian approaches to classification. In *Classification of Plant Communities,* ed. R. H. Whittaker, pp. 167–200. The Hague: Junk.

Ali, S. A., Gross, M. G., & Kishpaugh, J. R. L. (1975). Cluster analysis of marine sediments and waste deposits in New York Bight. *Environmental Geology,* 1, 143–8.

Allen, T. F. H. (1971). Multivariate approaches to the ecology of algae on terrestrial rock surfaces in North Wales. *Journal of Ecology,* 59, 803–26.

Allen, T. F. H., Bartell, S. M., & Koonce, J. F. (1977). Multiple stable configurations in ordination of phytoplankton community change rates. *Ecology,* 58, 1076–84.

Allen, T. F. H. & Koonce, J. F. (1973). Multivariate approaches to algal stratagems and tactics in systems analysis of phytoplankton. *Ecology,* 54, 1234–46.

Allen, T. F. H. & Skagen, S. (1973). Multivariate geometry as an approach to algal community analysis. *British Phycological Journal,* 8, 267–87.

Ambrose, J. D. (1980). A re-evaluation of the Melanthioideae (Liliaceae) using numerical analyses. In *Petaloid Monocotyledons,* eds. C. D. Brickell, D. F. Cutler & M. Gregory, pp. 65–82. Dorchester: Dorset Press.

Anderson, A. J. B. (1971). Ordination methods in ecology. *Journal of Ecology,* 59, 713–26.

Anderson, S. A. (1979). Habitat structure, succession and bird communities. In *Proceedings of the Workshop Management of Northcentral and North-*

eastern Forests for Nongame Birds, ed. R. M. DeGraaf, pp. 9–21. Orono, Me.: Northeastern Forest Experiment Station.

Andrewartha, H. G. (1961). *Introduction to the Study of Animal Populations.* London: Methuen.

Arabie, P. & Carroll, J. D. (1980). MAPCLUS: A mathematical programming approach to fitting the ADCLUS model. *Psychometrika,* 45, 211–35.

Archibold, O. W. (1978). Vegetation recovery following pollution control at Trail, British Columbia. *Canadian Journal of Botany,* 56, 1625–37.

Arfi, R., Champalbert, G., & Patriti, G. (1981). Système planctonique et pollution urbaine: Un aspect des populations zooplanctoniques. *Marine Biology,* 61, 133–41.

Aris, R. & Penn, M. (1980). The mere notion of a model. *Mathematical Modelling,* 1, 1–12.

Armitage, P. (1971). *Statistical Methods in Medical Research.* New York: Wiley.

Austin, M. P. (1968). An ordination study of a chalk grassland community. *Journal of Ecology,* 56, 739–57.

– (1972). Models and analysis of descriptive vegetation data. *Symposium of the British Ecological Society,* 12, 61–86.

– (1976*a*). On non-linear species response models in ordination. *Vegetatio,* 33, 33–41.

– (1976*b*). Performance of four ordination techniques assuming three different non-linear species response models. *Vegetatio,* 33, 43–9.

– (1977). Use of ordination and other multivariate descriptive methods to study succession. *Vegetatio,* 35, 165–75.

– (1978). Vegetation. In *Land Use on the South Coast of New South Wales,* ed. R. H. Gunn, vol. 2, pp. 44–67. Melbourne: Commonwealth Scientific and Industrial Research Organization.

– (1979). Current approaches to the non-linearity problem in vegetation analysis. In *Contemporary Quantitative Ecology and Related Ecometrics,* eds. G. P. Patil & M. Rosenzweig, pp. 197–210. Burtonsville, Md.: International Co-operative.

– (1980*a*). An exploratory analysis of grassland dynamics: An example of a lawn succession. *Vegetatio,* 43, 87–94.

– (1980*b*). Searching for a model for use in vegetation analysis. *Vegetatio,* 42, 11–21.

Austin, M. P. & Austin, B. O. (1980). Behavior of experimental plant communities along a nutrient gradient. *Journal of Ecology,* 68, 891–918.

Austin, M. P. & Greig-Smith, P. (1968). The application of quantitative methods to vegetation survey. II. Some methodological problems of data from rain forest. *Journal of Ecology,* 56, 827–44.

Austin, M. P. & Noy-Meir, I. (1971). The problem of non-linearity in ordination: Experiments with two-gradient models. *Journal of Ecology,* 59, 763–74.

Avedon, E. M. (editor) (1977). *Analysis Methods and Techniques for Recreation Research and Leisure Studies.* Waterloo, Ont.: University of Waterloo.

Avise, J. C. & Smith, M. H. (1977). Gene frequency comparisons between sunfish (Centrarchidae) populations at various stages of evolutionary divergence. *Systematic Zoology,* 26, 319–35.

Bailey, R. G., Pfister, R. D., & Henderson, J. A. (1978). Nature of land and resource classification – A review. *Journal of Forestry,* 76, 650–5.

Bakker, P. A. (1979). Vegetation science and nature conservation. In *The Study of Vegetation,* ed. M. J. A. Werger, pp. 247–88. The Hague: Junk.

Bakuzis, E. V. & Hansen, H. L. (1965). *Balsam Fir.* Minneapolis: University of Minnesota Press.

Balloch, D., Davies, C. E., & Jones, F. H. (1976). Biological assessment of water quality in three British rivers: The North Esk (Scotland), the Ivel (England) and the Taf (Wales). *Water Pollution Control,* 75, 92–114.

Bannister, P. (1966). The use of subjective estimates of cover-abundance as the basis for ordination. *Journal of Ecology,* 54, 665–74.

– (1968). An evaluation of some procedures used in simple ordinations. *Journal of Ecology,* 56, 27–34.

Barbour, M. G., Burk, J. H., & Pitts, W. D. (1980). *Terrestrial Plant Ecology.* Menlo Park, Calif.: Benjamin/Cummings.

Barkworth, M. E., McNeill, J., & Maze, J. (1979). A taxonomic study of *Stipa nelsonii* (Poaceae) with a key distinguishing it from related taxa in western North America. *Canadian Journal of Botany,* 57, 2539–53.

Barrett, H. C. & Rhodes, A. M. (1976). A numerical taxonomic study of affinity relationships in cultivated *Citrus* and its close relatives. *Systematic Botany,* 1, 105–36.

Barrett, N. S., Gerardi, G., & Hart, T. P. (1974). A factor analysis of quarterly price and wage behavior for U.S. manufacturing. *Quarterly Journal of Economics,* 88, 385–408.

Baum, B. R. (1978). Generic relationships in Triticeae based on computations of Jardine and Sibson B_k clusters. *Canadian Journal of Botany,* 56, 2948–54.

– (1980). Multivariate morphometric relationships between *Hordeum jubatum* and *Hordeum brachyantherum* in Canada and Alaska. *Canadian Journal of Botany,* 58, 604–23.

Baum, B. R. & Brach, E. J. (1975). Identification of oat cultivars by means of flourescence spectrography – A pilot study aimed at automatic identification of cultivars. *Canadian Journal of Botany,* 53, 305–9.

Baum, B. R. & Lefkovitch, L. P. (1972). A model for cultivar classification and identification with reference to oats (*Avena*). I. Establishment of the groupings by taximetric methods. *Canadian Journal of Botany,* 50, 121–30.

– (1973). A numerical taxonomic study of phylogenetic and phenetic relationships in some cultivated oats, using known pedigrees. *Systematic Zoology,* 22, 118–31.

Baum, B. R., Petruk, W., & Bailey, L. G. (1980). Assessment of the value of endospermic starch granules for the taxonomy of barley (*Hordeum*) species and cultivars with special emphasis on their identification, using the technique of image analysis. *Zeitschrift für Pflanzenzüchtung,* 85, 212–24.

Baum, B. R. & Thompson, B. K. (1976). Classification of Canadian oat cultivars by quantifying the size–shape of their "seeds": A step towards automatic identification. *Canadian Journal of Botany,* 54, 1472–80.

Bauzon, D., Ponge, J. F., & Dommergues, Y. (1974). Variations saisonniéres des caractéristiques chimiques et biologiques des sols forestiers interprétées par l'analyse factorielle des correspondances. *Revue d'écologie et de biologie du sol,* 11, 283–301.

Beals, E. W. (1960). Forest bird communities in the Apostle Islands of Wisconsin. *Wilson Bulletin,* 72, 156–81.

– (1973). Ordination: Mathematical elegance and ecological naïveté. *Journal of Ecology,* 61, 23–35.

Beaman, J. & Lindsay, S. (1976). Practical applications of cluster analysis. *Canadian Outdoor Recreation Demand Study, Parks Canada,* 2, 399–409.

Beatty, S. W. (1981). The Role of Treefalls and Forest Microtopography in Pattern Formation in Understory Communities. Ph.D. thesis, Cornell University, Ithaca.

Becking, R. W. (1957). The Zürich–Montpellier school of phytosociology. *Botanical Review,* 23, 411–88.

Benninghoff, W. S. & Southwood, W. C. (1964). Ordering of tabular arrays of phytosociological data by digital computer. *Tenth International Botanical Congress Abstracts,* pp. 331–2. Edinburgh: T. & A. Constable.

Benzécri, J.-P. (1969). Statistical analysis as a tool to make patterns emerge from data. In *Methodologies of Pattern Recognition,* ed. S. Watanabe, pp. 35–60. New York: Academic Press.

– (1973*a*). Algorithmes rapides d'agrégation. In *L'Analyse des données:* I. *La taxonomie,* eds. J.-P. Benzécri et al., pp. 288–319. Paris: Dunod.

– (1973*b*). *L'Analyse des données:* II. L'analyse des correspondances. Paris: Dunod.

Betters, D. R. & Rubingh, J. L. (1978). Suitability analysis and wildland classification: An approach. *Journal of Environmental Management,* 7, 59–72.

Bhatt, G. M. (1976). An application of multivariate analysis to selection for quality characters in wheat. *Australian Journal of Agricultural Research,* 27, 11–18.

Birks, H. J. B. (1973). *Past and Present Vegetation of the Isle of Skye: A Palaeoecological Study.* Cambridge University Press.

– (1974). Numerical zonations of Flandrian pollen data. *New Phytologist,* 73, 351–8.

– (1977). Modern pollen rain and vegetation of the St. Elias Mountains, Yukon Territory. *Canadian Journal of Botany,* 55, 2367–82.

– (1980). Modern pollen assemblages and vegetational history of the moraines of the Klutlan Glacier and its surroundings, Yukon Territory, Canada. *Quaternary Research,* 14, 101–29.

Birks, H. J. B. & Berglund, B. E. (1979). Holocene pollen stratigraphy of southern Sweden: A reappraisal using numerical methods. *Boreas,* 8, 257–79.

Birks, H. J. B. & Birks, H. H. (1980). *Quaternary Palaeoecology.* Baltimore: University Park Press.

Birks, H. J. B., Webb, T., & Berti, A. A. (1975). Numerical analysis of pollen samples from central Canada: A comparison of methods. *Review of Paleobotany and Palynology,* 20, 133–69.

Bissett, J. & Parkinson, D. (1979*a*). Functional relationships between soil fungi and environment in alpine tundra. *Canadian Journal of Botany,* 51, 1642–59.

– (1979*b*). Fungal community structure in some alpine soils. *Canadian Journal of Botany,* 57, 1630–41.

– (1979*c*). The distribution of fungi in some alpine soils. *Canadian Journal of Botany,* 57, 1609–29.

Black, R. A. & Bliss, L. C. (1978). Recovery sequence of *Picea mariana–Vaccinium uliginosum* forests after burning near Inuvik, Northwest Territories, Canada. *Canadian Journal of Botany,* 56, 2020–30.

Blanc, F., Blanc-Vernet, L., Laurec, A., Campion, J. Le, & Pastouret, L. (1976). Application paléoécologique de la méthode d'analyse factorielle en composantes principales: Interprétation des microfaunes de Foraminifères planctoniques quaternaires en Méditerranée. III. Les séquences paléoclimatiques. Conclusions générales. *Palaeogeography, Palaeoclimatology, Palaeoecology,* 20, 277–96.

Blashfield, R. K. & Aldenderfer, M. S. (1978). The literature on cluster analysis. *Multivariate Behavioral Research,* 13, 271–95.

Boesch, D. F. (1973). Classification and community structure of macrobenthos in the Hampton Roads area, Virginia. *Marine Biology,* 21, 226–44.

Bond, R. R. (1957). Ecological distribution of breeding birds in the upland forests of southern Wisconsin. *Ecological Monographs,* 27, 351–84.

Bonin, G. & Roux, M. (1978). Utilisation de l'analyse factorielle des correspondances dans l'étude phyto-écologique de quelques pelouses de l'Apennin lucano-calabrais. *Oecologia plantarum,* 13, 121–38.

Borowiec, S., Kutyna, I., & Skrzyczyńska, J. (1977). Occurrence of cropfield weed associations against environmental conditions in West Pomerania. *Ekologia Polska,* 25, 257–73.

Bouche, M.-B. & Beugnot, M. (1978). Action du chlorate de sodium sur le niveau des populations et l'activité biodégradatrice des lombriciens. *Phytiatrie–Phytopharmacie,* 27, 147–62.

Braithwaite, J. B. & Law, H. G. (1978). The structure of self-reported delinquency. *Applied Psychological Measurement,* 2, 221–38.

Bratton, S. P. (1975*a*). A comparison of the beta diversity functions of the overstory and herbaceous understory of a deciduous forest. *Bulletin of the Torrey Botanical Club,* 102, 55–60.

– (1975*b*). The effect of the European wild boar, *Sus scrofa,* on gray beech forest in the Great Smoky Mountains. *Ecology,* 56, 1356–66.

– (1976). Resource division in an understory herb community: Responses to temporal and microtopographic gradients. *American Naturalist,* 110, 679–93.

Bratton, S. P., Harmon, M. E., & White, P. S. (1981). Patterns of European wild boar rooting in the western Great Smoky Mountains. *Castanea* (in press).

Bratton, S. P., Hickler, M. G., & Graves, J. H. (1979). Trail erosion patterns in Great Smoky Mountains National Park. *Environmental Management,* 3, 431–45.

Braun-Blanquet, J. (1921). Prinzipien einer Systematik der Pflanzengesellschaften auf floristischer Grundlage. *St. Gallische Naturwissenschaftliche Gesellschaft,* 57, 305–51.

– (1928). *Pflanzensoziologie, Grundzüge der Vegetationskunde.* Berlin: Springer-Verlag.

– (1932). *Plant Sociology: The Study of Plant Communities,* trans. & ed. C. D. Fuller & H. S. Conrad. London: Hafner.

Bray, J. R. & Curtis, J. T. (1957). An ordination of the upland forest communities of southern Wisconsin. *Ecological Monographs,* 27, 325–49.

Briane, J.-P., Lazare, J.-J., & Salanon, R. (1977). *Le Traitement des trés grands ensembles de données en analyse factorielle des correspondances – Proposition d'une methodologie appliquée a la phytosociologie.* Nice Cedex, France: Parc Valrose.

Bridgewater, P. B. (1978). Coastal vegetation of East Gippsland, Victoria: A comparison of physiognomic and floristic classifications. *Phytocoenologia,* 4, 471–90.

Brock, T. D. (1966). *Principles of Microbial Ecology.* Englewood Cliffs, N. J.: Prentice-Hall.

Broich, S. L. & Palmer, R. G. (1980). A cluster analysis of wild and domesticated soybean phenotypes. *Euphytica,* 29, 23–32.

Broschat, T. K. (1979). Principal component analysis in horticultural research. *HortScience,* 14, 114–7.

Brown, A. L. (1978). *Ecology of Soil Organisms.* London: Heinemann.

Brown, D. (1954). *Methods of Surveying and Measuring Vegetation.* Bucks, England: Commonwealth Agricultural.

Brown, M. A., Frahm, R. R., Morrison, R. D., & McNew, R. W. (1979). Multivariate evaluation of phenotypic relationships between early performance and subsequent productivity in Hereford and Angus cows. *Journal of Animal Science,* 49, 378–90.

Bruno, M. G. & Lowe, R. L. (1980). Differences in the distribution of some bog diatoms: A cluster analysis. *American Midland Naturalist,* 104, 70–9.

Brush, G. S., Lenk, C., & Smith, J. (1980). The natural forests of Maryland: An explanation of the vegetation map of Maryland. *Ecological Monographs,* 50, 77–92.

Bruynooghe, M. (1978). Classification ascendante hiérarchique des grands ensembles de données: Un algorithme rapide fondé sur la construction des voisinages réducibles. *Cahiers analyse données,* 3, 7–33.

Bunce, R. G. H., Morrell, S. K., & Stel, H. E. (1975). An application of multivariate analysis to regional survey. *Journal of Environmental Management,* 3, 151–65.

Bunce, R. G. H. & Shaw, M. W. (1973). A standardized procedure for ecological survey. *Journal of Environmental Management,* 1, 239–58.

Bunce, R. G. H. & Smith, R. S. (1978). *An Ecological Survey of Cumbria.* Kendall, England: National Park Officer.

Burdon, J. J. (1980). Intra-specific diversity in a natural population of *Trifolium repens. Journal of Ecology,* 68, 717–35.

Burrough, P. A. & Webster, R. (1976). Improving a reconnaissance soil classification by multivariate methods. *Journal of Soil Science,* 27, 554–71.

Burt, R. L. (1976*a*). Description of introduced *Stylosanthes* material using morphological and performance attributes. In *Pattern Analysis in Agricultural Science,* ed. W. T. Williams, pp. 139–50. New York: Elsevier.

– (1976*b*). Sward performance of *Stylosanthes* accessions over a range of climates. In *Pattern Analysis in Agricultural Science,* ed. W. T. Williams, pp. 151–61. New York: Elsevier.

– (1976*c*). The climatic background of *Stylosanthes* accessions. In *Pattern Analysis in Agricultural Science,* ed. W. T. Williams, pp. 162–9. New York: Elsevier.

Burt, R. L., Edye, L. A., Williams, W. T., Gillard, P., Grof, B., Page, M., Shaw, N. H., Williams, R. J., & Wilson, G. P. M. (1974). Small-sward testing of *Stylosanthes* in Northern Australia: Preliminary considerations. *Australian Journal of Agricultural Research,* 25, 559–75.

Burt, R. L., Edye, L. A., Williams, W. T., Grof, B., & Nicholson, C. H. L. (1971). Numerical analysis of variation patterns in the genus *Stylosanthes* as an aid to plant introduction and assessment. *Australian Journal of Agricultural Research,* 22, 737–57.

Burt, R. L., Isbell, R. F., & Williams, W. T. (1979). Strategy of evaluation of a collection of tropical herbaceous legumes from Brazil and Venezuela. I. Ecological evaluation at the point of collection. *Agro-Ecosystems,* 5, 99–117.

Burt, R. L., Pengelly, B. C., & Williams, W. T. (1980). Network analysis of genetic resources data. III. The elucidation of plant/soil/climate relationships. *Agro-Ecosystems,* 6, 119–27.

Burt, R. L. & Reid, R. (1976). Exploration for, and utilization of, collections of tropical pasture legumes. III. The distribution of various *Stylosanthes* species with respect to climate and phytogeographic regions. *Agro-Ecosystems,* 2, 319–27.

Burt, R. L., Reid, R., & Williams, W. T. (1976). Exploration for, and utilization of, collections of tropical pasture legumes. I. The relationship between agronomic performance and climate of origin of introduced *Stylosanthes* spp. *Agro-Ecosystems,* 2, 293–307.

Burt, R. L. & Williams, W. T. (1979*a*). Strategy of evaluation of a collection of tropical herbaceous legumes from Brazil and Venezuela. II. Evaluation in the quarantine glasshouse. *Agro-Ecosystems,* 5, 119–34.

Burt, R. L. & Williams, W. T. (1979*b*). Strategy of evaluation of a collection of tropical herbaceous legumes from Brazil and Venezuela. III. The use of ordination techniques in evaluation. *Agro-Ecosystems,* 5, 135–46.

Buse, A. (1974). Habitats as recording units in ecological survey: A field trial in Caernarvonshire, North Wales. *Journal of Applied Ecology,* 11, 517–28.

Cain, S. A., & Castro, G. M. de Oliveira (1959). *Manual of Vegetation Analysis.* New York: Harper.

Cairns, J. (editor) (1977). *Aquatic Microbial Communities.* New York: Garland.

Campbell, G. S. (1977). *An Introduction to Environmental Biophysics.* New York: Springer-Verlag.

Campbell, L. G. & Lafever, H. N. (1977). Cultivar × environment interactions in soft red winter wheat yield tests. *Crop Science,* 17, 604–8.

Campbell, R. (1977). *Microbial Ecology.* New York: Wiley.

Capen, D. E. (editor) (1981). *The Use of Multivariate Statistics in Studies of Wildlife Habitat.* Ft. Collins, Colo.: Rocky Mountain Forest and Range Experiment Station.

Carleton, T. J. (1979). Floristic variation and zonation in the boreal forest south of James Bay: A cluster seeking approach. *Vegetatio,* 39, 147–60.

– (1980). Non-centered component analysis of vegetation data: A comparison of orthogonal and oblique rotation. *Vegetatio,* 42, 59–66.

Carleton, T. J. & Maycock, P. F. (1980). Vegetation of the boreal forests south of James Bay: Non-centered component analysis of the vascular flora. *Ecology,* 61, 1199–212.

Carpenter, J. A., Fitzhugh, H. A., Cartwright, T. C., Thomas, R. C., & Melton, A. A. (1978). Principal components for cow size and shape. *Journal of Animal Science,* 46, 370–5.

Caughley, G. (1977). *Analysis of Vertebrate Populations.* New York: Wiley.

Cavicchi, S. & Giorgi, G. (1976). Yield in tomato. II. Multivariate analysis on yield components. *Genetica Agraria,* 30, 315–26.

Češka, A. & Roemer, H. (1971). A computer program for identifying species-relevé groups in vegetaion studies. *Vegetatio,* 23, 255–77.

Chapman, S. B. (editor) (1976). *Methods in Plant Ecology.* London: Blackwell.

Chappell, V. G. (1976). Determining watershed sub-areas with principal component analysis. *Water Resources Bulletin,* 12, 1133–9.

Chardy, P., Glemarec, M., & Laurec, A. (1976). Application of inertia methods to benthic marine ecology: Practical implications of the basic options. *Estuarine and Coastal Marine Science,* 4, 179–205.

Christal, R. E. (1974). *The United States Air Force Occupational Research Project.* San Antonio, Tx.: Lackland Air Force Base.

Cipra, J. E., Bidwell, O. W., & Rohlf, F. J. (1970). Numerical taxonomy of soils from nine orders by cluster and centroid-component analyses. *Soil Science Society of America Proceedings,* 34, 281–7.

Cisne, J. L., Chandlee, G. O., Rabe, B. D., & Cohen, J. A. (1980). Geographic variation and episodic evolution in an Ordovician trilobite. *Science,* 209, 925–7.

Cisne, J. L., Molenock, J., & Rabe, B. D. (1980). Evolution in a cline: The trilobite *Triarthrus* along an Ordovician depth gradient. *Lethaia,* 13, 47–59.

Cisne, J. L. & Rabe, B. D. (1978). Coenocorrelation: Gradient analysis of fossil communities and its applications in stratigraphy. *Lethaia,* 11, 341–64.

Clarke, J. F. & Peterson, J. T. (1973). An empirical model using eigenvectors to calculate the temporal and spatial variations of the St. Louis heat island. *Journal of Applied Meteorology,* 12, 195–210.

Clausen, A. R. & Horn, C. E. Van (1977). How to analyze too many roll calls and related issues in dimensional analysis. *Political Methodology,* 4, 313–31.

Clifford, H. T. & Williams, W. T. (1976). Similarity measures. In *Pattern Analysis in Agricultural Science,* ed. W. T. Williams, pp. 37–46. New York: Elsevier.

Clint, M. & Jennings, A. (1970). The evaluation of eigenvalues and eigenvectors of real symmetric matrices by simultaneous iteration. *Computer Journal,* 13, 76–80.

Clover, R. C. (1979). Phenetic relationships among populations of *Podarcis sicula* and *P. melisellensis* (Sauria: Lancertidae) from islands in the Adriatic Sea. *Systematic Zoology,* 28, 284–98.

Clucas, R. D. & Ladiges, P. Y. (1979). Variations in populations of *Eucalyptus ovata* Labill., and the effects of waterlogging on seedling growth. *Australian Journal of Botany,* 27, 301–15.

Clymo, R. S. (1980). Preliminary survey of the peat-bog Hummell Knowe Moss using various numerical methods. *Vegetatio,* 42, 129–48.

Cody, M. L. (1974). *Competition and the Structure of Bird Communities.* Princeton, N. J.: Princeton University Press.

Coetzee, B. J. & Werger, M. J. A. (1975). On association-analysis and the classification of plant communities. *Vegetatio,* 30, 201–6.

Colquhoun, I. R. & Watson, E. M. (1975). A modified method of mapping hill vegetation in relation to animal distribution. *Journal of Applied Ecology,* 12, 343–8.

Colwell, R. K. & Fuentes, E. R. (1975). Experimental studies of the niche. *Annual Review of Ecology and Systematics,* 6, 281–310.

Colwell, R. R. & Morita, R. Y. (editors) (1974). *Effect of the Ocean Environment on Microbial Activities.* Baltimore: University Park Press.

Connell, J. H. (1978). Diversity in tropical rain forests and coral reefs. *Science,* 199, 1302–10.

– (1980). Diversity and the coevolution of competitors, or the ghost of competition past. *Oikos,* 35, 131–8.

Cormack, R. M. (1971). A review of classification. *Journal of the Royal Statistical Society, Series A,* 134, 321–67.

– (1979). Spatial aspects of competition between individuals. In *Spatial and Temporal Analysis in Ecology,* eds. R. M. Cormack & J. K. Ord, pp. 151–212. Burtonsville, Md.: International Co-operative.

Cottam, G. & Curtis, J. T. (1956). The use of distance measures in phytosociological sampling. *Ecology,* 37, 451–60.

Cottam, G., Goff, F. G., & Whittaker, R. H. (1978). Wisconsin comparative ordination. In *Ordination of Plant Communities,* ed. R. H. Whittaker, pp. 185–213. The Hague: Junk.

Couderc, H. (1977). Application de l'analyse factorielle des correspondances à l'étude systématique de l'*Anthyllis vulneraria* L. *Revue générale de botanique,* 84, 61–77.

– (1978). Etude comparée de populations françaises et islandaises de l'*Anthyllis vulneraria* L. ssp. *borealis* (Rouy) Jalas. *Bulletin Société Botanique de France,* 125, 73–88.

Crawford, H. S. & Titterington, R. W. (1979). Effects of silvicultural practices on bird communities in upland spruce–fir stands. In *Proceedings of the Workshop Management of Northcentral and Northeastern Forests for Nongame Birds,* ed. R. M. DeGraaf, pp. 110–9. Orono, Me.: Northeastern Forest Experiment Station.

Crawford, R. M. M. & Wishart, D. (1967). A rapid multivariate method for the detection and classification of groups of ecologically related species. *Journal of Ecology,* 55, 505–24.

– (1968). A rapid classification and ordination method and its application to vegetation mapping. *Journal of Ecology,* 56, 385–404.

Crovello, T. J. (1970). Analysis of character variation in ecology and systematics. *Annual Review of Ecology and Systematics,* 1, 55–98.

Crow, T. R. & Grigal, D. F. (1979). A numerical analysis of arborescent communities in the rain forest of the Luquillo Mountains, Puerto Rico. *Vegetatio,* 40, 135–46.

Crowe, T. M. (1979). Lots of weeds: Insular phytogeography of vacant urban lots. *Journal of Biogeography,* 6, 169–81.

Cuadras, J. & Pereira, F. (1977). Invertebrates associated with *Dardanus arrosor* (Anomura, Diogenidae). *Vie et milieu, série A,* 27, 301–10.

Cunningham, K. M. & Ogilvie, J. C. (1972). Evaluation of hierarchical grouping techniques: A preliminary study. *Computer Journal,* 15, 209–13.

Curtis, J. T. (1959). *The Vegetation of Wisconsin, An Ordination of Plant Communities.* Madison, Wisc.: University of Wisconsin Press.

Curtis, J. T. & McIntosh, R. P. (1951). An upland forest continuum in the prairie–forest border region of Wisconsin. *Ecology,* 32, 476–96.

Dagnelie, P. (1978). Factor analysis. In *Ordination of Plant Communities,* ed. R. H. Whittaker, pp. 215–38. The Hague: Junk.

Dahl, E. (1960). Some measures of uniformity in vegetation analysis. *Ecology,* 41, 805–8.

Dale, M. B. (1975). On objectives of methods of ordination. *Vegetatio,* 30, 15–32.

– (1976*a*). A comparison of some recently developed methods. In *Pattern Analysis in Agricultural Science,* ed. W. T. Williams, pp. 316–26. New York: Elsevier.

– (1976*b*). Ordination: Recent developments and future possibilities. In *Pattern Analysis in Agricultural Science,* ed. W. T. Williams, pp. 70–5. New York: Elsevier.

– (1980). A syntactic basis of classification. *Vegetatio,* 42, 93–8.

Dale, M. B. & Anderson, D. J. (1973). Inosculate analysis of vegetation data. *Australian Journal of Botany,* 21, 253–76.

Dale, M. B. & Clifford, H. T. (1976). On the effectiveness of higher taxonomic ranks for vegetation analysis. *Australian Journal of Ecology,* 1, 37–62.

Dale, M. B. & Quadraccia, L. (1973). Computer assisted tabular sorting of phytosociological data. *Vegetatio,* 28, 57–73.

Darom, E., Sharan, S., & Hertz-Lazarowitz, R. (1978). The development and validation of a multidimensional scale for assessment of teachers' attitudes toward small-group teaching. *Educational and Psychological Measurement,* 38, 1233–8.

Date, R. A., Burt, R. L., & Williams, W. T. (1979). Affinities between various *Stylosanthes* species as shown by rhizobial, soil pH and geographic relationships. *Agro-Ecosystems,* 5, 57–67.

Daubenmire, R. (1968). *Plant Communities.* New York: Harper.

David, P., Lepart, J., & Romane, F. (1979). Elements for a system of data processing in phytosociology and ecology. *Vegetatio,* 40, 115–23.

Davies, W. G. (1978). Cluster analysis applied to the classification of postures in the Chilean flamingo (*Phoenicopterus chilensis*). *Animal Behavior,* 26, 381–8.

Denis, J. C. & Adams, M. W. (1978). A factor analysis of plant variables related to yield in dry beans. I. Morphological traits. *Crop Science,* 18, 74–8.

Denys, E. (1980). A tentative phytogeographical division of tropical Africa based on a mathematical analysis of distribution maps. *Bulletin du Jardin Botanique National de Belgique,* 50, 465–504.

Diaz, L. R., Novo, F. G., & Merino, J. (1976). On the ecological interpretation of principal components in factor analysis. *Oecologia Plantarum,* 11, 137–41.

Diday, E. (1971). Une Nouvelle Méthode en classification automatique et reconnaissance de formes: La méthode des nuées dynamiques. *Revue de statistique appliquée,* 19, 19–33.

Dierschke, H. (1977). Waldrand-Gessellschaften als natürliches Modell für Schutzpflanzungen. In *Vegetation Science and Environmental Protection,* eds. A. Miyawaki & R. Tüxen, pp. 343–9. Tokyo: Maruzen.

Ditton, R. B. (1975). Clustering recreation participation data to identify recreation types. In *Indicators of Change in the Recreation Environment – A National Research Symposium,* ed. B. van der Smissen, pp. 75–86. University Park, Pa.: Pennsylvania State University Press.

Ditton, R. B., Goodale, T. L., & Johnsen, P. K. (1975). A cluster analysis of activity, frequency, and environment variables to identify water-based recreation types. *Journal of Leisure Research,* 7, 282–95.

Dix, R. L. & Smeins, F. E. (1967). The prairie, meadow, and marsh vegetation of Nelson County, North Dakota. *Canadian Journal of Botany,* 45, 21–58.

Dobben, W. H. van (1979). Autecology and vegetation science. In *The Study of Vegetation,* ed. M. J. A. Werger, pp. 1–10. The Hague: Junk.

Doncaster, C. P. (1981). The spatial distribution of ants' nests on Ramsey Island, South Wales. *Journal of Animal Ecology,* 50, 195–218.

Dor, I., Schechter, H., & Shuval, H. I. (1976). Biological and chemical succession in Nahal Soreq: A free-flowing wastewater stream. *Journal of Applied Ecology,* 13, 475–89.

Drake, D. W. (1980). Contrasting success of natural hybridization in two *Eucalyptus* species pairs. *Australian Journal of Botany,* 28, 167–91.

Drennan, R. D. (1976). A refinement of chronological seriation using nonmetric multidimensional scaling. *American Antiquity,* 41, 290–301.

Droop, M. R. & Jannasch, H. W. (editors) (1977). *Advances in Aquatic Microbiology,* vol. 1. New York: Academic Press.

– (editors) (1980). *Advances in Aquatic Microbiology,* vol. 2. New York: Academic Press.

Dume, G. (1978). Contribution à l'étude phytosociologique des forêts à Chêne et à Charme de la région parisienne. *Bulletin Société Botanique de France,* 125, 167–98.

Dumond, D. E. (1974). Some uses of *R*-mode analysis in archaeology. *American Antiquity,* 39, 253–70.

Duncan, D. J. (1978). Leisure types: Factor analyses of leisure profiles. *Journal of Leisure Research,* 10, 113–25.

Dye, P. J. & Walker, B. H. (1980). Vegetation–environment relations on sodic soils of Zimbabwe Rhodesia. *Journal of Ecology,* 68, 589–606.

Dyer, T. G. J. (1975). The assignment of rainfall stations into homogeneous groups: An application of principal component analysis. *Quarterly Journal of the Royal Meteorological Society,* 101, 1005–13.

Edye, L. A. (1976*a*). Analysis of agronomic data from *Stylosanthes* introductions. In *Pattern Analysis in Agricultural Science,* ed. W. T. Williams, pp. 170–80. New York: Elsevier.

– (1976*b*). Statistical and pattern analysis of a small-sward trial. In *Pattern Analysis in Agricultural Science,* ed. W. T. Williams, pp. 181–93. New York: Elsevier.

Edye, L. A., Williams, W. T., Anning, P., Holm, A. McR., Miller, C. P., Page, M. C., & Winter, W. H. (1975). Sward tests of some morphological–agronomic groups of *Stylosanthes* accessions in dry tropical environments. *Australian Journal of Agricultural Research,* 26, 481–96.

Edye, L. A., Williams, W. T., & Pritchard, A. J. (1970). A numerical analysis of variation patterns in Australian introductions of *Glycine wightii (G. javanica). Australian Journal of Agricultural Research,* 21, 57–69.

Egerton, F. N. (1976). Ecological studies and observations before 1900. In *Issues and Ideas in America,* eds. B. J. Taylor & T. J. White, pp. 311–51. Norman, Okla.: University of Oklahoma Press.

Ellenberg, H. (1948). Unkrautgesellschaften als Mass für den Säuregrad, die Verdichtung und andere Eigenschaften des Ackerbodens. *Berichte über Landtechnik, Kuratorium für Technik und Bauwesen in der Landwirtschaft,* 4, 130–46.

– (1952). *Landwirtschaftliche Pflanzensoziologie, Bd. II, Wiesen und Weiden und ihre Standortliche Bewertung*. Stuttgart: Ulmer.

– (1979). Man's influence on tropical mountain ecosystems in South America. *Journal of Ecology,* 67, 401–16.

Ellis, S. L., Fallat, C., Reece, N., & Riordan, C. (1977). *Guide to Land Cover and Use Classification Systems Employed by Western Governmental Agen-*

cies. Washington, D.C.: Office of Biological Services, Fish and Wildlife Service.

Elsasser, W. M. (1969). Acausal phenomena in physics and biology: A case for reconstruction. *American Scientist,* 57, 502–16.

Emmons, L. H. (1980). Ecology and resource partitioning among nine species of African rain forest squirrels. *Ecological Monographs,* 50, 31–54.

Essenwanger, O. (1976). *Applied Statistics in Atmospheric Science, Part A, Frequencies and Curve Fitting.* New York: Elsevier.

Etherington, J. R. (1975). *Environment and Plant Ecology.* New York: Wiley.

Everett, J. E. & Entrekin, L. V. (1980). Factor comparability and the advantages of multiple group factor analysis. *Multivariate Behavioral Research,* 15, 165–80.

Everitt, B. S. (1978). *Graphical Techniques for Multivariate Data.* New York: North-Holland.

Falbo, T. (1977). Multidimensional scaling of power strategies. *Journal of Personality and Social Psychology,* 8, 537–47.

Fasham, M. J. R. (1977). A comparison of nonmetric multidimensional scaling, principal components and reciprocal averaging for the ordination of simulated coenoclines and coenoplanes. *Ecology,* 58, 551–61.

Fekete, G. & Szőcs, Z. (1974). Studies on interspecific association processes in space. *Acta Botanica Academiae Scientiarum Hungaricae,* 20, 227–41.

Felley, J. D. & Avise, J. C. (1980). Genetic and morphological variation of bluegill populations in Florida lakes. *Transactions of the American Fisheries Society,* 109, 108–15.

Fenker, R. & Tees, S. (1976). Measuring the cognitive structures of pre-school children: A multidimensional scaling analysis of classification performance and similarity estimation. *Multivariate Behavioral Research,* 11, 339–52.

Feoli, E. & Feoli Chiapella, L. (1979). Relevé ranking based on a sum of squares criterion. *Vegetatio,* 39, 123–5.

– (1980). Evaluation of ordination methods through simulated coenoclines: Some comments. *Vegetatio,* 42, 35–41.

Feoli, E. & Lagonegro, M. (1979). Intersection analysis in phytosociology: Computer program and application. *Vegetatio,* 40, 55–9.

Feoli, E. & Orlóci, L. (1979). Analysis of concentration and detection of underlying factors in structured tables. *Vegetatio,* 40, 49–54.

Filip, M. (1970). Multidimenzionálna analýza v muzikológii. *Musicologica Slovaca,* 2, 17–49.

Fisher, R. A. (1940). The precision of discriminant functions. *Annals of Eugenics,* 10, 422–9.

Foin, T. C. & Jain, S. K. (1977). Ecosystems analysis and population biology: Lessons for the development of community ecology. *BioScience,* 27, 532–8.

Ford-Lloyd, B. V. & Williams, J. T. (1975). A revision of *Beta* section *Vulgares* (Chenopodiaceae), with new light on the origin of cultivated beets. *Botanical Journal of the Linnean Society,* 71, 89–102.

Forgas, J. P. (1976). The perception of social episodes: categorical and dimensional representations in two different social milieus. *Journal of Personality and Social Psychology,* 34, 199–209.

Forsythe, W. L. & Loucks, O. L. (1972). A transformation for species response to habitat factors. *Ecology,* 53, 1112–9.

Foster, P., Savidge, G., Foster, G. M., Hunt, D. T. E., & Pugh, K. B. (1976). Multivariate analysis of surface water characteristics in the summer regime of the western Irish Sea. *Journal of Experimental Marine Biology and Ecology,* 25, 171–85.

Fourt, D. F., Donald, D. G. M., Jeffers, J. N. R., & Binns, W. O. (1971). Corsican pine (*Pinus nigra* var. *maritima* (Ait.) Melville) in southern Britain. *Forestry,* 44, 189–207.

Francis, I. S. (1979). *A Comparative Review of Statistical Software.* Voorburg, The Netherlands: International Association for Statistical Computing.

Freedman, B. & Hutchinson, T. C. (1980). Long-term effects of smelter pollution at Sudbury, Ontario, on forest community composition. *Canadian Journal of Botany,* 58, 2123–40.

Frenkel, R. E. & Harrison, C. M. (1974). An assessment of the usefulness of phytosociological and numerical classificatory methods for the community biogeographer. *Journal of Biogeography,* 1, 27–56.

Frieze, A. M. (1980). Probabilistic analysis of some Euclidean clustering problems. *Discrete Applied Mathematics,* 2, 295–309.

Frijters, J. E. R. (1976). Evaluation of a texture profile for cooked chicken breast meat by principal component analysis. *Poultry Science,* 55, 229–34.

Fritschen, L. J. & Gay, L. W. (1979). *Environmental Instrumentation.* New York: Springer-Verlag.

Gandour, J. T. (1978). Perceived dimensions of 13 tones: A multidimensional scaling investigation. *Phonetica,* 35, 169–79.

Garten, C. T. (1978). Multivariate perspectives on the ecology of plant mineral element composition. *American Naturalist,* 112, 533–44.

Gauch, H. G. (1973*a*). A quantitative evaluation of the Bray–Curtis ordination. *Ecology,* 54, 829–36.

– (1973*b*). The relationship between sample similarity and ecological distance. *Ecology,* 54, 618–22.

– (1977). *ORDIFLEX–A Flexible Computer Program for Four Ordination Techniques: Weighted Averages, Polar Ordination, Principal Components Analysis, and Reciprocal Averaging, Release B.* Ithaca, N. Y: Cornell University.

– (1979). *COMPCLUS–A FORTRAN Program for Rapid Initial Clustering of Large Data Sets.* Ithaca, N. Y: Cornell University.

– (1980). Rapid initial clustering of large data sets. *Vegetatio,* 42, 103–11.

– (1981). Noise reduction by eigenvector ordination (manuscript).

Gauch, H. G., Chase, G. B., & Whittaker, R. H. (1974). Ordination of vegetation samples by Gaussian species distributions. *Ecology,* 55, 1382–90.

Gauch, H. G. & Scruggs, W. M. (1979). Variants of polar ordination. *Vegetatio,* 40, 147–53.

Gauch, H. G. & Stone, E. L. (1979). Vegetation and soil pattern in a mesophytic forest at Ithaca, New York. *American Midland Naturalist,* 102, 332–45.

Gauch, H. G. & Wentworth, T. R. (1976). Canonical correlation analysis as an ordination technique. *Vegetatio,* 33, 17–22.

Gauch, H. G. & Whittaker, R. H. (1972*a*). Coenocline simulation. *Ecology,* 53, 446–51.

– (1972*b*). Comparison of ordination techniques. *Ecology,* 53, 868–75.

– (1976). Simulation of community patterns. *Vegetatio,* 33, 13–16.

– (1981). Hierarchical classification of community data. *Journal of Ecology,* 69, 135–52.

Gauch, H. G., Whittaker, R. H., & Singer, S. B. (1981). A comparative study of nonmetric ordinations. *Journal of Ecology,* 69, 135–52.

Gauch, H. G., Whittaker, R. H., & Wentworth, T. R. (1977). A comparative study of reciprocal averaging and other ordination techniques. *Journal of Ecology,* 65, 157–74.

Gause, G. F. (1930). Studies on the ecology of the Orthoptera. *Ecology,* 11, 307–25.

Gauthreaux, S. A. (1978). The structure and organization of avian communities in forests. In *Proceedings of the Workshop Management of Southern Forests for Nongame Birds,* ed. R. M. DeGraaf, pp. 17–37. Asheville, N.C.: Southern Forest Experiment Station.

Geiger, R. (1965). *The Climate Near the Ground.* Cambridge, Mass.: Harvard University Press.

Gether, J. & Seip, H. M. (1979). Analysis of air pollution data by the combined use of interactive graphic presentation and a clustering technique. *Atmospheric Environment,* 13, 87–96.

Ghaderi, A., Everson, E. H., & Cress, C. E. (1980). Classification of environments and genotypes in wheat. *Crop Science,* 20, 707–10.

Ghaderi, A., Shishegar, M., Rezai, A., & Ehdaie, B. (1979). Multivariate analysis of genetic diversity for yield and its components in mung bean. *Journal of the American Society for Horticultural Science,* 104, 728–31.

Gilbert, M. L. & Curtis, J. T. (1953). Relation of the understory to the upland forest in the prairie–forest border region of Wisconsin. *Transactions of the Wisconsin Academy of Science, Arts and Letters,* 42, 183–95.

Gillard, P. (1976). Classification of sequential vegetation data: A three-dimensional approach. In *Pattern analysis in Agricultural Science,* ed. W. T. Williams, pp. 259–66. New York: Elsevier.

Gillen, B. & Sherman, R. C. (1980). Physical attractiveness and sex as determinants of trait attributions. *Multivariate Behavioral Research,* 15, 423–37.

Gittins, R. (1969). The application of ordination techniques. In *Ecological Aspects of the Mineral Nutrition of Plants,* ed. I. H. Rorison, pp. 37–66. Oxford: Blackwell.

– (1979). Ecological applications of canonical analysis. In *Multivariate Methods in Ecological Work,* eds. L. Orlóci, C. R. Rao, & W. M. Stiteler, pp. 309–535. Burtonsville, Md.: International Co-operative.

Gladfelter, W. B., Ogden, J. C., & Gladfelter, E. H. (1980). Similarity and diversity among coral reef fish communities: A comparison between tropical western Atlantic (Virgin Islands) and tropical central Pacific (Marshall Islands) patch reefs. *Ecology,* 61, 1156–68.

Gleason, H. A. (1926). The individualistic concept of the plant association. *Bulletin of the Torrey Botanical Club,* 53, 7–26.
Glenn-Lewin, D. C. (1979). Natural revegetation of acid coal spoils in southeast Iowa. In *Ecology and Coal Resource Development,* ed. M. K. Wali, vol. 2, pp. 568–75. New York: Pergamon.
– (1980). The individualistic nature of plant community development. *Vegetatio,* 43, 141–6.
Goff, F. G. (1968). Use of size stratification and differential weighting to measure forest trends. *American Midland Naturalist,* 79, 402–18.
– (1975). Comparison of species ordinations resulting from alternative indices of interspecific association and different numbers of included species. *Vegetatio,* 31, 1–14.
Goff, F. G. & Cottam, G. (1967). Gradient analysis: The use of species and synthetic indices. *Ecology,* 48, 793–806.
Goff, F. G. & Mitchell, R. (1975). A comparison of species ordination results from plot and stand data. *Vegetatio,* 31, 15–22.
Goff, F. G. & Zedler, P. H. (1968). Structural gradient analysis of upland forests in the western Great Lakes area. *Ecological Monographs,* 38, 65–86.
– (1972). Derivation of species succession vectors. *American Midland Naturalist,* 87, 397–412.
Goldsmith, F. B. & Harrison, C. M. (1976). Description and analysis of vegetation. In *Methods in Plant Ecology,* ed. S. B. Chapman, pp. 85–155. London: Blackwell.
Goodall, D. W. (1953). Objective methods for the classification of vegetation. I. The use of positive interspecific correlation. *Australian Journal of Botany,* 1, 39–63.
– (1954*a*). Objective methods for the classification of vegetation. III. An essay in the use of factor analysis. *Australian Journal of Botany,* 2, 304–24.
– (1954*b*). Vegetational classification and vegetational continua. *Festschrift für Erwin Aichinger,* 1, 168–82.
– (1962). Bibliography of statistical plant sociology. *Excerpta Botanica, Sectio B,* 4, 253–322.
– (1970). Statistical plant ecology. *Annual Review of Ecology and Systematics,* 1, 99–124.
– (1978*a*). Numerical classification. In *Classification of Plant Communities,* ed. R. H. Whittaker, pp. 247–86. The Hague: Junk.
– (1978*b*). Sample similarity and species correlation. In *Ordination of Plant Communities,* ed. R. H. Whittaker, pp. 99–149. The Hague: Junk.
Gordon, A. D. & Birks, H. J. B. (1972). Numerical methods in quaternary palaeoecology. I. Zonation of pollen diagrams. *New Phytologist,* 71, 961–79.
– (1974). Numerical methods in quaternary palaeoecology. *New Phytologist,* 73, 221–49.
Gordon, A. D. & Henderson, J. T. (1977). An algorithm for Euclidean sum of squares classification. *Biometrics,* 33, 355–62.
Gorham, E. (1953). Chemical studies on the soils and vegetation of water-

logged habitats in the English lake district. *Journal of Ecology*, 41, 345–60.

Gower, J. C. (1966). Some distance properties of latent root and vector methods used in multivariate analysis. *Biometrika*, 53, 325–38.

– (1967*a*). A comparison of some methods of cluster analysis. *Biometrics*, 23, 623–37.

– (1967*b*). Multivariate analysis and multivariate geometry. *The Statistician*, 17, 13–28.

– (1971). Statistical methods of comparing different multivariate analyses of the same data. In *Mathematics in the Archaeological and Historical Sciences*, eds. F. R. Hodson, D. G. Kendall, & P. Tartu, pp. 138–49. Edinburgh: Edinburgh University Press.

– (1974). Maximal predictive classification. *Biometrics*, 30, 643–54.

Grant, W. S. & Utter, F. M. (1980). Biochemical genetic variation in walleye pollock, *Theragra chalcogramma:* Population structure in the southeastern Bering Sea and the Gulf of Alaska. *Canadian Journal of Fisheries and Aquatic Science*, 37, 1093–100.

Green, B. F. (1977). Parameter sensitivity in multivariate methods. *Journal of Multivariate Behavioral Research*, 12, 163–87.

Green, P. E. (1978). *Analyzing Multivariate Data*. Hinsdale, Ill.: Dryden.

Green, P. E. & Rao, V. R. (1972). *Applied Multidimensional Scaling: A Comparison of Approaches and Algorithms*. New York: Holt.

– (1977). Nonmetric approaches to multivariate analysis in marketing. In *Multivariate Methods for Market and Survey Research*, ed. J. N. Sheth, pp. 237–53. Chicago: American Marketing Association.

Green, P. E., Rao, V. R., & DeSarbo, W. S. (1978). Incorporating group-level similarity judgments in conjoint analysis. *Journal of Consumer Research*, 5, 187–93.

Green, R. H. (1971). A multivariate statistical approach to the Hutchinsonian niche: Bivalve molluscs of central Canada. *Ecology*, 52, 543–56.

– (1979). *Sampling Design and Statistical Methods for Environmental Biologists*. New York: Wiley.

– (1980). Multivariate approaches in ecology: The assessment of ecologic similarity. *Annual Review of Ecology and Systematics*, 11, 1–14.

Green, R. H. & Vascotto, G. L. (1978). A method for the analysis of environmental factors controlling patterns of species composition in aquatic communities. *Water Research*, 12, 583–90.

Greig-Smith, P. (1964). *Quantitative Plant Ecology*, 2d. ed. London: Butterworths.

– (1971). Analysis of vegetation data: The user viewpoint. In *Statistical Ecology*, eds. G. P. Patil, E. C. Pielou, & W. E. Waters, vol. 3, pp. 149–66. University Park, Pa.: Pennsylvania State University Press.

– (1980). The development of numerical classification and ordination. *Vegetatio*, 42, 1–9.

Greszta, J., Braniewski, S., Marczyńska-Gałkowska, K., & Nosek, A. (1979). The effect of dusts emitted by non-ferrous metal smelters on the soil, soil microflora and selected tree species. *Ekologia Polska*, 27, 397–426.

Griffin, D. M. (1972). *Ecology of Soil Fungi.* London: Chapman & Hall.
Grigal, D. F. & Arneman, H. F. (1970). Quantitative relationships among vegetation and soil classifications from northeastern Minnesota. *Canadian Journal of Botany,* 48, 555–66.
Grime, J. P. (1979). *Plant Strategies and Vegetational Processes.* New York: Wiley.
Grodzińska, K. (1977). Changes in the forest environment in southern Poland as a result of steel mill emissions. In *Vegetation Science and Environmental Protection,* eds. A. Miyawaki & R. Tüxen, pp. 207–15. Tokyo: Maruzen.
Groenewoud, H. van (1965). Ordination and classification of Swiss and Canadian coniferous forests by various biometric and other methods. *Bericht Eidgenössische Technische Hochschule, Geobotanisches Institut, Stiftung Rübel, Zürich,* 36, 28–102.
– (1976). Theoretical considerations on the covariation of plant species along ecological gradients with regard to multivariate analysis. *Journal of Ecology,* 64, 837–47.
Guinochet, M. (1973). *Phytosociologie.* Paris: Masson.
Gupta, S. C., Wet, J. M. J. de, & Harlan, J. R. (1978). Morphology of Saccharum–Sorghum hybrid derivatives. *American Journal of Botany,* 65, 936–42.
Haedrich, R. L., Rowe, G. T., & Polloni, P. T. (1975). Zonation and faunal composition of epibenthic populations on the continental slope south of New England. *Journal of Marine Research,* 33, 191–212.
Hale, M. E. (1955). Phytosociology of corticolous cryptogams in the upland forests of southern Wisconsin. *Ecology,* 36, 45–63.
Hall, J. B. & Swaine, M. D. (1976). Classification and ecology of closed-canopy forest in Ghana. *Journal of Ecology,* 64, 913–51.
Hamer, A. D. & Soulsby, P. G. (1980). An approach to chemical and biological river monitoring systems. *Water Pollution Control,* 79, 56–69.
Hanks, S. & Fryxell, P. A. (1979). Palynological studies of *Gaya* and *Herissantia* (Malvaceae). *American Journal of Botany,* 66, 494–501.
Hanley, T. V. & Cox, D. L. (1979). Individual differences in visual discrimination of letters. *Perceptual and Motor Skills,* 48, 539–50.
Harada, H. (1980). Vertical distribution of oribatid mites in moss and lichen. II. Ecological studies on soil arthropods of Mt. Fujisan. *Japanese Journal of Ecology,* 30, 75–83.
Hardy, D. M. & Walton, J. J. (1978). Principal components analysis of vector wind measurements. *Journal of Applied Meteorology,* 17, 1153–62.
Hartigan, J. A. (1975). *Clustering Algorithms.* New York: Wiley.
Hatheway, W. H. (1971). Contingency-table analysis of rain forest vegetation. In *Statistical Ecology,* eds. G. P. Patil, E. C. Pielou, & W. E. Waters, vol. 3, pp. 271–313. University Park, Pa.: Pennsylvania State University Press.
Hayward, B. W. & Buzas, M. A. (1979). Taxonomy and paleoecology of early Miocene benthic foraminifera of northern New Zealand and the north Tasman Sea. *Smithsonian Contributions to Paleobiology,* 36, 1–145.
Hazel, J. E. (1977). Use of certain multivariate and other techniques in assemblage zonal biostratigraphy: Examples utilizing Cambrian, Cretaceous,

and tertiary benthic invertebrates. In *Concepts and Methods of Biostratigraphy,* eds. E. G. Kauffman & J. E. Hazel, pp. 187–212. Stroudsburg, Pa.: Dowden, Hutchinson & Ross.

Hecke, P. van, Impens, I., Goossens, R., & Hebrant, F. (1980). Multivariate analysis of multispectral remote sensing data on grasslands from different soil types. *Vegetatio,* 42, 165–70.

Hedges, D. A. (1976*a*). Some relationships between voluntary feed consumption and feed characteristics. In *Pattern Analysis in Agricultural Science,* ed. W. T. Williams, pp. 290–301. New York: Elsevier.

– (1976*b*). The efficiency of utilization of forage oats by sheep: Use of ordination and canonical coordinate procedures. In *Pattern Analysis in Agricultural Science,* ed. W. T. Williams, pp. 280–9. New York: Elsevier.

Henry, R. C. & Hidy, G. M. (1970). Multivariate analysis of particulate sulfate and other air quality variables by principal components. 1. Annual data from Los Angeles and New York. *Atmospheric Environment,* 13, 1581–96.

Hermy, M. & Stieperaera, H. (1981). An indirect gradient analysis of the ecological relationships between ancient and recent riverine woodlands to the south of Bruges (Flanders, Belgium). *Vegetatio,* 44, 43–9.

Hicks, D. J. (1980). Intrastand distribution patterns of southern Appalachian cove forest herbaceous species. *American Midland Naturalist,* 104, 209–23.

Hill, M. O. (1973). Reciprocal averaging: An eigenvector method of ordination. *Journal of Ecology,* 61, 237–49.

– (1974). Correspondence analysis: A neglected multivariate method. *Journal of the Royal Statistical Society, Series C,* 23, 340–54.

– (1977). Use of simple discriminant functions to classify quantitative phytosociological data. In *First International Symposium on Data Analysis and Informatics,* eds. E. Diday, L. Lebart, J. P. Pages, & R. Tomassone, vol. 1, pp. 181–99. Le Chesnay, France: Institut de Recherche d'Informatique et d'Automatique.

– (1979*a*). *DECORANA – A FORTRAN Program for Detrended Correspondence Analysis and Reciprocal Averaging.* Ithaca, N.Y.: Cornell University.

– (1979*b*). *TWINSPAN – A FORTRAN Program for Arranging Multivariate Data in an Ordered Two-Way Table by Classification of the Individuals and Attributes.* Ithaca, N.Y.: Cornell University.

Hill, M. O., Bunce, R. G. H., & Shaw, M. W. (1975). Indicator species analysis, a divisive polythetic method of classification, and its application to a survey of native pinewoods in Scotland. *Journal of Ecology,* 63, 597–613.

Hill, M. O. & Gauch, H. G. (1980). Detrended correspondence analysis, an improved ordination technique. *Vegetatio,* 42, 47–58.

Hill, M. O. & Smith, A. J. E. (1976). Principal component analysis of taxonomic data with multi-state discrete characters. *Taxon,* 25, 249–55.

Hirano, H., Inokuchi, T., & Nakajima, T. (1980). Relationships between amino acid contents and peroxidase isozymes in leaf blades of mulberry (*Morus* spp.). *Euphytica,* 29, 145–53.

Hirschfeld, H. O. (1935). A connection between correlation and contingency. *Proceedings of the Cambridge Philosophical Society,* 31, 520–4.

Hodson, F. R. (1970). Cluster analysis and archaeology: Some new developments and applications. *World Archaeology,* 1, 299–320.

Holman, E. W. (1972). The relation between hierarchical and Euclidean models for psychological distances. *Psychometrika,* 37, 417–23.

Holme, N. A. & McIntyre, A. D. (1971). *Methods for the Study of Marine Benthos: IBP Handbook Number 16.* Oxford: Blackwell.

Holmes, R. T., Bonney, R. E., & Pacala, S. W. (1979). Guild structure of the Hubbard Brook bird community: A multivariate approach. *Ecology,* 60, 512–20.

Holzner, W., Werger, M. J. A., & Ellenbroek, G. A. (1978). Automatic classification of phytosociological data on the basis of species groups. *Vegetatio,* 38, 157–64.

Hopke, P. K., Gladney, E. S., Gordon, G. E., Zoller, W. H., & Jones, A. G. (1976). The use of multivariate analysis to identify sources of selected elements in the Boston urban aerosol. *Atmospheric Environment,* 10, 1015–25.

Horst, P. (1935). Measuring complex attitudes. *Journal of Social Psychology,* 6, 369–74.

Hotelling, H. (1933). Analysis of a complex of statistical variables into principal components. *Journal of Educational Psychology,* 24, 417–41, 498–520.

Hubbell, S. P. (1979). Tree dispersion, abundance, and diversity in a tropical dry forest. *Science,* 203, 1299–1309.

Hughes, R. N. & Thomas, M. L. H. (1971). The classification and ordination of shallow-water benthic samples from Prince Edward Island, Canada. *Journal of Experimental Marine Biology and Ecology,* 7, 1–39.

Huntingford, F. A. (1976). An investigation of the territorial behavior of the three-spined stickleback (*Gasterosteus aculeatus*) using principal components analysis. *Animal Behavior,* 24, 822–34.

Huntley, B. (1979). The past and present vegetation of the Caenlochan National Nature Reserve, Scotland. I. Present vegetation. *New Phytologist,* 83, 215–83.

Huntley, B. & Birks, H. J. B. (1979*a*). The past and present vegetation of the Morrone Birkwoods National Nature Reserve, Scotland. I. A primary phytosociological survey. *Journal of Ecology,* 67, 417–46.

– (1979*b*). The past and present vegetation of the Morrone Birkwoods National Nature Reserve, Scotland. II. Woodland vegetation and soils. *Journal of Ecology,* 67, 447–67.

Huschle, G. & Hironaka, M. (1980). Classification and ordination of seral plant communities. *Journal of Range Management,* 33, 179–82.

Hussaini, S. H., Goodman, M. M., & Timothy, D. H. (1977). Multivariate analysis and the geographical distribution of the world collection of finger millet. *Crop Science,* 17, 257–63.

Ihm, P. & Groenewoud, H. van (1975). A multivariate ordering of vegetation data based on Gaussian type gradient response curves. *Journal of Ecology,* 63, 767–77.

Imbrie, J. & Newell, N. (1964). *Approaches to Paleoecology*. New York: Wiley.

Isebrands, J. G. & Crow, T. R. (1975). *Introduction to Uses and Interpretation of Principal Component Analysis in Forest Biology*. St. Paul: North Central Forest Experiment Station.

Itow, S. (1963). Grassland vegetation in uplands of western Honshu, Japan. II. Succession and grazing indicators. *Japanese Journal of Botany,* 18, 133–67.

Ivimey-Cook, R. B. & Proctor, M. C. F. (1966). The application of association-analysis to phytosociology. *Journal of Ecology,* 54, 179–92.

Jackson, D. M. (1969). Comparison of classifications. In *Numerical Taxonomy,* ed. A. J. Cole, pp. 91–113. New York: Academic Press.

Jacquot, M. & Arnaud, M. (1979). Classification numérique de variétés de riz. *Agronomie tropicale,* 34, 157–73.

James, F. C. (1971). Ordinations of habitat relationships among breeding birds. *Wilson Bulletin,* 83, 215–36.

Jancey, R. C. (1975). A new source of evidence for the polarized nucleus in maize. *Canadian Journal of Genetics and Cytology,* 17, 245–52.

– (1980). The minimisation of random events in the search for group structure. *Vegetatio,* 42, 99–101.

Janssen, C. R. (1979). The development of palynology in relation to vegetation science, especially in The Netherlands. In *The Study of Vegetation,* ed. M. J. A. Werger, pp. 229–46. The Hague: Junk.

Janssen, J. G. M. (1975). A simple clustering procedure for preliminary classification of very large sets of phytosociological results. *Vegetatio,* 30, 67–71.

Jardine, N. & Sibson, R. (1968). The construction of hierarchic and non-hierarchic classifications. *Computer Journal,* 11, 177–84.

– (1971). *Mathematical Taxonomy*. New York: Wiley.

Jeffree, R. A. & Williams, N. J. (1980). Mining pollution and the diet of the purple-striped gudgeon *Mogurnda mogurnda* Richardson (Eleotridae) in the Finniss River, Northern Territory, Australia. *Ecological Monographs,* 50, 457–85.

Jeglum, J. K. (1974). Relative influence of moisture-aeration and nutrients on vegetation and black spruce growth in northern Ontario. *Canadian Journal of Forest Research,* 4, 114–24.

Jeglum, J. K., Wehrhahn, C. P., & Swan, J. M. A. (1971). Comparisons of environmental ordinations with principal component vegetational ordinations for sets of data having different degrees of complexity. *Canadian Journal of Forest Research,* 1, 99–112.

Jeník, J., Bureš, L., & Burešová, Z. (1980). Syntaxonomic study of vegetation in Velká Kotlina Cirque, the Sudeten Mountains. *Folia geobotanica et phytotaxonomica,* 14, 337–448.

Jennings, A. (1967). A direct iteration method for obtaining the latent roots and vectors of a symmetric matrix. *Proceedings of the Cambridge Philosophical Society,* 63, 755–65.

Jensen, R. H. & Eshbaugh, W. H. (1976*a*). Numerical taxonomic studies of hybridization in *Quercus*. I. Populations of restricted areal distribution and low taxonomic diversity. *Systematic Botany,* 1, 1–10.
– (1976*b*). Numerical taxonomic studies of hybridization in *Quercus*. II. Populations with wide areal distributions and high taxonomic diversity. *Systematic Botany,* 1, 11–19.
Jensen, R. J. (1977*a*). A preliminary numerical analysis of the red oak complex in Michigan and Wisconsin. *Taxon,* 26, 399–407.
– (1977*b*). Numerical analysis of the scarlet oak complex (*Quercus* subgen. *Erythrobalanus*) in the eastern United States: Relationships above the species level. *Systematic Botany,* 2, 122–33.
Jensen, R. J., McLeod, M. J., Eshbaugh, W. H., & Guttman, S. I. (1979). Numerical taxonomic analyses of allozymic variation in *Capsicum* (Solanaceae). *Taxon,* 28, 315–27.
Jensen, S. (1978). Influences of transformation of cover values on classification and ordination of lake vegetation. *Vegetatio,* 37, 19–31.
– (1979). Classification of lakes in southern Sweden on the basis of their macrophyte composition by means of multivariate methods. *Vegetatio,* 39, 129–46.
Jensen, S. & Maarel, E. van der (1980). Numerical approaches to lake classification with special reference to macrophyte communities. *Vegetatio,* 42, 117–28.
Johnson, D. W. (1979). Air pollution and the distribution of corticolous lichens in Seattle, Washington. *Northwest Science,* 53, 257–63.
Johnson, E. A. (1977*a*). A multivariate analysis of the niches of plant populations in raised bogs. I. Niche dimensions. *Canadian Journal of Botany,* 55, 1201–10.
– (1977*b*). A multivariate analysis of the niches of plant populations in raised bogs. II. Niche width and overlap. *Canadian Journal of Botany,* 55, 1211–20.
– (1981). Vegetation organization and dynamics of lichen woodland communities in the Northwest Territories, Canada. *Ecology,* 62, 200–15.
Johnson, E. A. & Rowe, J. S. (1977). *Fire and Vegetation Change in the Western Subarctic.* Ottawa: Minister of Supply and Services.
Johnson, R. W. & Goodall, D. W. (1979). A maximum likelihood approach to non-linear ordination. *Vegetatio,* 41, 133–42.
Johnston, D. W. & Odum, E. P. (1956). Breeding bird populations in relation to plant succession on the piedmont of Georgia. *Ecology,* 37, 50–62.
Jonasson, S. (1981). Plant communities and species distribution of low alpine *Betula nana* heaths in northernmost Sweden. *Vegetatio,* 44, 51–64.
Jones, E. B. G. (1974). Aquatic fungi: Freshwater and marine. In *Biology of Plant Litter Decomposition,* eds. C. H. Dickinson & G. J. F. Pugh, pp. 337–83. New York: Academic Press.
Jones, J. H., Card, W., Chapman, M., Lennard-Jones, J. E., Morson, B. C., Sackin, M. J., & Sneath, P. H. A. (1970). Heterogeneity of disease. *Classification Society Bulletin,* 2, 33–8.

Jones, R. K. (1976). The use of classification to elucidate nutrient responses in *Stylosanthes.* In *Pattern Analysis in Agricultural Science,* ed. W. T. Williams, pp. 194–9. New York: Elsevier.

Jonescu, M. E. (1979). Natural revegetation of strip-mined land in the lignite coalfields of southern Saskatchewan. In *Ecology and Coal Resource Development,* ed. M. K. Wali, vol. 2, pp. 592–608. New York: Pergamon.

Jöreskog, K. G., Klovan, J. E., & Reyment, R. A. (1976). *Geological Factor Analysis.* New York: Elsevier.

Kachi, N. & Hirose, T. (1979*a*). Multivariate approaches to the plant communities related with edaphic factors in the dune system at Azigaura, Ibaraki Pref. I. Association-analysis. *Japanese Journal of Ecology,* 29, 17–27.

– (1979*b*). Multivariate approaches to the plant communities related with edaphic factors in the dune system at Azigaura, Ibaraki Pref. II. Ordination. *Japanese Journal of Ecology,* 29, 359–68.

Kaltsikes, P. J. (1973). Multivariate statistical analysis of yield, its components and characters above the flag leaf node in spring rye. *Theoretical and Applied Genetics,* 43, 88–90.

– (1974). Application of multivariate statistical techniques to yield and characters associated with it in hexaploid *Triticale. Zeitschrift für Pflanzenzüchtung,* 72, 252–9.

Karr, J. R. (1980). Geographic variation in the avifaunas of tropical forest undergrowth. *Auk,* 97, 283–98.

Karr, J. R. & Martin, T. E. (1981). Random numbers and principal components: Further searches for the unicorn. In *The Use of Multivariate Statistics in Studies of Wildlife Habitat,* ed. D. E. Capen. Ft. Collins, Colo.: Rocky Mountain Forest and Range Experiment Station (in press).

Kelsey, C. T., Goff, F. G., & Fields, D. (1976). *Theory and Analysis of Vegetation Pattern.* Oak Ridge, Tenn.: Oak Ridge National Laboratory.

Kendall, D. G. (1971). Seration from abundance matrices. In *Mathematics in the Archaeological and Historical Sciences,* eds. F. R. Hodson, D. G. Kendall, & P. Tautu, pp. 215–52. Edinburgh: Edinburgh University Press.

Kendeigh, S. C. (1961). *Animal Ecology.* Englewood Cliffs, N.J.: Prentice-Hall.

Kendrick, W. B. & Burges, A. (1962). Biological aspects of the decay of *Pinus sylvestris* leaf litter. *Nova Hedwigia,* 4, 313–42.

Kershaw, K. A. (1973). *Quantitative and Dynamic Plant Ecology.* 2d. ed. New York: Elsevier.

Kessell, S. R. (1976). Gradient modeling: A new approach to fire modeling and wilderness resource management. *Environmental Management,* 1, 39–48.

– (1979). *Gradient Modeling.* New York: Springer-Verlag.

Kessell, S. R. & Cattelino, P. J. (1978). Evaluation of a fire behavior information integration system for southern California chaparral wildlands. *Environmental Management,* 2, 135–59.

Kessell, S. R. & Whittaker, R. H. (1976). Comparisons of three ordination techniques. *Vegetatio,* 32, 21–9.

Kidson, J. W. (1975). Eigenvector analysis of monthly mean surface data. *Monthly Weather Review,* 103, 177–86.

Kinne, O. (editor) (1970). *Marine Ecology,* vol. I, *Environmental Factors.* New York: Wiley.

Knapp, R. R., Knapp, L., & Michael, W. B. (1979). The relationship of clustered interest measures and declared college major: Concurrent validity of the COPSystem interest inventory. *Educational and Psychological Measurement,* 39, 939–45.

Knight, D. H. & Loucks, O. L. (1969). A quantitative analysis of Wisconsin forest vegetation on the basis of plant function and gross morphology. *Ecology,* 50, 219–34.

Koeppl, J. W., Hoffmann, R. S., & Nadler, C. F. (1978). Pattern analysis of acoustical behavior in four species of ground squirrels. *Journal of Mammalogy,* 59, 677–96.

Komárkova, V. (1980). Classification and ordination in the Indian Peaks Area, Colorado Rocky Mountains. *Vegetatio,* 42, 149–63.

Kosina, R. (1980). Evaluation of the structure and caryopsis quality of some species and hybrids of spring wheat with application of multivariate analysis. *Zeitschrift für Pflanzenzüchtung,* 85, 294–307.

Krajina, V. J. (1975). Some observations on the three subalpine biogeoclimatic zones in British Columbia, Yukon and Mackenzie Districts. *Phytocoenologia,* 2, 396–400.

Krasilov, V. A. (1975). *Paleoecology of Terrestrial Plants,* trans. H. Hardin. New York: Wiley.

Krupa, S. V. & Dommergues, Y. R. (editors) (1979). *Ecology of Root Pathogens.* New York: Elsevier.

Kruskal, J. B. (1964*a*). Multidimensional scaling by optimizing goodness of fit to a nonmetric hypothesis. *Psychometrika,* 29, 1–27.

– (1964*b*). Nonmetric multidimensional scaling: A numerical method. *Psychometrika,* 29, 115–29.

– (1977). The relationship between multidimensional scaling and clustering. In *Clustering and Classification,* ed. J. Van Ryzin, pp. 17–44. New York: Academic Press.

Krzanowski, W. J. (1972). Techniques in multivariate analysis. In *The Way Ahead in Plant Breeding,* ed. F. G. H. Lupton, G. Jenkins, & R. Johnson, pp. 147–55. Thrumpington: Plant Breeding Institute.

Kuris, A. M. & Brody, M. S. (1976). Use of principal components analysis to describe the snail shell resource for hermit crabs. *Journal of Experimental Marine Biology and Ecology,* 22, 69–77.

Lacoste, A. (1975). La Végétation de l'étage subalpin du bassin supérieur de la Tinée (Alpes–Maritimes). *Phytocoenologia,* 3, 83–122.

Ladd, P. G. (1979). Past and present vegetation on the Delegate River in the highlands of eastern Victoria. I. Present vegetation. *Australian Journal of Botany,* 27, 167–84.

LaFrance, C. R. (1972). Sampling and ordination characteristics of computer-simulated individualistic communities. *Ecology,* 53, 387–97.

Lambert, J. M., Meacock, S. E., Barrs, J., & Smartt, P. F. M. (1973). AXOR and MONIT: Two new polythetic-divisive strategies for hierarchical classification. *Taxon,* 22, 173–6.

Lambert, J. M. & Williams, W. T. (1962). Multivariate methods in plant ecology. IV. Nodal analysis. *Journal of Ecology,* 50, 775–802.
– (1966). Multivariate methods in plant ecology. VI. Comparison of information-analysis and association-analysis. *Journal of Ecology,* 54, 635–64.
Lamoureux, G. & Lacoursière, E. (1976). Etude Préliminaire des groupements végétaux caractérisant quelques gîtes larvaires à moustiques dans la région de Trois-Rivières (Québec). *Canadian Journal of Botany,* 54, 177–90.
Lance, G. N. & Williams, W. T. (1965). Computer programs for monothetic classification ("Association analysis"). *Computer Journal,* 8, 246–9.
Lance, G. N. & Williams, W. T. (1966). A generalized sorting strategy for computer classifications. *Nature,* 212, 218.
– (1967). A general theory of classificatory sorting strategies. I. Hierarchical systems. *Computer Journal,* 9, 373–80.
Laperriere, A. J., Lent, P. C., Gassaway, W. C., & Nodler, F. A. (1980). Use of Landsat data for moose-habitat analyses in Alaska. *Journal of Wildlife Management,* 44, 881–7.
LaPorte, R. E. & Voss, J. F. (1979). Prose representation: A multidimensional scaling approach. *Multivariate Behavioral Research,* 14, 39–56.
Larson, D. W. (1980). Patterns of species distribution in an *Umbilicaria* dominated community. *Canadian Journal of Botany,* 58, 1269–79.
Lausi, D. & Feoli, E. (1979). Hierarchical classification of European salt marsh vegetation based on numerical methods. *Vegetatio,* 39, 171–84.
Lavrentiades, G. (1980). On the grain-field weeds of the American Farm School of Thessaloniki. *Phytocoenologia,* 7, 318–35.
Lebrun, J. (1977). Applications phytosociologiques à l'aménagement du territoire. *Vegetatio,* 35, 123–9.
Lee, R. (1978). *Forest Microclimatology.* New York: Columbia University Press.
Lepart, J. & Debussche, M. (1980). Information efficiency and regional constellation of environmental variables. *Vegetatio,* 42, 85–91.
Levin, M. H. & Tedrow, J. C. F. (1980). Environmental history of the lower Metedeconk River region, New Jersey pine barrens. *Bulletin of the New Jersey Academy of Science,* 25, 59.
Levine, D. M. (1977). Multivariate analysis of the visual information processing of numbers. *Journal of Multivariate Behavioral Research,* 12, 347–55.
Lewin, D. C. (1974). The vegetation of the ravines of the Southern Finger Lakes, New York Region. *American Midland Naturalist,* 91, 315–42.
Liddle, M. J. & Greig-Smith, P. (1975). A survey of tracks and paths in a sand dune ecosystem. II. Vegetation. *Journal of Applied Ecology,* 12, 909–30.
Lieth, H. & Moore, G. W. (1971). Computerized clustering of species in phytosociological tables and its utilization for field work. In *Spatial Patterns and Statistical Distributions, Statistical Ecology,* eds. G. P. Patil, E. C. Pielou, & W. E. Waters, vol. 1, pp. 403–22. University Park, Pa.: Pennsylvania State University Press.
Lind, O.T. (1979). *Handbook of Common Methods in Limnology.* 2d. ed. St. Louis: C.V. Mosby.
Lindsay, M. M. (1978). *The Vegetation of the Grassy Balds and Other High*

Elevation Disturbed Areas in Great Smoky Mountains National Park. Gatlinburg, Tenn.: Uplands Field Research Laboratory, Great Smoky Mountains National Park.

Lindsay, M. M. & Bratton, S. P. (1979). The vegetation of grassy balds and other high elevation disturbed areas in the Great Smoky Mountains National Park. *Bulletin of the Torrey Botanical Club,* 106, 264–75.

Longton, R. E. (1979). Vegetation ecology and classification in the Antarctic Zone. *Canadian Journal of Botany,* 57, 2264–78.

Lötschert, W. (1977). Bark of deciduous trees as an indicator for air pollution. In *Vegetation Science and Environmental Protection,* eds. A. Miyawaki & R. Tüxen, pp. 247–55. Tokyo: Maruzen.

Loucks, O. L. (1962). Ordinating forest communities by means of environmental scalars and phytosociological indices. *Ecological Monographs,* 32, 137–66.

Loucks, O. L. & Schnur, B. J. (1976). A gradient in understory shrub composition in southern Wisconsin. In *Central Hardwoods Forest Conference,* eds. J. S. Fralish, G. T. Weaver, & R. C. Schlesinger, pp. 99–117. St. Paul: North Central Forest Experiment Station.

Louppen, J. M. W. & Maarel, E. van der (1979). CLUSLA: A computer program for the clustering of large phytosociological data sets. *Vegetatio,* 40, 107–14.

Maarel, E. van der (1972). Ordination of plant communities on the basis of their plant genus, family and order relationships. In *Grundfragen und Methoden in der Pflanzensoziologie,* eds. E. van der Maarel & R. Tüxen, pp. 183–206. The Hague: Junk.

– (1974). The Working Group for Data-Processing of the International Society for Plant Geography and Ecology in 1972–1973. *Vegetatio,* 29, 63–7.

– (1975). The Braun-Blanquet approach in perspective. *Vegetatio,* 30, 213–19.

– (1979*a*). Multivariate methods in phytosociology, with reference to The Netherlands. In *The Study of Vegetation,* ed. M. J. A. Werger, pp. 161–225. The Hague: Junk.

– (1979*b*). Transformation of cover-abundance values in phytosociology and its effects on community similarity. *Vegetatio,* 39, 97–114.

– (1980*a*). On the interpretability of ordination diagrams. *Vegetatio,* 42, 43–5.

– (1980*b*). Vegetation development in a former orchard under different treatments: A preliminary report. *Vegetatio,* 43, 95–102.

Maarel, E. van der, Janssen, J. G. M., & Louppen, J. M. W. (1978). TABORD, a program for structuring phytosociological tables. *Vegetatio,* 38, 143–56.

Maarel, E. van der, Orlóci, L., & Pignatti, S. (1976). Data-processing in phytosociology, retrospect and anticipation. *Vegetatio,* 32, 65–72.

– (editors) (1980). *Data-processing in phytosociology.* The Hague: Junk.

Maarel, E. van der, Tüxen, R., & Westhoff, V. (1970). Bibliographie pflanzensoziologischer Lehrbücher und verwandter Schriften. *Excerpta Botanica, Section B,* 11, 86–160.

Maarel, E. van der & Werger, M. J. A. (1978). On the treatment of sucession data. *Phytocoenosis,* 7, 257–78.

Macan, T. T. & Worthington, E. B. (1951). *Life in Lakes and Rivers*. London: Collins.
Macfadyen, A. (1975). Some thoughts on the behavior of ecologists. *Journal of Applied Ecology,* 12, 351–63.
Madden, L. & Pennypacker, S. P. (1979). Principal component analysis of tomato early blight epidemics. *Phytopathologische Zeitschrift,* 95, 364–9.
Majer, J. D. (1976). The influence of ants and ant manipulation on the cocoa farm fauna. *Journal of Applied Ecology,* 13, 157–75.
Malhotra, K. C., Chakraborty, R., & Chakravarti, A. (1978). Gene differentiation among the Dhangar caste-cluster of Maharashtra, India. *Human Heredity,* 28, 26–36.
Manion, C. V. & Hassanein, K. (1980). A hospital use evaluation by numerical taxonomy. *Computers and Biomedical Research,* 13, 567–80.
Mannetje, L. 't. (1967). A comparison of eight numerical procedures applied to the classification of some African *Trifolium* taxa based on *Rhizobium* affinities. *Australian Journal of Botany,* 15, 521–8.
Marsh, J. E. & Nash, T. H. (1979). Lichens in relation to the Four Corners Power Plant in New Mexico. *Bryologist,* 82, 20–8.
Matson, R. G. & True, D. L. (1974). Site relationships at Quebrada Tarapaca, Chile: A comparison of clustering and scaling techniques. *American Antiquity,* 39, 51–74.
Matthews, J. A. (1979*a*). The vegetation of the Storbreen gletschervorfeld, Jotunheimen, Norway. I. Introduction and approaches involving classification. *Journal of Biogeography,* 6, 17–47.
– (1979*b*). The vegetation of the Storbreen gletschervorfeld, Jotunheimen, Norway. II. Approaches involving ordination and general conclusions. *Journal of Biogeography,* 6, 133–67.
Matveev, V. A. & Tikhomirova, A. L. (1975). Succession in the staphylinid fauna of spruce cuttings in Mari, Armenian SSR. *Soviet Journal of Ecology,* 6, 548–52.
McCullough, J. M., Martinsen, C. S., & Moinpour, R. (1978). Application of multidimensional scaling to the analysis of sensory evaluations of stimuli with known attribute structures. *Journal of Applied Psychology,* 1, 103–9.
McIntosh, R. P. (1962). Raunkiaer's "law of frequency." *Ecology,* 43, 533–5.
– (1970). Community, competition, and adaptation. *Quarterly Review of Biology,* 45, 259–80.
– (1975). H. A. Gleason – "Individualistic Ecologist" – 1882–1975: His contributions to ecological theory. *Bulletin of the Torrey Botanical Club,* 102, 253–73.
– (1976). Ecology since 1900. In *Issues and Ideas in America,* eds. B. J. Taylor & T. J. White, pp. 353–72. Norman, Okla.: University of Oklahoma Press.
– (1978). Matrix and plexus techniques. In *Ordination of Plant Communities,* ed. R. H. Whittaker, pp. 151–84. The Hague: Junk.
– (1980). The background and some current problems of theoretical ecology. *Synthese,* 43, 195–255.

McKerrow, W. S. (1978). *The Ecology of Fossils*. Cambridge, Mass.: MIT Press.

Mertz, D. B. & McCauley, D. E. (1980). The domain of laboratory ecology. *Synthese,* 43, 95–110.

Meulen, F. van der, Morris, J. W., & Westfall, R. (1978). A computer aid for the preparation of Braun-Blanquet tables. *Vegetatio,* 38, 129–34.

Meyers, J. M. & Johnson, A. S. (1978). Bird communities associated with succession and management of loblolly-shortleaf pine forests. In *Proceedings of the Workshop Management of Southern Forests for Nongame Birds,* ed. R. M. DeGraaf, pp. 50–65. Asheville, N.C.: Southeastern Forest Experiment Station.

Milne, P. W. (1976). The Canberra programs and their accession. In *Pattern Analysis in Agricultural Science,* ed. W. T. Williams, pp. 116–23. New York: Elsevier.

Mirkin, B. M. & Rozenberg, G. S. (1977). Experience of application of the method of principal components of vegetational variation. *Soviet Journal of Ecology,* 8, 403–10.

Mittelhammer, R. C., Young, D. L., Tasanasanta, D., & Donnelly, J. T. (1980). Mitigating the effects of multicollinearity using exact and stochastic restrictions: The case of an aggregate agricultural production function in Thailand. *American Journal of Agricultural Economics,* 62, 199–210.

Mohler, C. L. (1981). Effects of sample distribution on eigenvector ordination. *Vegetatio* (in press).

Molina-Cano, J. L. (1976). A numerical classification of some European barley cultivars (*Hordeum vulgare* L.s.l.). *Zeitschrift für Pflanzenzüchtung,* 76, 320–33.

Molina-Cano, J. L. & Elena Rosselló, J. M. (1978). A further contribution to the classification of barley cultivars: Use of numerical taxonomy and biochemical methods. *Seed Science and Technology,* 6, 593–615.

Moore, A. W. & Russell, J. S. (1976*a*). Ordination of soil data. In *Pattern Analysis in Agricultural Science,* ed. W. T. Williams, pp. 204–14. New York: Elsevier.

– (1976*b*). Problems in numerical classification of soil data. In *Pattern Analysis in Agricultural Science,* ed. W. T. Williams, pp. 215–23. New York: Elsevier.

Moore, A. W., Russell, J. S., & Ward, W. T. (1972). Numerical analysis of soils: A comparison of three soil profile models with field classification. *Journal of Soil Science,* 23, 193–209.

Moore, J. J. (1971). *Phyto – A suite of programs in Fortran IV for the manipulation of phytosociological tables according to the principles of Braun-Blanquet.* Dublin: University College.

– (1972). An outline of computer-based methods for the analysis of phytosociological data. In *Grundfragen und Methoden in der Pflanzensoziologie,* eds. E. van der Maarel & R. Tüxen, pp. 29–38. The Hague: Junk.

Moore, J. J., Fitzsimons, P., Lambe, E., & White, J. (1970). A comparison and evaluation of some phytosociological techniques. *Vegetatio,* 20, 1–20.

Moore, J. J. & O'Sullivan, A. (1970). A comparison between the results of the Braun-Blanquet method and those of cluster analysis. In *Gesellschaftsmorphologie,* ed. R. Tüxen, pp. 26–30. The Hague: Junk.

– (1978). A phytosociological survey of the Irish Molinio–Arrhenatheretea using computer techniques. *Vegetatio,* 38, 89–93.

Moral, R. del (1980). On selecting indirect ordination methods. *Vegetatio,* 42, 75–84.

Moral, R. del & Denton, M. F. (1977). Analysis and classification of vegetation based on family composition. *Vegetatio,* 34, 155–65.

Moral, R. del & Watson, A. F. (1978). Gradient structures of forest vegetation in the central Washington Cascades. *Vegetatio,* 38, 29–48.

Moravec, J. (1978). Application of constancy-species groups for numerical ordering of phytosociological tables–The synoptic table version. *Vegetatio,* 37, 33–42.

Morgan, B. J. T., Simpson, M. J. A., Hanby, J. P., & Hall-Craggs, J. (1976). Visualizing interaction and sequential data in animal behavior: Theory and application of cluster-analysis methods. *Behavior,* 56, 1–43.

Morgenstern, E. K. (1978). Range-wide genetic variation of black spruce. *Canadian Journal of Forest Research,* 8, 463–73.

Morin, G., Fortin, J.-P., Sochanska, W., & Lardeau, J.-P. (1979). Use of principal component analysis to identify homogeneous precipitation stations for optimal interpolation. *Water Resources Research,* 15, 1841–50.

Mosley, M. P. (1981). Delimitation of New Zealand hydrologic regions. *Journal of Hydrology,* 49, 173–92.

Moss, W. W. & Hendrickson, J. A. (1973). Numerical taxonomy. *Annual Review of Entomology,* 18, 227–58.

Mueller, W. H. & Reid, R. M. (1979). A multivariate analysis of fatness and relative fat patterning. *American Journal of Physical Anthropology,* 50, 199–208.

Mueller-Dombois, D. & Ellenberg, H. (1974). *Aims and Methods of Vegetation Ecology.* New York: Wiley.

Mueller-Dombois, D. & Spatz, G. (1975). Application of the relevé method to insular tropical vegetation for an environmental impact study. *Phytocoenologia,* 2, 417–29.

Muir, J. W., Hardie, H. G. M., Inkson, R. H. E., & Anderson, A. J. B. (1970). The classification of soil profiles by traditional and numerical methods. *Geoderma,* 4, 81–90.

Mungomery, V. E., Shorter, R., & Byth, D. E. (1974). Genotype x environment interactions and environmental adaptation. I. Pattern analysis–application to soya bean populations. *Australian Journal of Agricultural Research,* 25, 59–72.

Munn, R. E. (1970). *Biometeorological Methods.* New York: Academic Press.

Myers, J. G. (1977). Cluster analysis of marketing data. In *Multivariate Methods for Market and Survey Research,* ed. J. N. Sheth, pp. 163–85. Chicago: American Marketing Association.

Myers, W. L. & Shelton, R. L. (1980). *Survey Methods for Ecosystem Management.* New York: Wiley.

Nairn, A. E. M. (editor) (1964). *Problems in Palaeoclimatology*. New York: Wiley.

Nalepa, T. F. & Thomas, N. A. (1976). Distribution of macrobenthic species in Lake Ontario in relation to sources of pollution and sediment parameters. *Journal of Great Lakes Research,* 2, 150–63.

National Academy of Sciences (1971). *A Guide to Environmental Research on Animals.* Washington, D.C.: National Academy of Sciences.

Neff, N. A. & Marcus, L. F. (1980). *A Survey of Multivariate Methods for Systematics.* New York: American Museum of Natural History.

Nevers, J. V. (1972). Multidimensional Scaling Applications in Market Segmentation. Ph.D. thesis, Purdue University, Lafayette, Ind.

Newell, C. A. & Hymowitz, T. (1978). A reappraisal of the subgenus *Glycine. American Journal of Botany,* 65, 168–79.

Newnham, R. M. (1968). A classification of climate by principal component analysis and its relationship to tree species distribution. *Forest Science,* 14, 254–64.

Nichols, S. (1977). On the interpretation of principal components analysis in ecological contexts. *Vegetatio,* 34, 191–7.

Niemi, G. J. & Pfannmuller, L. (1979). Avian communities: Approaches to describing their habitat associations. In *Proceedings of the Workshop Management of Northcentral and Northeastern Forests for Nongame Birds,* ed. R. M. DeGraaf, pp. 154–78. Orono, Me.: Northeastern Forest Experiment Station.

Niklas, K. J. (1976). Chemical examinations of some non-vascular Paleozoic plants. *Brittonia,* 28, 113–37.

Niklas, K. J. & Gensel, P. G. (1976). Chemotaxonomy of some Paleozoic vascular plants. Part I: Chemical compositions and preliminary cluster analysis. *Brittonia,* 28, 353–78.

– (1977). Chemotaxonomy of some Paleozoic vascular plants. Part II: Chemical characterization of major plant groups. *Brittonia,* 29, 100–11.

– (1978). Chemotaxonomy of some Paleozoic vascular plants. Part III: Cluster configurations and their bearing on taxonomic relationships. *Brittonia,* 30, 216–32.

Nishisato, S. (1980). *Analysis of Categorical Data: Dual Scaling and Its Applications.* Toronto: University of Toronto Press.

Noble, I. R. & Slatyer, R. O. (1980). The use of vital attributes to predict successional changes in plant communities subject to recurrent disturbances. *Vegetatio,* 43, 5–21.

Noon, B. N. (1981). The distribution of an avian guild along a temperate elevational gradient: The importance and expression of competition. *Ecological Monographs,* 51, 105–24.

Noy-Meir, I. (1971). Multivariate analysis of the semi-arid vegetation in southeastern Australia: Nodal ordination by component analysis. *Proceedings of the Ecological Society of Australia,* 6, 159–93.

– (1973*a*). Data transformations in ecological ordination. I. Some advantages of non-centering. *Journal of Ecology,* 61, 329–41.

– (1973*b*). Divisive polythetic classification of vegetation data by optimized division on ordination components. *Journal of Ecology,* 61, 753–60.
– (1974*a*). Catenation: Quantitative methods for the definition of coenoclines. *Vegetatio,* 29, 89–99.
– (1974*b*). Multivariate analysis of the semiarid vegetation in south-eastern Australia. II. Vegetation catenae and environmental gradients. *Australian Journal of Botany,* 22, 115–40.
– (1979). Graphical models and methods in ecology. In *Contemporary Quantitative Ecology and Related Ecometrics,* eds. G. P. Patil & M. Rosenzweig, pp. 453–72. Burtonsville, Md.: International Co-operative.
Noy-Meir, I. & Austin, M. P. (1970). Principal component ordination and simulated vegetational data. *Ecology,* 61, 551–2.
Noy-Meir, I., Walker, D., & Williams, W. T. (1975). Data transformations in ecological ordination. II. On the meaning of data standardization. *Journal of Ecology,* 63, 779–800.
Noy-Meir, I. & Whittaker, R. H. (1977). Continuous multivariate methods in community analysis: Some problems and developments. *Vegetatio,* 33, 79–98.
Nygren, T. E. (1977). The relationship between the perceived risk and attractiveness of gambles: A multidimensional analysis. *Applied Psychological Measurement,* 4, 565–79.
Olsson, H. (1978). Vegetation of artificial habitats in northern Malmö and environs. *Vegetatio,* 36, 65–82.
Olsvig, L. S., Cryan, J. F., & Whittaker, R. H. (1979). Vegetational gradients of the pine plains and barrens of Long Island, New York. In *Pine Barrens – Ecosystem and Landscape,* ed. R. T. T. Forman, pp. 265–82. New York: Academic Press.
Onans, J. & Parsons, R. F. (1980). Regeneration of native plants on abandoned mallee farmland in south-eastern Australia. *Australian Journal of Botany,* 28, 479–93.
Onyekwelu, S. S. C. & Okafor, J. C. (1979). Ordination of a savanna woodland in Nigeria using woody and herbaceous species. *Vegetatio,* 40, 95–100.
Orlóci, L. (1966). Geometric models in ecology. I. The theory and application of some ordination methods. *Journal of Ecology,* 54, 193–215.
– (1967). An agglomerative method for the classification of plant communities. *Journal of Ecology,* 55, 193–206.
– (1968). Definitions of structure in multivariate phytosociological samples. *Vegetatio,* 15, 281–91.
– (1972). On objective functions of phytosociological resemblance. *American Midland Naturalist,* 88, 28–55.
– (1973). An algorithm for cluster seeking in ecological collections. *Vegetatio,* 27, 339–45.
– (1974*a*). On information flow in ordination. *Vegetatio,* 29, 11–16.
– (1974*b*). Revisions for the Bray and Curtis ordination. *Canadian Journal of Botany,* 52, 1773–6.
– (1975). Measurement of redundancy in species collections. *Vegetatio,* 31, 65–7.

– (1978*a*). *Multivariate Analysis in Vegetation Research.* 2d. ed. The Hague: Junk.
– (1978*b*). Ordination by resemblance matrices. In *Ordination of Plant Communities,* ed. R. H. Whittaker, pp. 239–75. The Hague: Junk.
– (1980). An algorithm for predictive ordination. *Vegetatio,* 42, 23–5.
Orlóci, L., Feoli, E., & Fewster, P. (1977). *Multivariate Analysis of Vegetation Data.* London, Ont.: University of Western Ontario.
Orlóci, L. & Mukkattu, M. M. (1973). The effect of species number and type of data on the resemblance structure of a phytosociological collection. *Journal of Ecology,* 61, 37–46.
Orlóci, L. & Stanek, W. (1979). Vegetation survey of the Alaska Highway, Yukon Territory: Types and gradients. *Vegetatio,* 41, 1–56.
Orr, D. M. (1980). Effects of sheep grazing *Astrebla* grassland in central western Queensland. I. Effects of grazing pressure and livestock distribution. *Australian Journal of Agricultural Research,* 31, 797–806.
Orsay, J.-C. K. (1979). Application de l'analyse factorielle des correspondances à l'étude phytosociologique de l'étage alpin des Pyrénées centrales. *Phytocoenologia,* 5, 125–88.
Osmond, C. B., Björkman, O., & Anderson, D. J. (1980). *Physiological Processes in Plant Ecology.* New York: Springer-Verlag.
Ozimek, T. (1978). Effect of municipal sewage on the submerged macrophytes of a lake littoral. *Ekologia Polska,* 26, 3–39.
Pakarinen, P. (1976). Agglomerative clustering and factor analysis of south Finnish mire types. *Annales Botanici Fennici,* 13, 35–41.
Pakarinen, P. & Ruuhijärvi, R. (1978). Ordination of northern Finnish peatland vegetation with factor analysis and reciprocal averaging. *Annales Botanici Fennici,* 15, 147–57.
Parker, W. H., Bradfield, G. E., Maze, J., & Lin, S.-C. (1979). Analysis of variation in leaf and twig characters of *Abies lasiocarpa* and *A. amabilis* from north-coastal British Columbia. *Canadian Journal of Botany,* 57, 1354–66.
Parren, S. G., Thompson, F. R., & Capen, D. E. (1980). *A Selected Bibliography: The Use of Multivariate Statistics in Studies of Wildlife Habitat.* Burlington, Vt.: University of Vermont.
Passer, M. W., Kelley, H. H., & Michela, J. L. (1978). Multidimensional scaling of the causes of negative interpersonal behavior. *Journal of Personality and Social Psychology,* 36, 951–62.
Paterson, J. G., Goodchild, N. A., & Boyd, W. J. R. (1978). Classifying environments for sampling purposes using a principal component analysis of climatic data. *Agricultural Meteorology,* 19, 349–62.
Pearce, K. & Lester, R. N. (1979). Chemotaxonomy of the cultivated eggplant – A new look at the taxonomic relationships of *Solanum melongena* L. In *The Biology and Taxonomy of the Solanaceae,* eds. J. G. Hawkes, R. N. Lester, & A. D. Skelding, pp. 615–27. New York: Academic Press.
Pearson, K. (1901). On lines and planes of closest fit to systems of points in space. *Philosophical Magazine, Sixth Series,* 2, 559–72.

Peet, R. K. (1978). Forest vegetation of the Colorado Front Range: Patterns of species diversity. *Vegetatio,* 37, 65–78.

– (1980). Ordination as a tool for analyzing complex data sets. *Vegetatio,* 42, 171–4.

Peet, R. K. & Christensen, N. L. (1980). Succession: A population process. *Vegetatio,* 43, 131–40.

Peet, R. K. & Loucks, O. L. (1977). A gradient analysis of southern Wisconsin forests. *Ecology,* 58, 485–99.

Pemadasa, M. A. & Mueller-Dombois, D. (1979). An ordination study of montane grasslands of Sri Lanka. *Journal of Ecology,* 67, 1009–23.

Pentecost, A. (1980). The lichens and bryophytes of rhyolite and pumice-tuff rock outcrops in Snowdonia, and some factors affecting their distribution. *Journal of Ecology,* 68, 251–67.

Persson, S. (1980). Succession in a south Swedish deciduous wood: A numerical approach. *Vegetatio,* 43, 103–22.

– (1981). Ecological indicator values as an aid in the interpretation of ordination diagrams. *Journal of Ecology,* 69, 71–84.

Peterson, J. T. (1970). Distribution of sulfur dioxide over metropolitan St. Louis, as described by empirical eigenvectors, and its relation to meteorological parameters. *Atmospheric Environment,* 4, 501–18.

Pfister, R. D. & Arno, S. F. (1980). Classifying forest habitat types based on potential climax vegetation. *Forest Science,* 26, 52–70.

Phalen, W. J. (1975). *Comprehensive Occupational Data Analysis Programs (CODAP): Ordering of Hierarchically Grouped Case Data (KPATH) and Print (PRKPTH) Programs.* San Antonio, Tx.: Lackland Air Force Base.

Phillips, D. L. (1978). Polynomial ordination: Field and computer simulation testing of a new method. *Vegetatio,* 37, 129–40.

Pickersgill, B., Heiser, C. B., & McNeill, J. (1979). Numerical taxonomic studies on variation and domestication in some species of *Capsicum.* In *The Biology and Taxonomy of the Solanaceae,* eds. J. G. Hawkes, R. N. Lester, & A. D. Skelding, pp. 679–700. New York: Academic Press.

Pielou, E. C. (1974). *Population and Community Ecology.* New York: Gordon & Breach.

– (1977). *Mathematical Ecology.* 2d. ed. New York: Wiley.

– (1979). Interpretation of paleoecological similarity matrices. *Paleobiology,* 5, 435–43.

Pignatti, S. (1980). Reflections on the phytosociological approach and the epistemological basis of vegetation science. *Vegetatio,* 42, 181–5.

Platts, W. S. (1980). A plea for fishery habitat classification. *Fisheries,* 5, 2–6.

Poag, C. W. (1977). Biostratigraphy in Gulf Coast petroleum exploration. In *Concepts and Methods of Biostratigraphy,* eds. E. G. Kauffman & J. E. Hazel, pp. 213–33. Stroudsburg, Pa.: Dowden, Hutchinson & Ross.

Podani, J. (1979). Association-analysis based on the use of mutual information. *Acta Botanica Academiae Scientiarum Hungaricae,* 25, 125–30.

– (1980). *SYN-TAX: Számitógépes Programcsomag Ökológiai, Cönoógiai és Taxonómiai Osztályozások Végrehajtására.* Vácrátót, Hungary: Botanikai Kutató Intézete.

Poore, G. C. B. & Mobley, M. C. (1980). Canonical correlation analysis of marine macrobenthos survey data. *Journal of Experimental Marine Biology and Ecology,* 45, 37–50.

Poore, M. E. D. (1955). The use of phytosociological methods in ecological investigations. I. The Braun-Blanquet system. *Journal of Ecology,* 43, 226–44.

– (1956). The use of phytosociological methods in ecological investigations. IV. General discussion of phytosociological problems. *Journal of Ecology,* 44, 28–50.

– (1962). The method of successive approximation in descriptive ecology. *Advances in Ecological Research,* 1, 35–68.

Prentice, I. C. (1977). Non-metric ordination methods in ecology. *Journal of Ecology,* 65, 85–94.

– (1980*a*). Multidimensional scaling as a research tool in quaternary palynology: A review of theory and methods. *Review of Palaeobotany and Palynology,* 31, 71–104.

– (1980*b*). Vegetation analysis and order invariant gradient models. *Vegetatio,* 42, 27–34.

Preston, F. W. (1948). The commonness, and rarity, of species. *Ecology,* 29, 254–83.

– (1980). Noncanonical distributions of commonness and rarity. *Ecology,* 61, 88–97.

Pritchard, N. M. & Anderson, A. J. B. (1971). Observations on the use of cluster analysis in botany with an ecological example. *Journal of Ecology,* 59, 727–47.

Puckett, K. J. & Finegan, E. J. (1980). An analysis of the element content of lichens from the Northwest Territories, Canada. *Canadian Journal of Botany,* 58, 2073–89.

Rabe, B. D. & Cisne, J. L. (1980). Chronostratigraphic accuracy of ecostratigraphic correlation in the Trenton Group (Ordovician, New York). *Lethaia,* 13, 109–18.

Ramensky, L. G. (1930). Zur Methodik der vergleichenden Bearbeitung und Ordnung von Pflanzenlisten und anderen Objekten, die durch mehrere, verschiedenartig wirkende Faktoren bestimmt werden. *Beiträge zur Biologie der Pflanzen,* 18, 269–304.

Rand, W. M. (1971). Objective criteria for the evaluation of clustering methods. *Journal of the American Statistical Association, Theory and Methods Section,* 66, 846–50.

Rao, V. R. (1975). Taxonomy of television programs based on viewing behavior. *Journal of Marketing Research,* 12, 355–8.

– (1977). Conjoint measurement in marketing analysis. In *Multivariate Methods for Market and Survey Research,* ed. J. N. Sheth, pp. 257–86. Chicago: American Marketing Association.

– (1980). Books on quantitative methods for consumer research. *Journal of Consumer Research,* 7, 198–210.

Raunkaier, C. (1934). *The Life Forms of Plants and Statistical Plant Geography.* Oxford: Clarendon.

Rawls, W. J. & Jackson, T. J. (1979). Pattern recognition analysis of snowdrifts. *Nordic Hydrology,* 10, 251–60.

Refseth, D. (1980). Ecological analyses of carabid communities – Potential use in biological classification for nature conservation. *Biological Conservation,* 17, 131–41.

Regnéll, G. (1979). Vegetationsförändringar vid upphörande bete i en shånsk kalkfuktäng. Utvärdering och tolkningsmöjligheter med datateknik. *Svensk botanisk tidskrift,* 73, 139–59.

– (1980). A numerical study of successions in an abandoned, damp calcareous meadow in S. Sweden. *Vegetatio,* 43, 123–30.

Reid, R. (1973). A numerical classification of sown tropical pasture regions based on the performance of sown pasture species. *Tropical Grasslands,* 7, 331–40.

– (1976). Delimitation of world pasture types by numerical methods. In *Pattern Analysis in Agricultural Science,* ed. W. T. Williams, pp. 242–8. New York: Elsevier.

Reid, R., Ryan, D. M., & Burt, R. L. (1976). Exploration for, and utilization of, collections of tropical pasture legumes. II. The Papakakis system of climatic classification applied to testing areas in northern Australia. *Agro-Ecosystems,* 2, 309–18.

Reyment, R. A. (1980). *Morphometric Methods in Biostratigraphy.* New York: Academic Press.

Reynolds, J. F. & Crawford, D. J. (1980). A quantitative study of variation in the *Chenopodium atrovirens–desiccatum–pratericola* complex. *American Journal of Botany,* 67, 1380–90.

Rhodes, A. M., Bemis, W. P., Whitaker, T. W., & Carmer, S. G. (1968). A numerical taxonomic study of *Cucurbita. Brittonia,* 20, 251–66.

Rhodes, A. M., Campbell, C., Malo, S. E., & Carmer, S. G. (1970). A numerical taxonomic study of the mango *Mangifera indica* L. *Journal of the American Society for Horticultural Science,* 95, 252–6.

Rhodes, A. M., Carmer, S. G., & Courter, J. W. (1969). Measurement and classification of genetic variability in horseradish. *Journal of the American Society for Horticultural Science,* 94, 98–102.

Rhodes, A. M., Malo, S. E., Campbell, C. W., & Cramer, S. G. (1971). A numerical taxonomic study of the avocado (*Persea americana* Mill.). *Journal of the American Society for Horticultural Science,* 96, 391–5.

Richards, B. N. (1974). *Introduction to the Soil Ecosystem.* Essex: Longman.

Richardson, M. W. & Kuder, G. F. (1933). Making a rating scale that measures. *Personnel Journal,* 12, 36–40.

Richens, R. H. & Jeffers, J. N. R. (1978). Multivariate analysis of the elms of northern France. II. Pooled analysis of the elm populations of northern France and England. *Silvae Genetica,* 27, 85–95.

Ricklefs, R. E. & Lau, M. (1980). Bias and dispersion of overlap indices: Results of some Monte Carlo simulations. *Ecology,* 61, 1019–24.

Riechert, S. E. (1978). Games spiders play: Behavioral variability and territorial disputes. *Behavioral Ecology and Sociobiology,* 3, 135–62.

Riggs, R. D., Hamblen, M. L., & Rakes, L. (1981). Infra-species variation in reactions to hosts in *Heterodera glycines* populations. *Journal of Nematology,* 13, 171–9.

Ritchie, J. C. & Yarranton, G. A. (1978*a*). Pattern of change in the late-quaternary vegetation of the western interior of Canada. *Canadian Journal of Botany,* 56, 2177–83.

– (1978*b*). The late-quaternary history of the boreal forest of central Canada, based on standard pollen stratigraphy and principal components analysis. *Journal of Ecology,* 66, 199–212.

Roberts, A. V. (1977). Relationships between species in the genus *Rosa,* section *Pimpinellifoliae. Botanical Journal of the Linnean Society,* 74, 309–28.

Roberts, R. D., Marrs, R. H., Skeffington, R. A., & Bradshaw, A. D. (1981). Ecosystem development on naturally-colonized china clay wastes. I. Vegetation changes and overall accumulation of organic matter and nutrients. *Journal of Ecology,* 69, 153–61.

Robertson, P. A. (1978). Comparison of techniques for ordinating and classifying old-growth floodplain forests in southern Illinois. *Vegetatio,* 37, 43–51.

– (1979). Comparisons among three hierarchical classification techniques using simulated coenoplanes. *Vegetatio,* 40, 175–83.

Robinson, P. J. (1976). Analysis of electrophoretic data using pattern-analysis techniques. In *Pattern Analysis in Agricultural Science,* ed. W. T. Williams, pp. 200–3. New York: Elsevier.

Robinson, P. J., Burt, R. L., & Williams, W. T. (1980). Network analysis of genetic resources data. II. The use of isozyme data in elucidating geographic relationships. *Agro-Ecosystems,* 6, 111–8.

Rohlf, F. J. (1974). Methods of comparing classifications. *Annual Review of Ecology and Systematics,* 5, 101–13.

Rohlf, F. J., Kishpaugh, J., & Kirk, D. (1972). *NT-SYS Numerical Taxonomy System of Multivariate Statistical Programs.* Stony Brook, N.Y.: State University of New York.

Rohlf, F. J., Rodman, T. C., & Flehinger, B. J. (1980). The use of nonmetric multidimensional scaling for the analysis of chromosomal associations. *Computers and Biomedical Research,* 13, 19–35.

Roi, G. H. La & Hnatiuk, R. J. (1980). The *Pinus contorta* forests of Banff and Jasper National Parks: A study in comparative synecology and syntaxonomy. *Ecological Monographs,* 50, 1–29.

Romesburg, H. C. (1979). Use of cluster analysis in leisure research. *Journal of Leisure Research,* 11, 144–53.

Rösler, F. (1978). Cortical potential correlates of selective attention in multidimensional scaling. *Biological Physiology,* 7, 223–38.

Ross, D. J., Speir, T. W., Giltrap, D. J., McNeilly, B. A., & Molloy, L. F. (1975). A principal components analysis of some biochemical activities in a climosequence of soils. *Soil Biology and Biochemistry,* 7, 349–55.

Ross, G. J. S., Lauckner, F. B., & Hawkins, D. (1976). *CLASPX Classification Program.* Harpenden, England: Rothamsted Experimental Station.

Rotenberry, J. T. & Wiens, J. A. (1980). Habitat structure, patchiness, and avian communities in North American steppe vegetation: A multivariate analysis. *Ecology,* 61, 1228–50.

Roux, G. & Roux, M. (1967). A Propos de Quelques Méthodes de classification en phytosociologie. *Revue de statistique appliquée,* 15, 59–72.

Rowe, J. S. (1956). Uses of undergrowth plant species in forestry. *Ecology,* 37, 461–73.

Russell, J. S. & Moore, A. W. (1970). Detection of homoclimates by numerical analysis with reference to the Brigalow Region (Eastern Australia). *Agricultural Meteorology,* 7, 455–79.

– (1976). Classification of climatic data for Northern Australia. In *Pattern Analysis in Agricultural Science,* ed. W. T. Williams, pp. 231–41. New York: Elsevier.

Sabo, S. R. (1980). Niche and habitat relations in subalpine bird communities of the White Mountains of New Hampshire. *Ecological Monographs,* 50, 241–59.

Sabo, S. R. & Whittaker, R. H. (1979). Bird niches in a subalpine forest: An indirect ordination. *Proceedings of the National Academy of Sciences,* 76, 1338–42.

Sachan, K. S. & Sharma, J. R. (1971). Multivariate analysis of genetic divergence in tomatos. *Indian Journal of Genetics and Plant Breeding,* 31, 86–93.

Saeki, M., Kunii, K., Seki, T., Sugiyama, K., Suzuki, T., & Shishido, S. (1977). Metal burden of urban lichens. *Environmental Research,* 13, 256–66.

Salbu, B., Pappas, A. C., & Steinnes, E. (1979). Elemental composition of Norwegian rivers. *Nordic Hydrology,* 10, 115–40.

Salton, G. (1975). *Dynamic Information and Library Processing.* Englewood Cliffs, N. J.: Prentice-Hall.

Salton, G. & Wong, A. (1978). Generation and search of cluster files. *Association for Computing Machinery Transactions on Database Systems,* 3, 321–46.

Sanson-Fisher, R. W. & Mulligan, B. (1977). The validity of a behavioral rating scale: Applications of a psychophysical technique. *Journal of Multivariate Behavioral Research,* 12, 357–72.

Schäfer, W. (1972). *Ecology and Palaeoecology of Marine Environments,* trans. I. Oertel, ed. G. Y. Craig. Edinburgh: Oliver and Boyd.

Schiffman, S. S., Reilly, D. A., & Clark, T. B. (1979). Quantitative differences among sweeteners. *Physiology and Behavior,* 23, 1–9.

Schlegel, C. C. (1981). Development, equity, and level of living in peninsular Malaysia. *Journal of Developing Areas,* 15, 297–316.

Schlieper, C. (1972). *Research Methods in Marine Biology,* trans. E. Drucker. London: Sidgwick & Jackson.

Schubert, R. (1977). Lichens as bioindicators for SO_2 atmospheric-pollution in cities and industrial areas. In *Vegetation Science and Environmental Protection,* eds. A. Miyawaki & R. Tüxen, pp. 225–34. Tokyo: Maruzen.

Schultz, V., Eberhardt, L. L., Thomas, J. M., & Cochran, M. I. (1976). *A Bibliography of Quantitative Ecology.* Stroudsburg, Pa.: Dowden, Hutchinson & Ross.

Scott, D. (1974). Description of relationships between plants and environment. In *Vegetation and Environment,* eds. B. R. Strain & W. D. Billings, pp. 49–69. The Hague: Junk.
Seeman, J., Chirkov, Y. I., Lomas, J., & Primault, B. (1979). *Agrometeorology.* New York: Springer-Verlag.
Seligson, M. A. (1977). Prestige among peasants: A multidimensional analysis of preference data. *American Journal of Sociology,* 83, 632–52.
Shannon, C. E. & Weaver, W. (1949). *The Mathematical Theory of Communication.* Urbana: University of Illinois Press.
Sharp, D. (1976). A phytosociological study of weed communities on the southwestern coastal plain of North Carolina. *Vegetatio,* 31, 103–36.
Shaw, R. H. (editor) (1967). *Ground Level Climatology.* Washington, D.C.: American Association for the Advancement of Science.
Shepard, R. N. (1962). The analysis of proximities: Multidimensional scaling with an unknown distance function. *Psychometrika,* 27, 125–39, 219–46.
Shepard, R. N. & Carroll, J. D. (1966). Parametric representation of nonlinear data structures. In *Multivariate Analysis,* ed. P. R. Krishnaiah, pp. 561–92. New York: Academic Press.
Sherman, C. R. (1977). *A Multidimensional Model of Medical School Similarities.* Washington, D. C.: Bureau of Health Manpower, U. S. Department of Health, Education, and Welfare.
Shewry, P. R., Woolhouse, H. W., & Thompson, K. (1979). Relationships of vegetation to copper and cobalt in the copper clearings of Haut Shaba, Zaïre. *Botanical Journal of the Linnean Society,* 79, 1–35.
Shimwell, D. W. (1971). *The Description and Classification of Vegetation.* London: Sidgwick & Jackson.
Shmida, A. & Whittaker, R. H. (1981). Pattern and biological microsite effects in two shrub communities, southern California. *Ecology* 62, 234–51.
Shorter, R., Byth, D. E., & Mungomery, V. E. (1977). Genotype × environment interactions and environmental adaptation. II. Assessment of environmental contributions. *Australian Journal of Agricultural Research,* 28, 223–35.
Sibson, R. (1972). Order invariant methods for data analysis. *Journal of the Royal Statistical Society, Series B,* 34, 311–49.
Simberloff, D. (1978). Using island biogeographic distributions to determine if colonization is stochastic. *American Naturalist,* 112, 713–26.
– (1980). A succession of paradigms in ecology: Essentialism to materialism and probability. *Synthese,* 43, 3–39.
Simon, H. A. (1962). The architecture of complexity. *Proceedings of the American Philosophical Society,* 106, 467–82.
– (1981). Studying human intelligence by creating artificial intelligence. *American Scientist,* 69, 300–9.
Singer, S. B. (1980). *DATAEDIT–A FORTRAN Program for Editing Data Matrices.* Ithaca, N.Y.: Cornell University.
Singh, S. & Murry, T. (1978). Multidimensional classification of normal voice qualities. *Journal of the Acoustical Society of America,* 64, 81–7.
Singh, S. P., Singh, H. N., & Rai, J. N. (1980). Multivariate analysis in relation to breeding system in okra [*Abelmoschus esculentus* (L.) Moench]. *Zeitschrift für Pflanzenzüchtung,* 84, 57–62.

Small, E. (1978*a*). A numerical and nomenclatural analysis of morpho-geographic taxa of *Humulus. Systematic Botany,* 3, 37–76.
– (1978*b*). A numerical taxonomic analysis of the *Daucus carota* complex. *Canadian Journal of Botany,* 56, 248–76.
– (1980). The relationships of hop cultivars and wild variants of *Humulus lupulus. Canadian Journal of Botany,* 58, 676–86.
– (1981). A numerical analysis of morpho-geographic groups of cultivars of *Humulus lupulus* based on samples of cones. *Canadian Journal of Botany,* 59, 311–24.
Smartt, P. F. M., Meacock, S. E., & Lambert, J. M. (1974). Investigations into the properties of quantitative vegetational data. *Journal of Ecology,* 62, 735–59.
Smith, A. J. E. & Hill, M. O. (1975). A taxonomic investigation of *Ulota bruchii* Hornsch. ex Brid., *U. crispa* (Hedw.) Brid. and *U. crispula* Brid. I. European material. *Journal of Bryology,* 8, 423–33.
Sneath, P. H. A. (1969). Evaluation of clustering methods. In *Numerical Taxonomy,* ed. A. J. Cole, pp. 257–71. New York: Academic Press.
Sneath, P. H. A. & Sokal, R. R. (1973). *Numerical Taxonomy.* San Francisco: W. H. Freeman.
Snedecor, G. W. & Cochran, W. G. (1967). *Statistical Methods.* 6th ed. Ames: Iowa State University Press.
Sobolev, L. N. (1975). Ecology and typology of agricultural lands. *Soviet Journal of Ecology,* 6, 307–14.
Sobolev, L. N. & Utekhin, V. D. (1978). Russian (Ramensky) approaches to community systematization. In *Ordination of Plant Communities,* ed. R. H. Whittaker, pp. 71–97. The Hague: Junk.
Sokal, R. R. (1974). Classification: Purposes, principles, progress, prospects. *Science,* 185, 1115–23.
– (1977). Clustering and classification: Background and current direction. In *Classification and Clustering,* ed. J. Van Ryzin, pp. 1–15. New York: Academic Press.
Sokal, R. R. & Michener, C. D. (1958). A statistical method for evaluating systematic relationships. *University of Kansas Science Bulletin,* 38, 1409–38.
Solomon, D. L. (1979). On a paradigm for mathematical modeling. In *Contemporary Quantitative Ecology and Related Ecometrics,* eds. G. P. Patil & M. Rosenzweig, pp. 231–50. Burtonsville, Md.: International Co-operative.
Sondheim, M. W., Singleton, G. A., & Lavkulich, L. M. (1981). Numerical analysis of a chronosequence, including the development of a chronofunction. *Soil Science Society of America Proceedings,* 45, 538–63.
Sousa, W. P. (1979). Experimental investigations of disturbance and ecological succession in a rocky intertidal algal community. *Ecological Monographs,* 49, 227–54.
Southwood, T. R. E. (1978). *Ecological Methods: With Particular Reference to the Study of Insect Populations.* New York: Wiley.
– (1980). Ecology – A mixture of pattern and probabilism. *Synthese,* 43, 111–22.

Sparling, D. W. & Williams, J. D. (1978). Multivariate analysis of avian vocalizations. *Journal of Theoretical Biology,* 74, 83–107.
Spatz, G. (1969). Elektronische Datenverarbeitung bei pflanzensoziologischer Tabellenarbeit. *Naturwissenschaften,* 56, 470–1.
– (1972). Eine Möglichkeit zum Einsatz der elektronischen Datenverarbeitung bei der pflanzensoziologischen Tabellenarbeit. In *Grundfragen und Methoden in der Pflanzensoziologie,* eds. E. van der Maarel & R. Tüxen, pp. 251–8. The Hague: Junk.
Stalker, H. T., Harlan, J. R., & Wet, J. M. J. de. (1977). Observations on introgression of *Tripsacum* into maize. *American Journal of Botany,* 64, 1162–9.
Stanek, W. (1973). A comparison of Braun-Blanquet's method with sum-of-squares agglomeration for vegetation classification. *Vegetatio,* 27, 323–38.
Steele, J. H. (1977). *Spatial Patterns in Plankton Communities.* New York: Plenum.
Steubing, L. (1977). The value of lichens as indicators of immission load. In *Vegetation Science and Environmental Protection,* eds. A. Miyawaki & R. Tüxen, pp. 235–46. Tokyo: Maruzen.
Stevenson, L. H. & Colwell, R. R. (editors) (1973). *Estuarine Microbial Ecology.* Columbia, S.C.: University of South Carolina Press.
Stitt, F. W., Frane, M., & Frane, J. W. (1977). Mood change in rheumatoid arthritis: Factor analysis as a tool in clinical research. *Journal of Chronic Diseases,* 30, 135–45.
Stocker, M., Gilbert, F. F., & Smith, D. W. (1977). Vegetation and deer habitat relations in southern Ontario: Classification of habitat types. *Journal of Applied Ecology,* 14, 419–32.
Stocum, A. S. (1980). Natural Vegetation and Its Relationship to the Environment of Selected Abandoned Coal Surface Mines in the Cumberland Mountains of Tennessee. Ph.D. thesis, University of Tennessee, Knoxville.
Stopher, P. R. & Meyburg, A. H. (1979). *Survey Sampling and Multivariate Analysis for Social Scientists and Engineers.* Lexington, Mass.: D.C. Heath.
Streibig, J. C. (1979). Numerical methods illustrating the phytosociology of crops in relation to weed flora. *Journal of Applied Ecology,* 16, 577–87.
Strong, D. R. (1980). Null hypotheses in ecology. *Synthese,* 43, 271–85.
Swain, P. H. (1978). Fundamentals of pattern recognition in remote sensing. In *Remote Sensing: The Quantitative Approach,* eds. P. H. Swain & S. M. Davis, pp. 136–87. New York: McGraw-Hill.
Swain, P. H. & Davis, S. M. (editors) (1978). *Remote Sensing: The Quantitative Approach.* New York: McGraw-Hill.
Swaine, M. D. & Greig-Smith, P. (1980). An application of principal components analysis to vegetation change in permanent plots. *Journal of Ecology,* 68, 33–41.
Swaine, M. D. & Hall, J. B. (1976). An application of ordination to the identification of forest types. *Vegetatio,* 32, 83–6.
Swan, F. R. (1961). Lianas in Southern Wisconsin Forests. M.S. thesis, University of Wisconsin, Madison.

Swan, J. M. A. (1970). An examination of some ordination problems by use of simulated vegetational data. *Ecology,* 51, 89–102.

Szocs, Z. (1973). On the botanical application of some multivariate analyses. II. General characterization. (Magyar with English summary.) *Botanikai Kozlemenyek,* 60, 29–34.

Tai, G. C. C. & Jong, H. De (1980). Multivariate analyses of potato hybrids. I. Discrimination between tetraploid–diploid hybrid families and their relationship to cultivars. *Canadian Journal of Genetics and Cytology,* 22, 227–35.

Tai, G. C. C. & Tarn, T. R. (1980). Multivariate analyses of potato hybrids. II. Discrimination between *Tuberosum–Andigena* hybrid families and their relationship to their parents. *Canadian Journal of Genetics and Cytology,* 22, 279–86.

Tanner, J. T. (1978). *Guide to the Study of Animal Populations.* Knoxville: University of Tennesee Press.

Taylor, D. W. (1977). Floristic relationships along the Cascade–Sierran axis. *American Midland Naturalist,* 97, 333–49.

Tazaki, T. & Ushijima, T. (1977). The vegetation in the neighbourhood of smelting factories and the amount of heavy metals absorbed and accumulated by various species. In *Vegetation Science and Environmental Protection,* eds. A. Miyawaki & R. Tüxen, pp. 217–24. Tokyo: Maruzen.

Teil, H. & Cheminee, J. L. (1975). Application of correspondence factor analysis to the study of major and trace elements in the Erta Ale Chain (Afar, Ethiopia). *Journal of the International Association for Mathematical Geology,* 7, 13–30.

Terborgh, J. (1971). Distribution on environmental gradients: Theory and a preliminary interpretation of distributional patterns in the avifauna of the Cordillera Vilcabamba, Peru. *Ecology,* 52, 23–40.

Thayer, D. W. (editor) (1975). *Microbial Interaction with the Physical Environment.* Stroudsburg, Pa.: Dowden, Hutchinson & Ross.

Thompson, D. C. (1980). A classification of the vegetation of Boothia Peninsula and the Northern District of Keewatin, N. W. T. *Arctic,* 33, 73–99.

Thompson, J. P., Lloyd, D. L., & Moore, A. W. (1976). Pattern analysis of nitrogen mineralization potentials in soil from a ley-pasture crop rotation experiment. In *Pattern Analysis in Agricultural Science,* ed. W. T. Williams, pp. 249–58. New York: Elsevier.

Thurstone, L. L. & Chave, E. J. (1929). *The Measurement of Attitude.* Chicago: University of Chicago Press.

Tomlinson, R. (1981). A rapid sampling technique suitable for expedition use, with reference to the vegetation of the Faroe Islands. *Biological Conservation,* 20, 69–81.

Tothill, J. C. (1976). Classification of gilgai data. In *Pattern Analysis in Agricultural Science,* ed. W. T. Williams, pp. 274–9. New York: Elsevier.

Trenbath, B. R. (1974). Neighbour effects in the genus *Avena.* II. Comparison of weed species. *Journal of Applied Ecology,* 11, 111–25.

Trenbath, B. R. & Harper, J. L. (1973). Neighbour effects in the genus *Avena.* I. Comparison of crop species. *Journal of Applied Ecology,* 10, 379–400.

Tresner, H. D., Bakus, M. P., & Curtis, J. T. (1954). Soil microfungi in relation to the hardwood forest continuum in southern Wisconsin. *Mycologia,* 46, 314–33.

Tsunewaki, K., Mukai, Y., Ryu Endo, T., Tsuji, S., & Murata, M. (1976). Genetic diversity of the cytoplasm in *Triticum* and *Aegilops.* V. Classification of 23 cytoplasms into eight plasma types. *Japanese Journal of Genetics,* 51, 175–91.

Tulloch, A. P., Baum, B. R., & Hoffman, L. L. (1980). A survey of epicuticular waxes among genera of Triticeae. 2. Chemistry. *Canadian Journal of Botany,* 58, 2602–15.

Vanneman, R. (1977). The occupational composition of American classes: Results from cluster analysis. *American Journal of Sociology,* 82, 783–807.

Veblen, T. T. & Ashton, D. H. (1979). Successional pattern above timberline in south-central Chile. *Vegetatio,* 40, 39–47.

Verma, M. M., Murty, B. R., Jain, O. P., & Rao, U. M. B. (1974). Factor analysis of diversity in soybean. *Genetica Agraria,* 28, 142–9.

Verneaux, J. (1976*a*). Biotypologie de l'écosystème "eau courante." La structure biotypologique. *Comptes rendus des séances de l'Académie des Sciences, série D,* 283, 1663–6.

– (1976*b*). Biotypologie de l'écosystème "eau courante." Les groupements socio-écologiques. *Comptes rendus des séances de l'Académie des Sciences, série D,* 283, 1791–3.

Visser, S. & Parkinson, D. (1975). Fungal succession on aspen poplar leaf litter. *Canadian Journal of Botany,* 53, 1640–51.

Vitt, D. H. & Slack, N. G. (1975). An analysis of the vegetation of *Sphagnum*-dominated kettle-hole bogs in relation to environmental gradients. *Canadian Journal of Botany,* 53, 332–59.

Volland, L. A. & Connelly, M. (1978). *Computer Analysis of Ecological Data: Methods and Programs.* Portland, Ore.: Forest Service.

Walker, B. H. & Wehrhahn, C. F. (1971). Relationships between derived vegetation gradients and measured environmental variables in Saskatchewan wetlands. *Ecology,* 52, 86–95.

Walton, P. D. (1971). The use of factor analysis in determining characters for yield selection in wheat. *Euphytica,* 20, 416–21.

Ward, J. H. (1963). Hierarchical grouping to optimize an objective function. *American Statistical Association Journal,* 58, 236–44.

Waring, R. H. & Major, J. (1964). Some vegetation of the California coastal redwood region in relation to gradients of moisture, nutrients, light and temperature. *Ecological Monographs,* 34, 167–215.

Watanabe, R. & Miyai, S. (1978). Studies on vegetation in Kiyosumi region. I. Numerical classification of forest vegetation. *Japanese Journal of Ecology,* 28, 281–90.

Watts, W. A. (1980). The late quaternary vegetation history of the southeastern United States. *Annual Review of Ecology and Systematics,* 11, 387–409.

Webb, L. J., Tracey, J. G., Williams, W. T., & Lance, G. N. (1967). Studies in the numerical analysis of complex rain forest communities. II. The problem of species sampling. *Journal of Ecology,* 55, 525–38.

Webb, L. J., Tracey, J. G., Williams, W. T., & Lance, G. N. (1971). Prediction of agricultural potential from intact forest vegetation. *Journal of Applied Ecology,* 8, 99–121.

Webb, T., Laseski, R. A., & Bernabo, J. C. (1978). Sensing vegetational patterns with pollen data: Choosing the data. *Ecology,* 59, 1151–63.

Webster, R. & Burrough, P. A. (1972*a*). Computer-based soil mapping of small areas from sample data. I. Multivariate classification and ordination. *Journal of Soil Science,* 23, 210–21.

– (1972*b*). Computer-based soil mapping of small areas from sample data. II. Classification smoothing. *Journal of Soil Science,* 23, 222–34.

Webster, R. & Butler, B. E. (1976). Soil classification and survey studies at Ginninderra. *Australian Journal of Soil Research,* 14, 1–24.

Weimarck, G. (1981). Numerical analysis of the floristic composition of localities including *Hierochloë* (Poaceae) species in northern Europe. *Vegetatio,* 44, 101–35.

Weisberg, H. F. (1980). A multidimensional conceptualization of party identification. *Political Behavior,* 2, 33–60.

Wentworth, T. R. (1981). Vegetation on limestone and granite in the Mule Mountains, Arizona. *Ecology,* 62, 469–82.

Westhoff, V. (1979). Phytosociology in The Netherlands: History, present state, future. In *The Study of Vegetation,* ed. M. J. A. Werger, pp. 81–121. The Hague: Junk.

Westhoff, V. & Maarel, E. van der (1978). The Braun-Blanquet approach. In *Classification of Plant Communities,* ed. R. H. Whittaker, pp. 287–399. The Hague: Junk.

Westman, W. E. (1980). Gaussian analysis: Identifying environmental factors influencing bell-shaped species distributions. *Ecology,* 61, 733–9.

– (1981). Factors influencing the distribution of species of Californian coastal sage scrub. *Ecology,* 62, 439–55.

Wetzel, R. G. & Likens, G. E. (1979). *Limnological Analyses.* Philadelphia: W. B. Saunders.

Wheeler, B. D. (1980*a*). Plant communities of rich-fen systems in England and Wales. I. Introduction. Tall sedge and reed communities. *Journal of Ecology,* 68, 365–95.

– (1980*b*). Plant communities of rich-fen systems in England and Wales. II. Communities of calcareous mires. *Journal of Ecology,* 68, 405–20.

White, C. W., Lockhead, G. R., & Evans, N. J. (1977). Multidimensional scaling of subjective colors by color-blind observers. *Perception and Psychophysics,* 6, 522–6.

White, R. F. & Lewinson, T. M. (1977). Probabilistic clustering for attributes of mixed type with biopharmaceutical applications. *Journal of the American Statistical Association,* 72, 271–7.

Whitney, G. G. & Adams, S. D. (1980). Man as a maker of new plant communities. *Journal of Applied Ecology,* 17, 431–48.

Whittaker, R. H. (1948). A Vegetation Analysis of the Great Smoky Mountains. Ph.D. thesis, University of Illinois, Urbana.

– (1952). A study of summer foliage insect communities in the Great Smoky Mountains. *Ecological Monographs,* 22, 1–44.

– (1956). Vegetation of the Great Smoky Mountains. *Ecological Monographs,* 26, 1–80.

– (1960). Vegetation of the Siskiyou Mountains, Oregon and California. *Ecological Monographs,* 30, 279–338.

– (1962). Classification of natural communities. *Botanical Review,* 28, 1–239.

– (1966). Forest dimensions and production in the Great Smoky Mountains. *Ecology,* 47, 103–21.

– (1967). Gradient analysis of vegetation. *Biological Reviews,* 42, 207–64.

– (1970). The population structure of vegetation. In *Gesellschaftsmorphologie,* ed. R. Tüxen, pp. 39–62. The Hague: Junk.

– (1975). The design and stability of plant communities. In *Unifying Concepts in Ecology,* eds. W. H. van Dobben & R. H. Lowe-McConnell, pp. 169–81. The Hague: Junk.

– (editor) (1978*a*). *Classification of Plant Communities.* The Hague: Junk.

– (1978*b*). Direct gradient analysis. In *Ordination of Plant Communities,* ed. R. H. Whittaker, pp. 7–50. The Hague: Junk.

– (editor) (1978*c*). *Ordination of Plant Communities.* The Hague: Junk.

Whittaker, R. H. & Gauch, H. G. (1978). Evaluation of ordination techniques. In *Ordination of Plant Communities,* ed. R. H. Whittaker, pp. 277–336. The Hague: Junk.

Whittaker, R. H., Gilbert, L. E., & Connell, J. H. (1979). Analysis of two-phase pattern in a mesquite grassland, Texas. *Journal of Ecology,* 67, 935–52.

Whittaker, R. H. & Levin, S. A. (editors) (1975). *Niche Theory and Application.* Stroudsburg, Pa.: Dowden, Hutchinson & Ross.

– (1977). The role of mosaic phenomena in natural communities. *Theoretical Population Biology,* 12, 117–39.

Whittaker, R. H., Levin, S. A., & Root, R. B. (1973). Niche, habitat, and ecotope. *American Naturalist,* 107, 321–38.

Whittaker, R. H. & Naveh, Z. (1979). Analysis of two-phase patterns. In *Contemporary Quantitative Ecology and Related Ecometrics,* eds. G. P. Patil & M. L. Rosenzweig, pp. 157–65. Burtonsville, Md.: International Co-operative.

Whittaker, R. H. & Niering, W. A. (1964). Vegetation of the Santa Catalina Mounains, Arizona. I. Ecological classification and distribution of species. *Journal of the Arizona Academy of Science,* 3, 9–34.

Whittaker, R. H., Niering, W. A., & Crisp, M. D. (1979). Structure, pattern, and diversity of a mallee community in New South Wales. *Vegetatio,* 39, 65–76.

Wiegleb, G. (1980). Some applications of principal components analysis in vegetation: Ecological research of aquatic communities. *Vegetatio,* 42, 67–73.

Wiens, J. A. & Rotenberry, J. T. (1981). Habitat associations and community structure of birds in shrubsteppe environments. *Ecological Monographs,* 51, 21–41.

Wikum, D. A. & Shanholtzer, G. F. (1978). Application of the Braun-Blanquet cover-abundance scale for vegetation analysis in land development studies. *Environmental Management,* 2, 323–9.

Wild, H. (1968). Geobotanical anomalies in Rhodesia. 1. The vegetation of copper bearing soils. *Kirkia,* 7, 1–72.

– (1970). Geobotanical anomalies in Rhodesia. 3. The vegetation of nickel bearing soils. *Kirkia,* 7 (Supplement), 1–62.

– (1974). Geobotanical anomalies in Rhodesia. 4. The vegetation of arsenical soils. *Kirkia,* 9, 243–64.

Wildi, O. (1979). *GRID–A* space density analysis for recognition of noda in vegetation samples. *Vegetatio,* 41, 95–100.

– (1980). Management and multivariate analysis of large data sets in vegetation research. *Vegetatio,* 42, 175–80.

Wildi, O. & Orlóci, L. (1980). *Management and Multivariate Analysis of Vegetation Data.* Birmensdorf: Swiss Federal Institute of Forestry Research.

Wilkes, R. E. & Uhr, E. B. (1978). An empirical application of multidimensional scaling to advertising pretesting. *Omega,* 6, 173–81.

Wilkinson, J. H. & Reinsch, C. (1971). *Handbook for Automatic Computation, Linear Algebra,* vol. 2, pp. 212–48. New York: Springer-Verlag.

Williams, W. T. (1962). Computers as botanists. *Proceedings of the Royal Institute,* 39, 306–12.

– (1971*a*). Principles of clustering. *Annual Review of Ecology and Systematics,* 2, 303–26.

– (1971*b*). Strategy and tactics in the acquisition of ecological data. *Proceedings of the Ecological Society of Australia,* 6, 57–62.

– (editor) (1976). *Pattern Analysis in Agricultural Science.* New York: Elsevier.

Williams, W. T., Burt, R. L., Pengelly, B. C., & Robinson, P. J. (1980). Network analysis of genetic resources data. I. Geographical relationships. *Agro-Ecosystems,* 6, 99–109.

Williams, W. T. & Edye, L. A. (1974). A new method for the analysis of three-dimensional data matrices in agricultural experimentation. *Australian Journal of Agricultural Research,* 25, 803–12.

Williams, W. T., Edye, L. A., Burt, R. L., & Grof, B. (1973). The use of ordination techniques in the preliminary evaluation of *Stylosanthes* accessions. *Australian Journal of Agricultural Research,* 24, 715–31.

Williams, W. T. & Gillard, P. (1971). Pattern analysis of a grazing experiment. *Australian Journal of Agricultural Research,* 22, 245–60.

Williams, W. T. & Lambert, J. M. (1959). Multivariate methods in plant ecology. I. Association-analysis in plant communities. *Journal of Ecology,* 47, 83–101.

– (1960). Multivariate methods in plant ecology. II. The use of an electronic digital computer for association-analysis. *Journal of Ecology,* 48, 689–710.

– (1961). Multivariate methods in plant ecology. III. Inverse association-analysis. *Journal of Ecology,* 49, 717–29.

Williams, W. T., Lambert, J. M., & Lance, G. N. (1966). Multivariate methods in plant ecology. V. Similarity analyses and information-analysis. *Journal of Ecology,* 54, 427–45.

Williams, W. T. & Lance, G. N. (1968). Choice of strategy in the analysis of complex data. *Statistician,* 18, 31–43.

Williams, W. T., Lance, G. N., Webb, L. J., & Tracey, J. G. (1973). Studies in the numerical analysis of complex rain-forest communities. VI. Models for the classification of quantitative data. *Journal of Ecology,* 61, 47–70.

Williams, W. T. & Stephenson, W. (1973). The analysis of three-dimensional data (sites × species × times) in marine ecology. *Journal of Experimental Marine Biology and Ecology,* 11, 207–27.

Williamson, C. J. & Killick, R. J. (1978). Multivariate methods as an aid in identifying *Poa ampla* × *P. pratensis* hybrids from maternal-type offspring. *Heredity,* 41, 215–25.

Will-Wolf, S. (1980). Structure of corticolous lichen communities before and after exposure to emissions from a "clean" coal-fired generating station. *Bryologist,* 83, 281–95.

Wilson, M. V. (1981). A statistical test of the accuracy and consistency of ordinations. *Ecology* 62, 8–12.

Wilson, M. V. & Mohler, C. L. (1981). *GRADBETA – A FORTRAN Program for Measuring Compositional Change Along Gradients.* Ithaca, N.Y.: Cornell University.

Wimsatt, W. C. (1980). Randomness and perceived-randomness in evolutionary biology. *Synthese,* 43, 287–329.

Winner, W. E. & Bewley, J. D. (1978*a*). Contrasts between bryophyte and vascular plant synecological responses in an SO_2-stressed white spruce association in central Alberta. *Oecologia,* 33, 311–25.

– (1978*b*). Terrestrial mosses as bioindicators of SO_2 pollution stress. *Oecologia,* 35, 221–30.

Winter, W. H. (1976). Classification of sheep body composition data. In *Pattern Analysis in Agricultural Science,* ed. W. T. Williams, pp. 302–9. New York: Elsevier.

Wishart, D. (1978). *Clustan Users Manual,* 3d. ed. Edinburgh: Edinburgh University.

Wishart, D. & Leach, S. V. (1970). A multivariate analysis of Platonic prose rhythm. *Computer Studies in the Humanities and Verbal Behavior,* 3, 90–9.

Wolff, J. D. (1980). The role of habitat patchiness in the population dynamics of snowshoe hares. *Ecological Monographs,* 50, 111–30.

Wood, R. D. (1975). *Hydrobotanical Methods.* Baltimore: University Park Press.

Yates, T. E., Brooks, R. R., & Boswell, C. R. (1974). Factor analysis in botanical methods of exploration. *Journal of Applied Ecology,* 11, 563–74.

Young, F. W., Bertoli, F., & Bertoli, S. (1981). Rural poverty and ecological problems: Results of a new type of baseline study. *Social Indicators Research* (in press).

Young, F. W., Freebairn, D. K., & Snipper, R. (1979). The structural context of rural poverty in Mexico: A cross-state comparison. *Economic Development and Cultural Change,* 27, 669–86.

Young, F. W. & Lewyckyj, R. (1979). *ALSCAL-4 User's Guide.* Carrboro, N. C.: Data Analysis and Theory Associates.

Young, L. D., Johnson, R. K., & Omtvedt, I. T. (1977). An analysis of the dependency structure between a gilt's prebreeding and reproductive traits. II. Principal component analysis. *Journal of Animal Science,* 44, 565–70.
Young, R. C. (1976). The structural context of Caribbean agriculture: A comparative study. *Journal of Developing Areas,* 10, 425–44.
Zajicek, G., Maayan, C., & Rosenmann, E. (1977). An application of cluster analysis to glomerular histopathology. *Computers and Biomedical Research,* 10, 471–81.
Zedler, P. H. & Goff, F. G. (1973). Size-association analysis of forest successional trends in Wisconsin. *Ecological Monographs,* 43, 79–94.
Ziegler, A. M. (1965). Silurian marine communities and their environmental significance. *Nature,* 207, 207–2.
Ziegler, A. M., Cocks, L. R. M., & Bambach, R. K. (1968). The composition and structure of Lower Siberian marine communities. *Lethaia,* 1, 1–27.
Zottoli, R. (1978). *Introduction to Marine Environments,* 2d. ed. St. Louis: C. V. Mosby.

Index